Lecture Notes in Computer Science

Vol. 192: Automata on Infinite Words. Proceedings, 1984. Edited by M. Nivat and D. Perrin. V, 216 pages.1985.

Vol. 193: Logics of Programs. Proceedings, 1985. Edited by R. Parikh. VI, 424 pages. 1985.

Vol. 194: Automata, Languages and Programming. Proceedings, 1985. Edited by W. Brauer. IX, 520 pages. 1985.

Vol. 195: H. J. Stüttgen, A Hierarchical Associative Processing System. XII, 273 pages. 1985.

Vol. 196: Advances in Cryptology. Proceedings of CRYPTO '84. Edited by G. R. Blakley and D. Chaum. IX, 491 pages.1985.

Vol. 197: Seminar on Concurrency. Proceedings, 1984. Edited by S. D. Brookes, A. W. Roscoe and G. Winskel. X, 523 pages. 1985.

Vol. 198: A. Businger, PORTAL Language Description. VIII, 186 pages. 1985.

Vol. 199: Fundamentals of Computation Theory. Proceedings, 1985. Edited by L. Budach. XII, 533 pages. 1985.

Vol. 200: J. L. A. van de Snepscheut, Trace Theory and VLSI Design. VI, 0–140 pages. 1985.

Vol. 201: Functional Programming Languages and Computer Architecture. Proceedings, 1985. Edited by J.-P. Jouannaud. VI, 413 pages. 1985.

Vol. 202: Rewriting Techniques and Applications. Edited by J.-P. Jouannaud. VI, 441 pages. 1985.

Vol. 203: EUROCAL '85. Proceedings Vol. 1, 1985. Edited by B. Buchberger. V, 233 pages. 1985.

Vol. 204: EUROCAL '85. Proceedings Vol. 2, 1985. Edited by B. F. Caviness. XVI, 650 pages. 1985.

Vol. 205: P. Klint, A Study in String Processing Languages. VIII, 165 pages. 1985.

Vol. 206: Foundations of Software Technology and Theoretical Computer Science. Proceedings, 1985. Edited by S. N. Maheshwari. IX, 522 pages. 1985.

Vol. 207: The Analysis of Concurrent Systems. Proceedings, 1983. Edited by B. T. Denvir, W. T. Harwood, M. I. Jackson and M. J. Wray. VII, 398 pages. 1985.

Vol. 208: Computation Theory. Proceedings, 1984. Edited by A. Skowron. VII, 397 pages. 1985.

Vol. 209: Advances in Cryptology. Proceedings of EUROCRYPT '84. Edited by T. Beth, N. Cot and I. Ingemarsson. VII, 491 pages. 1985.

Vol. 210: STACS 86. Proceedings, 1986. Edited by B. Monien and G. Vidal-Naquet. IX, 368 pages. 1986.

Vol. 211: U. Schöning, Complexity and Structure. V, 99 pages. 1986.

Vol. 212: Interval Mathematics 1985. Proceedings, 1985. Edited by K. Nickel. VI, 227 pages. 1986.

Vol. 213: ESOP 86. Proceedings, 1986. Edited by B. Robinet and R. Wilhelm. VI, 374 pages. 1986.

Vol. 214: CAAP '86. 11th Colloquium on Trees in Algebra and Programming. Proceedings, 1986. Edited by P. Franchi-Zannettacci. VI, 306 pages. 1986.

Vol. 215: Mathematical Methods of Specification and Synthesis of Software Systems '85. Proceedings, 1985. Edited by W. Bibel and K. P. Jantke. 245 pages. 1986.

Vol. 216: Ch. Fernstrom, I. Kruzela, B. Svensson, LUCAS Associative Array Processor: Design, Programming and Application Studies. XII, 323 pages. 1986.

Vol. 217: Programs as Data Objects. Proceedings, 1985. Edited by H. Ganzinger and N. D. Jones. X, 324 pages. 1986.

Vol. 218: Advances in Cryptology – CRYPTO '85. Proceedings, 1985. Edited by H. C. Williams. X, 548 pages. 1986.

Vol. 219: Advances in Cryptology – EUROCRYPT '85. Proceedings, 1985. Edited by F. Pichler. IX, 281 pages. 1986.

Vol. 220: RIMS Symposia on Software Science and Engineering II. Proceedings, 1983 and 1984. Edited by E. Goto, K. Araki and T. Yuasa. XI, 323 pages. 1986.

Vol. 221: Logic Programming '85. Proceedings, 1985. Edited by E. Wada. IX, 311 pages. 1986.

Vol. 222: Advances in Petri Nets 1985. Edited by G. Rozenberg. VI, 498 pages. 1986.

Vol. 223: Structure in Complexity Theory. Proceedings, 1986. Edited by A. L. Selman. VI, 401 pages. 1986.

Vol. 224: Current Trends in Concurrency. Overviews and Tutorials. Edited by J. W. de Bakker, W.-P. de Roever and G. Rozenberg. XII, 716 pages. 1986.

Vol. 225: Third International Conference on Logic Programming. Proceedings, 1986. Edited by E. Shapiro. IX, 720 pages. 1986.

Vol. 226: Automata, Languages and Programming. Proceedings, 1986. Edited by L. Kott. IX, 474 pages. 1986.

Vol. 227: VLSI Algorithms and Architectures – AWOC 86. Proceedings, 1986. Edited by F. Makedon, K. Mehlhorn, T. Papatheodorou and P. Spirakis. VIII, 328 pages. 1986.

Vol. 228: Applied Algebra, Algorithmics and Error-Correcting Codes. AAECC-2. Proceedings, 1984. Edited by A. Poli. VI, 265 pages. 1986.

Vol. 229: Algebraic Algorithms and Error-Correcting Codes. AAECC-3. Proceedings, 1985. Edited by J. Calmet. VII, 416 pages. 1986.

Vol. 230: 8th International Conference on Automated Deduction. Proceedings, 1986. Edited by J. H. Siekmann. X, 708 pages. 1986.

Vol. 231: R. Hausser, NEWCAT: Parsing Natural Language Using Left-Associative Grammar. II, 540 pages. 1986.

Vol. 232: Fundamentals of Artificial Intelligence. Edited by W. Bibel and Ph. Jorrand. VII, 313 pages. 1986.

Vol. 233: Mathematical Foundations of Computer Science 1986. Proceedings, 1986. Edited by J. Gruska, B. Rovan and J. Wiedermann. IX, 650 pages. 1986.

Vol. 234: Concepts in User Interfaces: A Reference Model for Command and Response Languages. By Members of IFIP Working Group 2.7. Edited by D. Beech. X, 116 pages. 1986.

Vol. 235: Accurate Scientific Computations. Proceedings, 1985. Edited by W. L. Miranker and R. A. Toupin. XIII, 205 pages. 1986.

Vol. 236: TEX for Scientific Documentation. Proceedings, 1986. Edited by J. Désarménien. VI, 204 pages. 1986.

Vol. 237: CONPAR 86. Proceedings, 1986. Edited by W. Händler, D. Haupt, R. Jeltsch, W. Juling and O. Lange. X, 418 pages. 1986.

Vol. 238: L. Naish, Negation and Control in Prolog. IX, 119 pages. 1986.

Vol. 239: Mathematical Foundations of Programming Semantics. Proceedings, 1985. Edited by A. Melton. VI, 395 pages. 1986.

Vol. 240: Category Theory and Computer Programming. Proceedings, 1985. Edited by D. Pitt, S. Abramsky, A. Poigné and D. Rydeheard. VII, 519 pages. 1986.

Vol. 241: Foundations of Software Technology and Theoretical Computer Science. Proceedings, 1986. Edited by K.V. Nori. XII, 519 pages. 1986.

Vol. 242: Combinators and Functional Programming Languages. Proceedings, 1985. Edited by G. Cousineau, P.-L. Curien and B. Robinet. V, 208 pages. 1986.

Vol. 243: ICDT '86. Proceedings, 1986. Edited by G. Ausiello and P. Atzeni. VI, 444 pages. 1986.

Vol. 244: Advanced Programming Environments. Proceedings, 1986. Edited by R. Conradi, T.M. Didriksen and D.H. Wanvik. VII, 604 pages. 1986

Lecture Notes in Computer Science

Edited by G. Goos and J. Hartmanis

283

D. H. Pitt A. Poigné D. E. Rydeheard (Eds.)

Category Theory and Computer Science

Edinburgh, U.K., September 7–9, 1987
Proceedings

Springer-Verlag
Berlin Heidelberg New York London Paris Tokyo

Editors

David H. Pitt
Department of Mathematics, University of Surrey
Guildford, Surrey GU2 5XH, U.K.

Axel Poigné
Department of Computer Science
Imperial College of Science and Technology, University of London
180 Queen's Gate, London SW7 2BZ, U.K.

David E. Rydeheard
Department of Computer Science, The University
Manchester M13 9PL, U.K.

CR Subject Classification (1987): D.2.1, D.3.1, D.3.3, F.3, F.4.1

ISBN 3-540-18508-9 Springer-Verlag Berlin Heidelberg New York
ISBN 0-387-18508-9 Springer-Verlag New York Berlin Heidelberg

Printing and binding: Druckhaus Beltz, Hemsbach/Bergstr.
2145/3140-543210

PREFACE

Category theory arose in the 1940s as an attempt to give a unified treatment to constructs which appear in various guises in algebra, set theory and topology. More recently, logical aspects of category theory have come to the fore. Also in the 1940s were the first attempts at programming electronic computers using sequences of machine instructions. Despite the apparent distance between these two subjects, category theory has become of increasing importance in understanding the process of computer programming.

This volume is a collection of research papers describing some of the links being established between category theory and computer programming. It is the proceedings of a conference held at the University of Edinburgh, 7th-9th September 1987. This conference was arranged as a sequel to that held at the University of Surrey in 1985, whose proceedings are published as number 240 in this series. For those interested in this topic, mention should also be made of the American Mathematical Society conference on Categories in Computer Science and Logic, held in Boulder, Colorado in June 1987.

D.H.Pitt
A.Poigné
D.E.Rydeheard

Organising and Program Committee

S.Abramsky, P.Dybjer, P.L.Curien, H.D.Ehrich, M.Fourman, D.H.Pitt, A.Poigné, D.E.Rydeheard, D.T.Sannella, E.Wagner.

Referees

S.Abramsky, P.Degano, P.Dybjer, P.Ciuffotti, P.L.Curien, R.Dyckhoff, H.D.Ehrich, M.Fourman, W.Harwood, Y.Lafont, G.Longo, V.Manca, I.Mason, D.MacQueen, E.Moggi, B.Monahan, I.C.Phillips, D.H.Pitt, A.J.Pitts, A.Poigné, D.E.Rydeheard, M.Sadler, D.T.Sannella, R.Shaw, P.Taylor, E.Wagner.

The Organising Committee would like to thank the University of Edinburgh for hosting the conference and in particular D.T.Sannella, J.M.Ratcliff and G.L.Cleland for their indispensable contribution.

CONTENTS

CATEGORIES AND EFFECTIVE COMPUTATIONS

G. ROSOLINI[1]

Department of Pure Mathematics and Mathematical Statistics
16 Mill Lane, Cambridge CB2 1SB, England,
and Dipartimento di Matematica
Università degli Studi, 43100 Parma, Italy

We are convinced that in a treatment of partial functions and convergence of computations it is convenient to have an intuitionist's mind. This conviction comes from the development of the abstract theory for categories of partial maps. It was begun by Eilenberg & Kelly [1966], but it laid forgotten for a long while. The idea was then brought back to life by the new approach to Scott domains of Plotkin [1986] which concurred with a renewed approach to the categorical theory for partial maps (see those references listed in the bibliography with either "partial" or "dominical" in their title). In particular, the author's D.Phil. thesis (Rosolini [1986]) shows how to represent any (essentially small) category of partial maps in a suitable topos.

We think that topos theory is the best expression of the link between category theory and intuitionism. Category theory has to do with closure properties of families of maps and with the characterisation of maps and spaces in terms of mapping properties. Definability of these maps is often important. Intuitionistic logic treats existence in terms of explicit constructions; hence definability is again important. The conservative translation of category theory into intuitionistic logic makes harmony between these theories precise. Moreover intuitionistic logic have models in which maps on the natural numbers are always computable (again a kind of definability), so we combine under one point of view both algebraic explicitness and computational notions.

We shall not treat here the algebraic theory for categories of partial maps (and simply refer the reader to Robinson & Rosolini [1986] for detailed information). Rather we prefer to concentrate our efforts on a particular instance of relevance for modelling computations. Applying topos-theoretical methods the author showed that the properties of the topological structure of a countable algebraic lattice depended upon those subobjects classified by a certain subobject Σ of truth values. We shall move from this remark and consider those objects which are completely determined by their Σ-classified subobjects and prove some of the properties they enjoy.

A first draft of this paper sketching ideas which developed from the author's D.Phil. thesis was written while he was at Carnegie Mellon University. He wishes to thank his supervisor Dana Scott for the enlightening suggestions and many helpful discussions.

1 Partiality in a topos

Suppose one wants to define a category of partial maps on a given category **A** with products: say continuous mappings of topological spaces defined on an open subspace. One realises immediately

[1]Research partly supported by a grant from Carnegie Mellon University and by a grant from S.E.R.C.

that, besides the category A, a class $\mathcal{M}$ of subobjects must be given collecting the "acceptable" domains of definition for the partial maps. In order to secure a properly defined category $\mathcal{M}$-Ptl(A), the class $\mathcal{M}$ must satisfies the following closure properties:

a) $\mathcal{M}$ contains all maximal subobjects

b) pullbacks of a subobject in $\mathcal{M}$ exist in A and they always are in $\mathcal{M}$

c) if both $[m\colon A \rightarrowtail B]$ and $[n\colon B \rightarrowtail C]$ are in $\mathcal{M}$, then $[nm\colon A \rightarrowtail C]$ is in $\mathcal{M}$.

In case the given category is a topos $\mathcal{E}$ and a lot more definability power is at hand, a class $\mathcal{M}_\Sigma$ of subobjects can be defined by considering a subobject $\Sigma \rightarrowtail \Omega$ of the subobject classifier and by letting $m\colon A \rightarrowtail X$ be in $\mathcal{M}_\Sigma$ exactly when its characteristic map factors through $\Sigma \rightarrowtail \Omega$. Then $\mathcal{M}_\Sigma$ obviously satisfies (b), and it satisfies (a) if and only if

1) $\top \in \Sigma$.

Moreover one can prove easily that (c) is equivalent to

2) $p \in \Sigma \ \wedge\ [p = \top \longrightarrow q \in \Sigma] \longrightarrow (p \wedge q) \in \Sigma$

by noticing that the categorical interpretation of the formula is as follows. Given maps $p, q\colon A \to \Omega$ classifying subobjects $P \rightarrowtail A$ and $Q \rightarrowtail A$ respectively, if $P \leq q^{-1}(\Sigma)$ then $P \wedge Q \rightarrowtail A$ is in $\mathcal{M}_\Sigma$. Therefore $\mathcal{M}_\Sigma$-Ptl($\mathcal{E}$) is a category if and only if $\Sigma \rightarrowtail \Omega$ satisfies conditions (1) and (2) above. Such a subobject of Ω is called a dominance in Rosolini [1986] where it is also shown that any category $\mathcal{M}$-Ptl(A) can be fully embedded in one of the form $\mathcal{M}_\Sigma$-Ptl($\mathcal{E}$) when A is essentially small.

In the topos $\mathcal{S}$ of sets there are only two trivial choices for Σ which yield the category $\mathcal{S}$ itself and the category of all partial functions between sets as the corresponding categories of partial maps. A more encouraging example is found in the recursive topos $\mathcal{R}$ introduced by Mulry [1982] where a subobject Ω_{re} is defined which internalises a notion of recursively enumerable subset. We shall be interested in a similar case when in the topos $\mathcal{E}$ the natural number object N satisfies some particularly nice properties with respect to computability.

In the sequel we suppose that a topos $\mathcal{E}$ is fixed with a natural number object N, and the following properties hold in the internal logic of $\mathcal{E}$:

the N-*Axiom of Choice* :
$\forall R \subset \mathsf{N} \times X\ [\forall n \in \mathsf{N}\ \exists x \in X.\ \langle n, x\rangle \in R \longrightarrow \exists f \in X^{\mathsf{N}}\ \forall n \in \mathsf{N}.\ \langle x, f(x)\rangle \in R]$

Church's Thesis :
$\forall f \in \mathsf{N}^{\mathsf{N}}\ \exists e \in \mathsf{N}\ \forall n \in \mathsf{N}\ \exists m \in \mathsf{N}\ [T(e, n, m)\ \wedge\ Um = fn]$,
where T is Kleene's predicate and U the output function

Markov's Principle :
$\forall R \subset \mathsf{N}\ [\forall n \in \mathsf{N}\ [n \in R\ \vee\ n \notin R]\ \wedge\ \neg\neg\exists m \in \mathsf{N}.\ m \in R \longrightarrow \exists m \in \mathsf{N}. m \in R]$.

The leading example of a topos satisfying the above is the effective topos (*cf.* Hyland [1982]).

In the topos $\mathcal{E}$ it is possible to define a dominance $\Sigma \rightarrowtail \Omega$ consisting of the *r.e. truth values* as

$$\Sigma = \{p \in \Omega \mid \exists f \in \mathsf{N}^{\mathsf{N}}\ [p \leftrightarrow \exists n \in \mathsf{N}.\ f(n) = 0]\}.$$

Apply (N-AC) to prove that $\Sigma^{\mathsf{N}} \rightarrowtail \Omega^{\mathsf{N}}$ is the subobject of r.e. subsets of N, thus comforting the idea that Σ consists of the r.e. truth values. In tune with this we shall reduce the phrase "subobject classified by Σ" to simply "Σ-subobject". Moreover we have the following

1.1 PROPOSITION Σ is a sublattice of Ω closed under N-sups. Moreover it satisfies conditions (1) and (2) above. Hence the partial maps in $\mathcal{E}$ defined on Σ-subobjects form a category $\mathcal{M}_\Sigma$-Ptl($\mathcal{E}$).

Proof. The first part is obvious. And condition (1) holds trivially. To prove (2) suppose that $p \leftrightarrow \exists n \in \mathsf{N}\ .\ f(n) = 0$ for some $f \in \mathsf{N}^{\mathsf{N}}$ and

$$\exists n \in \mathsf{N}\ .\ f(n) = 0 \longrightarrow \exists g \in \mathsf{N}^{\mathsf{N}}\ [q \leftrightarrow \exists m \in \mathsf{N}.\ g(m) = 0].$$

This is equivalent to

$$\forall n \in \mathsf{N}\ \exists g \in \mathsf{N}^{\mathsf{N}}\ [f(n) = 0 \longrightarrow [q \leftrightarrow \exists m \in \mathsf{N}.\ g(m) = 0]]$$

since "$f(n) = 0$" is decidable. Apply (N-AC) to get an $h \in \mathsf{N}^{\mathsf{N}\times\mathsf{N}}$ such that

$$\forall n \in \mathsf{N}\ [f(n) = 0 \longrightarrow [q \leftrightarrow \exists m \in \mathsf{N}.\ h(m,n) = 0]]$$

from which it follows that

$$p \wedge q \longleftrightarrow \exists n, m \in \mathsf{N}.\ f(n) + h(n,m) = 0$$

proving that $p \wedge q \in \Sigma$. □

By simple logical means it is possible to define the object $[X \rightharpoonup Y]$ of partial maps from X to Y defined on a Σ-subset as

$$\{\phi \in \tilde{Y}^X \mid \mathrm{dom}\,\phi \in \Sigma^X\} \rightarrowtail \tilde{Y}^X.$$

By (CT) it follows from the previous remark about Σ^{N} that $[\mathsf{N} \rightharpoonup \mathsf{N}]$ is the object of partial recursive functions on N.

2 The category of σ-sets

The category-theoretic approach to computability for a definition of a suitable category of "effective" objects dates back to the work of Eršov [1973]. There the description was too complicated for category theory to be of substantial help. The object $\Omega_{\mathrm{r.e.}}$ defined by Mulry is not used in order to develop arguments about partial recursive maps. The first time we saw an internal definition of Σ was at a lecture delivered by Dana Scott in Pisa in 1982. The major stress then was on the effective topos as a general ground from which to renew the attack to "effective objects". It was when we proved a characterisation of Σ-subsets of effective algebraic posets in terms of their Scott topology that he suggested to look at those objects which are completely determined by their Σ-subsets, and pointed out the notion of *σ-set*.

In order to define σ-sets and discuss some of their categorical properties it is useful to study a certain adjunction induced by Σ. To do this we need to introduce the category H of σ-algebras in $\mathcal{E}$. Define a *σ-algebra* $(H, \leq)$ as a partially ordered object with finite meets and N-joins satisfying

$$s \wedge \bigvee_n t_n = \bigvee_n (s \wedge t_n)$$

and a *σ-homomorphism* between σ-algebras as a map which preserves the operations. Let H be the category of σ-algebras and σ-homomorphisms. Clearly Σ is a σ-algebra.

2.1 THEOREM There is an adjunction of categories $\Sigma^{(-)} \dashv \mathrm{Hom}(-, \Sigma): \mathsf{H}^{\mathrm{op}} \to \mathcal{E}$.

Proof. Is straightforward. A homomorphism $f: H \to \Sigma^X$ is completely determined by the request on its adjoint $f': X \times H \to \Sigma$ that it satisfies

$$\forall x \in X\ [f'(s,x)\ \wedge\ f'(t,x) \leftrightarrow f'(s \wedge t, x)]$$
$$\forall x \in X\ [\exists n \in \mathbb{N}.\ f'(t_n, x) \leftrightarrow f'(\bigvee_n t_n, x)].$$

In turn the two conditions state that the other adjoint $f'': X \to \Sigma^X$ takes values into $\mathrm{Hom}(H, \Sigma)$. The assertion is now obvious. □

NOTATION We write $[X]$ for Σ^X and also $[H]$ for $\mathrm{Hom}(H, \Sigma)$ making sure that the context will always make clear which of the two function spaces is intended. We also write $U \subset_\Sigma X$ to mean $U \in \Sigma^X$ as well as $A \rightarrowtail_\Sigma X$ when $A \rightarrowtail X$ is classified by Σ. Finally recall that an object X is *countable* if there is an epimorphism $\mathbb{N} \twoheadrightarrow X + 1$—this amounts to X being the image of a decidable (=recursive) subset of N. We may use the same notion in the internal logic for subsets $U \in \Omega^X$ with the obvious meaning.

From the adjunction it follows that there are canonical maps from X into its dual $[[X]]$ given by

$$\sigma_X: x \mapsto \{U \subset_\Sigma X \mid x \in U\}: X \to [[X]]$$

and similarly from a σ-algebra H into $[[H]]$ defined by

$$s_H: h \mapsto \{F \in \mathrm{Hom}(H, \Sigma) \mid F(h) = \top\}: H \to [[H]].$$

DEFINITION An object X in $\mathcal{E}$ is a *σ-set* if the map $\sigma_X: X \to [[X]]$ is an isomorphism.

By definition one has $[[X]] \rightarrowtail \Sigma^{[X]}$, and this induces a partial order on $[[X]]$; moreover the join in $\Sigma^{[X]}$ of any N-chain of σ-homomorphisms is a σ-homomorphism. Therefore in case X is a σ-set there is a canonical partial order defined on it which has sups of N-chains. This seems to support the idea that in a suitably powerful intuitionistic universe there are special objects which carry a canonical effective structure. Notice that it follows immediately that *any* map between σ-sets in the topos is monotone and preserves sups of N-chains so that we can take the full subcategory of $\mathcal{E}$ on the σ-sets and be sure that functions have nice continuity properties.

A large family of σ-sets is supplied by the *effective algebraic cpo's* in the topos: to waste no time we stick to the simplest presentation from our point of view and leave to the reader to check that, under the prescribed assumptions on N, in the topos $\mathcal{E}$ our definition includes any other similar in the extant literature. Given a poset $(P, \sqsubseteq)$ let $\mathrm{Idl}(P)$ be the object of countable ideals of P (= countable, downward closed, upward directed subsets). An *effective algebraic cpo* (or, simply, *algebraic cpo*) is a poset (isomorphic to one) of the form $\mathrm{Idl}(P)$ with P countable and such that $\sqsubseteq \rightarrowtail_\Sigma P^2$. Clearly $\mathrm{Idl}(P)$ is closed under unions of N-chains.

In order to treat Σ-families of Σ-subsets we need a lemma which can be proved by a constructive Rice-Shapiro argument. It is convenient to have a notation like $\mathrm{Up}_\Sigma(P)$ for the object of upward closed Σ-subsets of P (= monotonic maps into Σ).

2.2 LEMMA Suppose $(P, \sqsubseteq)$ is a poset and has sups of N-chains. Given $U \subset_\Sigma P$ and an N-chain (x_n) in P, one has

$$\bigsqcup_n x_n \in U \longleftrightarrow \exists n \in \mathsf{N}.\ x_n \in U.$$

Proof. Suppose $x = \bigsqcup_n x_n \in U$ and consider the following family of chains:

$$c_n^{(m)} = \begin{cases} x_n & \forall z \leq n.\ \neg T(m,m,z) \\ x_{n'} & \exists z \leq n.\ T(m,m,z) \ \wedge \ \text{"}n'\text{ is the first such }z\text{"} \end{cases}$$

which is well-defined because the clauses are exhaustive and disjoint. Taking sups of chains defines a map $h : m \mapsto \bigsqcup_n c_n^{(m)} : \mathsf{N} \to P$ and

$$\{m\}(m)\uparrow \longrightarrow h(m) = x$$
$$\{m\}(m)\downarrow \longrightarrow \exists n \in \mathsf{N}.\ h(m) = x_n.$$

If $\neg\exists n \in \mathsf{N}.\ x_n \in U$, then $h^{-1}(U)$ is the complement of the halting set and a Σ-subset as inverse image of one such. This is a contradiction, thus $\neg\neg\exists n \in \mathsf{N}.\ x_n \in U$, and the conclusion follows by (MP). The converse is proved similarly. □

Finally we can give a family of examples of σ-sets.

2.3 THEOREM Suppose P is a countable poset such that $\sqsubseteq \rightarrowtail_\Sigma P^2$. Then the following hold in $\mathcal{E}$ for $\mathrm{Idl}(P)$

(i) the σ-algebra $[\mathrm{Idl}(P)]$ is isomorphic to $\mathrm{Up}_\Sigma(P)$,

(ii) $\mathrm{Idl}(P) \cong \mathrm{Hom}(\mathrm{Up}_\Sigma(P), \Sigma)$,

(iii) $\mathrm{Idl}(P)$ is a σ-set.

Proof. (i) Follows from Lemma 2.2 since every ideal is the union of a suitable chain of principal ideals by means of (N-AC). (ii) Another application of the Lemma. (iii) Follows immediately from (i) and (ii). □

Note that 2.3 can be read as stating that there are objects of the topos $\mathcal{E}$ which come endowed with an effective structure by their own definitional nature. Also the theorem provides us with many natural examples of σ-sets: all basic partial recursive types such as N, [N ⇀ 1] ($\cong \Sigma^{\mathsf{N}}$), [1 ⇀ N], [N ⇀ N] are σ-sets. We shall see later that all partial types are σ-sets as it will follow from general properties.

In case $\mathcal{E}$ is the effective topos, it is easy to prove that the category of effective cpo's as defined in Scott [1982] with effective continuous functions is a full subcategory of σ-sets extending a result about effective information systems of McCarty [1984]. The proof is achieved by using the results in Hyland [1982] about effective objects and Lemma 2.2 above.

2.4 THEOREM   The category of effective cpo's and effective continuous functions is a full subcategory of the category of σ-sets in the effective topos.

3 Properties of σ-sets

In this sections we shall extend some of the basic constructions on cpo's to σ-sets using the internal set-theory and exploiting the fact that the order structure on the σ-sets is naturally inherited by their very effective nature.

In the following proposition we give some closure and algebraic properties of σ-sets. We should notice that, although simple, none of these follows from abstract considerations about the adjunction in 2.1.

3.1 PROPOSITION   Suppose X is a σ-set. Then the following hold

(i) X is a $\neg\neg$-separated object of $\mathcal{E}$,

(ii) if $A \rightarrowtail_{\Sigma} X$, then A is a σ-set,

(iii) if $R \rightarrowtail X$ is a retract, then R is a σ-set.

Proof. (i) By definition $X \rightarrowtail \Sigma^{(\Sigma^X)}$: Σ is $\neg\neg$-separated, and so is any of its powers. (ii) The inclusion $A \rightarrowtail_{\Sigma} X$ induces a retraction pair $\Sigma^A \overset{\alpha}{\leftrightarrows} \Sigma^X$ with α a surjective homomorphism. Thus, given any σ-homomorphism $F : \Sigma^A \to \Sigma$, the composite $F\alpha$ is $F\alpha(U) = U(x)$ for a unique $x \in X$. But $\top = F(A) = F\alpha(A) = A(x)$. Hence x is in A and A is a σ-set. (ii) Similar to the previous one. □

For the next proposition we need to recall the notion of an *internal A-indexed family* as simply a map $f : Z \to A$. To see it as a family of objects is a straightforward generalisation from the case of the topos S of sets where f is completely determined by its fibres $(f^{-1}\{a\})_{a\in A}$. Indeed trying to maintain the similarity as close as possible we shall employ the same notation in the topos $\mathcal{E}$ as a shorthand.

3.2 PROPOSITION   Suppose $(X_n)_{n\in\mathsf{N}}$ is a internal family of σ-sets. Then their sum $\coprod_n X_n$ is a σ-set.

Proof. All $X_m \subset \coprod_n X_n$ are decidable, hence Σ. So any homomorphism $F : \coprod_n X_n \to \Sigma$ has value $\top$ on exactly one X_m. The conclusion is now easy. □

The main result of this section is that given an internal family of σ-sets $(X_n)_{n \in \mathsf{N}}$ and projections the limit of the sequence of retractions is isomorphic to the colimit of the sequence of injections and is a σ-set. It can be seen as a constructive version of the main theorem of Plotkin & Smyth [1982] for the sequence is *internally* indexed by N. The result can be used to define a subclass of "profinite" σ-sets which would then be closed under products and exponentials, and to generate solutions of fixed-point domain equations.

3.3 THEOREM Suppose

$$X_0 \underset{i_0}{\overset{p_0}{\leftrightarrows}} X_1 \underset{i_1}{\overset{p_1}{\leftrightarrows}} X_2 \underset{i_2}{\overset{p_2}{\leftrightarrows}} \dots \underset{i_{n-1}}{\overset{p_{n-1}}{\leftrightarrows}} X_n \underset{i_n}{\overset{p_n}{\leftrightarrows}} \dots$$

is a sequence of σ-sets and retraction pair satisfying the conditions $p_n i_n = \mathrm{id}$ and $i_n p_n \sqsubseteq \mathrm{id}$ with respect to the canonical order of the σ-sets. Then

$$\varprojlim_n (X_n, p_n) \cong [[\varinjlim_n (X_n, i_n)]]$$

and they are both σ-sets.

Proof. An element of the inverse limit is a sequence (x_n) such that $x_n \in X_n$ and $x_n = p_n(x_{n+1})$. The canonical maps $\pi_n : \varprojlim X_n \to X_n$ from the limits are defined by projection. Clearly there is an order on $\varprojlim X_n$ induced pointwise by the canonical orders on the X_n's, and $\varprojlim X_n$ has sups of N-chains with respect to that order. Each projection π_n has a left inverse $\iota_n : X_n \to \varprojlim X_n$ given by

$$\iota_n(x) = \langle \dots, p_{n-2}(p_{n-1}(x)), p_{n-1}(x), x, i_n(x), \dots \rangle.$$

To complete the proof we need a lemma.

3.4 LEMMA With the notation introduced above

(i) $\iota_n \pi_n \sqsubseteq \mathrm{id}$ for each n,

(ii) every z in $\varprojlim X_n$ is a sup of a chain $(\iota_n \pi_n(z))_{n \in \mathsf{N}}$.

Proof. (i) If $m \le n$ then the mth components of z and $\iota_n \pi_n(z)$ coincide as $z_m = p_m \circ \dots \circ p_{n_1}(z_n)$. If $m > n$ then $z_m = p_n \circ \dots \circ p_m(z_m)$. In this case the mth component of $\iota_n \pi_n(z_n)$ is equal to

$$i_m \circ \dots \circ i_n(z_n) = i_m \circ \dots \circ i_n \circ p_n \circ \dots \circ p_m(z_m) \sqsubseteq z_m.$$

(ii) In the proof of (i) we showed that $\iota_n \pi_n(z)$ agreees with z on its first n components. The conclusion now follows easily. □

By 2.2 and (ii) in the lemma,

$$z \in U \longleftrightarrow \exists n \in \mathsf{N}.\ \iota_n \pi_n(z) \in U$$

for z in $\varprojlim_n X_n$ and $U \subset_\Sigma \varprojlim_n X_n$. Thus $U = \bigcup_n \pi_n^{-1}\iota_n^{-1}(U)$, and from this it is easy to see that $[\varprojlim_n X_n]$ is a limit for the sequence

$$\ldots \xrightarrow{[i_n]} [X_n] \xrightarrow{[i_{n-1}]} \ldots \xrightarrow{[i_1]} [X_1] \xrightarrow{[i_0]} [X_0].$$

Hence $[\varprojlim X_n] \cong \varprojlim [X_n] \cong [\varinjlim X_n]$ so that

$$[[\varprojlim_n X_n]] \cong [[\varinjlim_n X_n]].$$

Finally an argument like that in Plotkin & Smyth [1982] shows that $\varprojlim[X_n] \cong \varinjlim[X_n]$. Hence

$$\varprojlim[[X_n]] \cong [\varinjlim[X_n]] \cong [\varprojlim[X_n]] \cong [[\varinjlim X_n]] \cong [[\varprojlim X_n]],$$

and this yields the assertion. □

4 Scott domains as σ-sets

Notice that in general nothing practical is known about the collection $[X]$ of Σ-subsets of an arbitrary σ-set X although we have property 2.2. Rather, because of this, singletons are very seldom Σ-subsets. So for X a fixed σ-set consider the canonical order $\sqsubseteq$ induced by the isomorphism, and for $b \in X$ let $\uparrow(b) = \{x \in X \mid b \sqsubseteq x\}$: call b *basic* if $\uparrow(b) \subset_\Sigma X$. Notice then that a basic element b is compact, in the sense that any countable directed set in X whose sup is above b has already one element above it. Theorem 2.3 can be improved to give a characterisation of algebraic cpo's in $\mathcal{E}$ in terms of their σ-set structure.

4.1 PROPOSITION Suppose X is a σ-set and let B be the subset of basic elements of X. Then X is an algebraic cpo if and only if B is countable and every element of X is the sup of a chain in B.

Proof. Suppose $X \cong \mathrm{Idl}(P)$ is an algebraic cpo. Every ideal in X is the union of an $\mathbb{N}$-chain of principal ideals. So $b \in X$ is basic if and only if it is a principal ideal. Hence $B \cong P$ is countable, and every element in X is a union of an $\mathbb{N}$-chain in B. Conversely, by definition the order on B is Σ. Moreover consider the map

$$x \mapsto \{b \in B \mid b \sqsubseteq x\} : X \to \mathrm{Idl}(B)$$

which is well-defined because every x is a sup of a chain in B. It is then easy to prove that it is iso. Hence the assertion follows. □

If every element of the σ-set X is the sup of an $\mathbb{N}$-chain of basic elements, we shall say that B is a *basis* for X (indeed unique up to isomorphism). Closure properties of countable cpo's can be obtained as an easy corollary to 4.1.

4.2 COROLLARY Suppose X, Y and Z_n ($n \in \mathbb{N}$), are σ-sets with countable bases. Then

(i) the product $X \times Y$ and the sum $\coprod_n Z_n$ are σ-sets with countable bases;

(ii) if $U \rightarrowtail_\Sigma X$, then U is a σ-set with countable basis;

(iii) if $A \overset{i}{\underset{p}{\leftrightarrows}} X$ is a retraction such that $ip \sqsubseteq \mathrm{id}$, then A is a σ-set;

(iv) if the basis on Y is an upper-semilattice, then the partial function space $[X \rightharpoonup Y]$ is a σ-set with countable basis;

(v) $\Sigma^X \cong [X \rightharpoonup 1]$ is a σ-set with a countable basis.

Exponentials are still missing from the picture. We think that partial function spaces are more appropriate to work with computations than total function spaces. It is possible to prove that, given σ-sets X and Y with countable bases B and C respectively, the partial functions $[X \rightharpoonup Y]$ form a σ-set when the order relation on B is decidable (= Δ-classified) and B and C are bounded complete in the sense that every pair of compatible elements has a join. Call the σ-set X a *σ-domain* if it has a bounded complete basis with a decidable order.

4.3 THEOREM The σ-domains are closed under all finite products, $\mathbb{N}$-coproducts, partial function spaces and limits of projections. Moreover, a σ-domain X has a fixed-point operator $F:[X \rightharpoonup X] \to X$ if and only if X has a least element, in which case a least fixed-point operator is definable.

Proof. The first part is easy. As to the second, recall that a fixed-point operator on the partial functions is a total functional F such that

$$\forall \phi \in [X \rightharpoonup X]\ [F(\phi)\downarrow \longrightarrow \phi(F(\phi)) = F(\phi)].$$

Consider then the map $r:[1 \rightharpoonup X] \to X$ which takes ς to $F(\varsigma \circ !)$ where $!$ is the unique (total) map into the terminal object. This is a total left inverse to the inclusion $X \rightarrowtail [1 \rightharpoonup X]$. As they are both σ-sets, the surjection r is monotone increasing. Hence the image under r of the least element in $[1 \rightharpoonup X]$ is the least element in X. The usual least-fixed-point operator is thus definable as all maps preserves sup of $\mathbb{N}$-chains. □

This theorem provides us with more examples of σ-sets in the hierarchy of the effective partial functionals. As 1 and $\mathbb{N}$ are σ-domains, all the hierarchy of partial functionals consists of σ-domains. And to define it one just has to use products and partial function spaces in the topos with no further inductive conditions to require (remember that the universe has some sort of effectiveness built in). In his D.Phil. thesis the author showed how arguments about the hierarchy of functionals can be extremely simplified by this.

So it looks like the σ-domains form a nice, simple basis for a synthetic theory of effective computability. Indeed one should note that all we have done was performed in a topos which is a universe of sets and the σ-sets were not structured sets there, but simply sets. Because a notion of recursiveness is inherent to the universe itself, these come endowed with a further canonical structure and one just has to recognise it, not to impose it.

We should also recall that a particular case of our work is that of the effective Scott domains (the embedding of the effective cpo's and continuous functions into the full subcategory of σ-sets

in the effective topos preserves all the constructions we considered). The synthetic approach has the great advantage of freeing us completely from the burdensome problem of dealing with enumerations all the time. In particular, it should sufficiently clear how to extend the results of this last section to retracts of σ-domains and how to treat effective sfp-domains.

REFERENCES

CURIEN, P.-L. & OBTUŁOWICZ, A.

[1986] *Partiality and cartesian closedness*, typescript, 1986

DIPAOLA, R. & HELLER, A.

[1986] *Dominical categories*, to appear in Journ. Symb. Logic, 1986

EILENBERG, S. & KELLY, G.M.

[1966] *Closed categories*, in **Proceedings of the Conference on Categorical Algebra** (edited by S. Eilenberg, D.K. Harrison, S. MacLane & H. Röhrl), Springer-Verlag, Berlin (1966) 421-562

ERŠOV, JU.L.

[1973] *Theorie der Numerierungen I*, in Zeitschrift für Math. Log. (4) **19** (1973) 289-388

HELLER, A.

[1985] *Dominical categories and recursion theory*, in **Atti della Scuola di Logica 2**, Università di Siena (1985) 339-344

HOEHNKE, H.J.

[1977] *On partial algebras*, in Col. Math. Soc. J. Bolyai **29** (1977) 373-412

HYLAND, J.M.E.

[1982] *The effective topos*, in **The L.E.J. Brouwer Centenary Symposium** (edited by A.S. Troelstra & D. van Dalen), North-Holland Publishing Company, Amsterdam (1982) 165-216

LONGO, G. & MOGGI, E.

[1984] *Cartesian closed categories and partial morphisms for effective type structures*, in **International Symposium on Semantics of Data Types** (edited by G. Kahn, D.B. McQueen & G. Plotkin), Lecture Notes in Computer Science 173, Springer-Verlag, Berlin (1984) 235-255

MCCARTY, D.C.

[1984] **Realizability and Recursive Mathematics**, D.Phil. thesis, University of Oxford, 1984

MULRY, P.

[1981] *Generalised Banach-Mazur functionals in the topos of recursive sets*, in J. Pure Appl. Alg. **26** (1981) 71-83

OBTUŁOWICZ, A.

[1986] *The logic of categories of partial functions and its applications*, in Diss. Math. **141** (1986)

PLOTKIN, G.

[1985] **Denotational Semantics with Partial Functions**, Lectures at the C.S.L.I. Summer School, Stanford, July1985

PLOTKIN, G.D. & SMYTH, M.B.

[1982] *The category-theoretic solution of recursive domain equations*, in SIAM J. Comp. **11** (1982) 761-783

ROBINSON, E.P. & ROSOLINI, G.

[1986] **Categories of partial maps**, Quaderno del Dipartimento di Matematica 18, Università di Parma, 1986

ROSOLINI, G.

[1986] **Continuity and effectiveness in topoi**, D.Phil. thesis, University of Oxford, 1986

SCOTT, D.S.

[1982] *Domains for denotational sematics*, in **Automata, Languages and Programming**, Ninth Colloquium, Aarhus, Denmark (edited by M. Nielsen & E.M. Schmidt), Lecture Notes in Computer Science 140, Springer-Verlag, Berlin (1982) 677-718

Polymorphism is Set Theoretic, Constructively

A.M.Pitts
Mathematics Division, University of Sussex,
Brighton BN1 9QH, England.

0. Introduction

The title of this paper ought really to be "Polymorphism *can* be set theoretic, constructively", but the obvious reference to Reynolds' paper "Polymorphism is not set theoretic" [R2] was too tempting. The purpose of the paper is to prove particular kinds of completeness and full embedding theorems for the polymorphic lambda calculus. In order to explain the theorems we need to review some of the recent history of type polymorphism:

In *loc.cit.*, it is shown that the standard interpretation of the first order typed lambda calculus in the category of sets cannot be extended to a model of the second order typed lambda calculus. Reynolds works with quite a general notion of what constitutes a second order model extending a standard first order model, but the naive idea would be to have a set U of sets closed under finite cartesian products, exponentiation and U-indexed cartesian products: the types would be interpreted by sets which are elements of U, with function types and polymorphic product types *both* being interpreted in a standard way using set-theoretic exponentiation and product. Putting this in more category-theoretic terms, one would have a small category **U** and a full and faithful functor from **U** to the category of sets, $G:\mathbf{U}\hookrightarrow\mathbf{Set}$, so that **U** has and G preserves finite products, exponentiation and products indexed by the set U of objects of **U**. A simple cardinality argument shows that *there are no non-trivial such* **U**,G (in the sense that any such U can only contain sets with at most one element). If one drops the requirement of having the structure preserving full embedding into **Set**, there are of course many, highly non-trivial, small cartesian closed categories **U** with the above limit closure properties — for example Lambek's *C-monoids*, studied in Part I of [LS], correspond to cartesian closed **U** with just two objects (a terminal object $\top$ and an object $X\cong(X\times X)\cong(X\to X)$), and which therefore certainly have U-indexed products. All such **U** give rise to models of the second order typed lambda calculus, but not ones extending the standard set theoretic interpretation of the first order calculus.

Recent work of Hyland, Moggi, Robinson, Rosolini [HRR], Carboni, Freyd and Scedrov [CFS] shows that this non-existence of standard models is due to the non-constructive nature of the category of sets. They have demonstrated that it is possible for elementary toposes (which in general model a particular kind of higher order constructive logic — see [LS]) to contain *internal full subcategories* whch are closed under the operations of taking *any* limits or colimits in the topos (and so in particular also are closed under taking exponentials since these are given by powers, i.e. by internal products). The particular topos they consider is Hyland's *effective topos* [H] and its internal small full subcategory of *subcountable* objects, i.e. those objects which are the quotient of the natural numbers object by a partial equivalence relation. (The related internal subcategory given by $\neg\neg$-*closed* partial equivalence relations also has these strong closure properties in the effective topos; the objects in this subcategory have been termed the *modest sets* by Dana Scott. See also [FS].)

Thus the effective topos is a model of higher order constructive logic in which there is a "set" U of "sets" whose elements are closed under all the operations (indeed, far more) needed to model the second order typed lambda calculus in a completely standard way: the polymorphic types (of one free type variable) are interpreted as arbitrary U-indexed collections of "sets" in U; function types are interpreted as full function exponentials in the topos; and the product type of a polymorphic type is interpreted as the actual product in the topos of the corresponding indexed collection (which product is again in U despite the size of the indexing "set"). There is no contradiction with Reynolds' result since in the effective topos (and in toposes in general) the classically valid Law of Excluded Middle needed to carry out his argument, is not valid.

Generalizing the naive idea of a standard set theoretic model of the second order typed lambda calculus discussed above, by a *topos model* we shall mean the following: a topos $\mathbf{E}$ equipped with an internal category $\mathbf{U}$ and a full and faithful diagram G of type $\mathbf{U}$ in $\mathbf{E}$, so that $\mathbf{U}$ has and G preserves finite products, exponentials and products indexed by the object of objects of $\mathbf{U}$. (We will spell out explicitly what such data amount to in section 3.) Much as indicated above, these properties of the internal subcategory are just what are needed to interpret the language of the second order typed lambda calculus (by which we shall mean Girard's system F [G1, G2] augmented with finite product types: see section 1), in such a way that the equations of β-conversion, η-conversion and surjective pairing and the ξ-rules of extensionality are always satisfied. Conversely we shall prove that there are enough topos models so that ***the only equational consequences of these axioms and rules are those which are satisfied by all topos models.***

More generally one can consider *$2T\lambda C$-theories* $\mathbf{T}=(L,A)$, consisting of a suitable language L (with symbols for constant types, type constructors and constant terms) and a set A of equations between terms in the second order typed lambda calculus built up from the language. A *topos model* of **T** will then mean a topos model in the sense of the previous paragraph, together with an interpretation of the language with the property that all the equations of A are satisfied. (Precise definitions of the notions of theory and satisfaction are given in sections 1 and 2.) We prove:

Theorem B. (Completeness of topos models.) *Let* **T**=(L,A) *be a* $2T\lambda C$*-theory. The equations between second order typed lambda calculus terms built up from* L *which are provable consequences in equational logic of the axioms* A *of* **T** *together with the axioms of* β*-conversion,* η*-conversion, surjective pairing and the* ξ*-rules, are just those satisfied by all topos models of* **T**. *In fact for each* **T**, *there is a* single *topos model whose valid equations are exactly those derivable in* **T**.

In fact our proof of this theorem is via an even stronger result which shows there are very many topos models:

Theorem A. (Full embedding in topos models.) *Every* hyperdoctrine model *of the second order typed lambda calculus fully embeds into a topos model.*

The category theoretic notion of *hyperdoctrine* was introduced by Lawvere [L] in his seminal work on the connexions between categories and logic. The correspondence between theories in a higher order (extensional) typed lambda calculus and an appropriate kind of hyperdoctrine has been developed by R.A.G.Seely [Se]. This correspondence readily specializes to one between $2T\lambda C$-theories and the kind of hyperdoctrines described in section 2 below. The passage from theory to hyperdoctrine is the familiar term-model type construction in categorical logic of organizing the syntax, suitably quotiented by provability in the theory, into a category with suitable extra categorical structure or properties. In this way Theorem B is deduced from Theorem A.

The proof of Theorem A proceeds by first using the *Grothendieck construction* to obtain the total category of the fibration corresponding to the given hyperdoctrine; and then one takes the functor category of contravariant set-valued functors on this total category. Thus the topos constructed is simply a *presheaf topos* and the original hyperdoctrine embeds into it essentially via the Yoneda embedding. The method of proof is therefore very similar to that used by Dana Scott [Sc1] in showing that a model of the untyped λ-calculus can be realized as a reflexive "set" inside a presheaf topos. The fact that the Yoneda embedding preserves any existing exponentials is a key ingredient of his proof, and here we use this and more — namely that the Yoneda embedding preserves any existing *local* exponentials and instances of right adjoints to pulling back. Part of the data which specifies the original hyperdoctrine naturally gives rise to an internal full subcategory in the constructed presheaf topos. The main part of the proof of Theorem A then resides in showing that this has the requisite limit closure properties.

The significance of Theorem A is that, *as long as our arguments can be carried out in intuitionistic higher order logic* (for a description of which see [LS, Part I] for example), *then we may reason about the second order typed lambda calculus as though polymorphic types were suitably indexed collections of sets, product types of polymorphic types were actual cartesian products of sets, function types were full exponentials of sets, etc.* Our method of proof in fact extends to the case of *higher order* typed lambda calculus (optionally augmented with *sum types* of polymorphic types). The corresponding kind of hyperdoctrine has a cartesian closed base category (optionally augmented with stable left adjoints to substitution along projections): see [Se]. The topos model constructed in the proof of Theorem A will in this case have the further property that the internal subcategory is closed in the topos under products indexed by any finite type on its object of objects (and optionally closed under similarly indexed coproducts). It is important to note that these closure properties are still much less than those enjoyed by the internal categories of the effective topos mentioned above: they are closed with respect to *all* internal limits and (hence) colimits. The question of whether Theorems A and B can be strengthened by restricting to such topos models is not addressed here.

Having read the above description of what will be proved in this paper, one might wonder why it was not entitled "Polymorphism is topos theoretic". What have these results to do with *constructive set theory* ? The kind of intuitionistic higher order logic which toposes model can certainly be regarded as a constructive theory of sets. But if one wishes to model the *untyped* ($\in$,=)-language of set theory, then elementary toposes in general are capable of modelling only a restricted form of Zermelo-Fraenkel set theory, with *bounded* separation and collection axioms: see [J, Chapter 9]. However, the toposes which arise in the proofs of Theorems A and B are *Grothendieck* toposes, and these M.Fourman [F] has shown admit the interpretation of a full intuitionistic Zermelo-Fraenkel set theory with atoms, *IZFA*. (Indeed, the toposes considered here are just *presheaf* toposes, for which the Fourman interpretation of *IZFA* takes a particularly simple form, as Dana Scott shows in [Sc2].) Thus Theorems A and B could be rephased in terms of models of the second order typed lambda calculus in *IZFA*. (It is interesting to note that even though Hyland's effective topos is definitely not a Grothendieck topos, nevertheless it does support an interpretation of *IZFA* — one that is entirely analogous to the standard Heyting-valued models. The resulting model of *IZFA* is presented by C.McCarty in [McC].)

Acknowledgements. I have had many helpful discussions about type polymorphism with J.M.E.Hyland, E.Robinson, G.Rosolini and R.A.G.Seely (who also suggested the title of this paper). My understanding of fibred category theory in general and internal full subcategories in particular has been very greatly enhanced by the work of J.Bénabou in this area. This research was partly carried out while visiting the Centre en Etudes Catégoriques in Montréal. I also gratefully acknowledge the financial support of the Royal Society.

1 Second order equational theories

In this section we review the syntax of equational theories in a system, $2T\lambda C$, of ***second order typed lambda calculus***. The type system is that of the Girard-Reynolds polymorphic lambda calculus [G1, G2, R1, R2], augmented with terminal and binary product types. (There are several good reasons for including these extra, first order types, not least of which is their presence in cartesian closed categories.) The equational logic we consider over the type system contains axioms for β-conversion, η-conversion, extensionality and surjective pairing.

1.1. Types. Let us fix a countably infinite set TV whose elements we shall call ***type variables*** and denote by $X,Y,Z,\ldots$. Given a multiset $ar:C \longrightarrow \mathbb{N}$, the set $Types(C)$ of $2T\lambda C$ ***types over*** C (denoted $\Phi,\Psi,\ldots$) is defined inductively by the clauses given below; simultaneously we define the finite subset $FTV(\Phi) \subseteq TV(C)$ of ***free type variables*** of a type Φ:

- if $X \in TV$, then $X \in Types(C)$ and $FTV(X)=\{X\}$;
- if $F \in C$ with $ar(F)=n$ say, and if $\Phi_0,\ldots,\Phi_{n-1} \in Types(C)$, then $F(\Phi_0,\ldots,\Phi_{n-1}) \in Types(C)$ and $FTV(F(\Phi_0,\ldots,\Phi_{n-1}))=FTV(\Phi_0)\cup\ldots\cup FTV(\Phi_{n-1})$;
- $\mathrm{T} \in Types(C)$ and $FTV(\mathrm{T})=\emptyset$;
- if $\Phi,\Psi \in Types(C)$, then for $* = \times$ or $\rightarrow$, $(\Phi * \Psi) \in Types(C)$ and $FTV(\Phi * \Psi)=FTV(\Phi)\cup FTV(\Psi)$;
- if $\Phi \in Types(C)$ and $X \in TV$, then $\prod X.\Phi \in Types(C)$ and $FTV(\prod X.\Phi)=FTV(\Phi)\setminus\{X\}$.

If $F \in C$ and $ar(F)=n$, we shall call F a ***type constructor symbol of arity*** n; in the special when case $n=0$, F is a ***type constant*** and we write F rather than $F()$ for the corresponding type. A type Φ is *closed* if $FTV(\Phi)=\emptyset$. The type variables occurring in a type Φ which are not in $FTV(\Phi)$ are the ***bound*** type variables of Φ (X is bound in $\prod X.\Phi$). As usual, we identify types up to α-equivalence, i.e. up to change of bound type variables. If $\Phi,\Psi \in Types(C)$ and $X \in FTV(\Phi)$, then $[X:=\Psi]\Phi$ will denote the result (defined up to α-equivalence) of substituting Ψ for X throughout Φ, avoiding variable capture.

1.2. Terms. Fix a countably infinite set IV of *individual variables* $x,y,z,...$. Elements of the product $IV\times Types(C)$ will called *typed* individual variables and a typical such pair will be denoted x^{Φ}. By a *language* L we will mean a pair $L=(C,K)$, where C is a multiset of type constructor symbols and K is a set of *individual constants* $(a,b,c,...)$ together with a specification of their types $(a{:}\Phi,...)$, which we will always assume to be closed types. Given such an L, the set $Terms(L)$ of $2T\lambda C$ *terms* over L (denoted $s,t,...$) is defined inductively by the clauses given below; simultaneously we define the type of each term (denoted $s{:}\Phi$) and the finite subsets $FIV(s)$, $FTV(s)$ of *free individual and type variables* of a term s:

- if $x^{\Phi}\in IV\times Types(C)$, then $x^{\Phi}\in Terms(L)$ with $x^{\Phi}{:}\Phi$, $FIV(x^{\Phi})=\{x^{\Phi}\}$ and $FTV(x^{\Phi})=FTV(\Phi)$;
- if $a\in K$, then $a\in Terms(L)$ with closed type as specified and $FIV(a)=FTV(a)=\emptyset$;
- $\langle\rangle\in Terms(L)$ with $\langle\rangle{:}\mathrm{T}$ and $FIV(\langle\rangle)=FTV(\langle\rangle)=\emptyset$;
- $\mathsf{Fst}\in Terms(L)$ with $\mathsf{Fst}{:}\prod X.\prod Y.(X\times Y)\to X$ and $FIV(\mathsf{Fst})=FTV(\mathsf{Fst})=\emptyset$;
- $\mathsf{Snd}\in Terms(L)$ with $\mathsf{Snd}{:}\prod X.\prod Y.(X\times Y)\to Y$ and $FIV(\mathsf{Snd})=FTV(\mathsf{Snd})=\emptyset$;
- if $s{:}\Phi$ and $t{:}\Psi$, then $\langle s,t\rangle{:}\Phi\times\Psi$ with $FIV(st)=FIV(s)\cup FIV(t)$ and $FTV(st)=FTV(s)\cup FTV(t)$;
- if $s{:}(\Phi\to\Psi)$ and $t{:}\Phi$, then $st{:}\Psi$ with $FIV(st)=FIV(s)\cup FIV(t)$ and $FTV(st)=FTV(s)\cup FTV(t)$;
- if $s{:}\prod X.\Phi$ and $\Psi\in Types(C)$, then $s\Psi{:}[X{:=}\Psi]\Phi$ with $FIV(s\Psi)=FIV(s)$ and $FTV(s\Psi)=FTV(s)\cup FTV(\Psi)$;
- if $s{:}\Psi$, then $\lambda x^{\Phi}.s : \Phi\to\Psi$ with $FIV(\lambda x^{\Phi}.s)=FIV(s)\setminus\{x^{\Phi}\}$ and $FTV(\lambda x^{\Phi}.s)= FTV(\Phi)\cup FTV(s)$;
- if $s{:}\Phi$, then $\lambda X.s : \prod X.\Phi$ with $FIV(\lambda X.s)=FIV(s)$ and $FTV(\lambda X.s)=FTV(s)\setminus\{X\}$.

A term s is *closed* if $FIV(s)=\emptyset$. The individual and type variables occurring in s but not in the sets $FIV(s)$, $FTV(s)$ are the *bound* variables of s (x^{Φ} is bound in $\lambda x^{\Phi}.s$ and X is bound in $\lambda X.s$). As for types, we identify terms up to α-equivalence, i.e. up to change of bound variables. If $s\in Terms(L)$ and $\Psi\in Types(C)$, then $[X{:=}\Psi]s$ denotes the result of substituting Ψ for a type variable X throughout s, avoiding variable capture; evidently $[X{:=}\Psi]s : [X{:=}\Psi]\Phi$ when $s{:}\Phi$. Similarly, if $s{:}\Phi$ and $t{:}\Psi$, then $[x^{\Psi}{:=}t]s{:}\Phi$ denotes the result of substituting t for the individual variable x^{Ψ} throughout s, avoiding variable capture.

1.3. Theories. Given a language $L = (C,K)$ as above, we consider judgements which assert the equality of two $2T\lambda C$ terms of equal type. Specifically, an *equality judgement* will take the form:

$$\vdash_{\underline{X},\underline{x}} s=t{:}\Phi$$

where $\underline{X}=X_0,...,X_{n-1}$ is a finite list of distinct type variables, $\underline{x}=x_0^{\Phi_0},...,x_{m-1}^{\Phi_{m-1}}$ is a finite list of distinct individual variables of various types over C, s and t are terms over L of type Φ whose free individual variables are contained in $\underline{x}$, and such that the free type variables of s, t, Φ and the Φ_i are contained in $\underline{X}$. Then a $2T\lambda C$-*theory* $\mathbf{T}$ is specified by a language L and a set A of equality judgements over L, called the *axioms of* $\mathbf{T}$.

1.4. Equational logic over $2T\lambda C$. We next give the basic logical axioms and rules for deriving equality judgements in $2T\lambda C$. In particular, if $\mathbf{T} = (L,A)$ is a $2T\lambda C$-theory, then the *theorems* of $\mathbf{T}$ comprise the least set of equality judgements over L containing A and the axioms below, and closed under the rules given below:

- *weakening* $$\frac{\vdash_{\underline{X},\underline{x}} s=t:\Phi}{\vdash_{\underline{X}',\underline{x}'} s=t:\Phi} \quad (\underline{X}\subseteq\underline{X}' \text{ and } \underline{x}\subseteq\underline{x}')$$
- *reflexivity* $$\vdash_{\underline{X},\underline{x}} s=s:\Phi$$
- *symmetry* $$\frac{\vdash_{\underline{X},\underline{x}} s=t:\Phi}{\vdash_{\underline{X},\underline{x}} t=s:\Phi}$$
- *transitivity* $$\frac{\vdash_{\underline{X},\underline{x}} r=s:\Phi \qquad \vdash_{\underline{X},\underline{x}} s=t:\Phi}{\vdash_{\underline{X},\underline{x}} r=t:\Phi}$$
- *terminal* $$\vdash_{\langle\rangle,\langle\rangle} t=\langle\rangle:\mathrm{T}$$
- *pairing* $$\frac{\vdash_{\underline{X},\underline{x}} s=s':\Phi \qquad \vdash_{\underline{X},\underline{x}} t=t':\Psi}{\vdash_{\underline{X},\underline{x}} \langle s,t\rangle=\langle s',t'\rangle:\Phi\times\Psi}$$
- *projections* $$\vdash_{\underline{X},\underline{x}} \mathrm{Fst}\,\Phi\Psi\langle s,t\rangle=s:\Phi \quad , \qquad \vdash_{\underline{X},\underline{x}} \mathrm{Snd}\Phi\Psi\langle s,t\rangle=t:\Psi$$
- *surjectivity* $$\vdash_{\underline{X},\underline{x}} \langle \mathrm{Fst}\Phi\Psi r,\mathrm{Snd}\Phi\Psi r\rangle=r:\Phi\times\Psi$$
- *individual ξ* $$\frac{\vdash_{\underline{X},\underline{x},x^\Phi} t=t':\Psi}{\vdash_{\underline{X},\underline{x}} \lambda x^\Phi.t=\lambda x^\Phi.t':\Phi\to\Psi} \quad (x^\Phi\notin\underline{x})$$
- *individual application* $$\frac{\vdash_{\underline{X},\underline{x}} s=s':\Phi\to\Psi \quad \vdash_{\underline{X},\underline{x}} t=t':\Phi}{\vdash_{\underline{X},\underline{x}} st=s't':\Psi}$$
- *individual β* $$\vdash_{\underline{X},\underline{x}} (\lambda x^\Phi.t)s=[x:=s]t:\Psi$$
- *individual η* $$\vdash_{\underline{X},\underline{x}} \lambda x^\Phi.(tx)=t:\Phi\to\Psi \qquad (x^\Phi\notin FIV(t))$$
- *type ξ* $$\frac{\vdash_{\underline{X},X,\underline{x}} t=t':\Phi}{\vdash_{\underline{X},\underline{x}} \lambda X.t=\lambda X.t'} \quad (X\notin\underline{X})$$
- *type application* $$\frac{\vdash_{\underline{X},\underline{x}} s=s':\prod X.\Phi}{\vdash_{\underline{X},\underline{x}} s\Psi=s'\Psi:[X:=\Psi]\Phi}$$
- *type β* $$\vdash_{\underline{X},\underline{x}} (\lambda X.t)\Psi=[X:=\Psi]t:[X:=\Psi]\Phi$$
- *type η* $$\vdash_{\underline{X},\underline{x}} \lambda X.(tX)=t:\prod X.\Phi \quad (X\notin\underline{X}) \ .$$

We shall write $\mathbf{T}\vdash_{\underline{X},\underline{x}} s=t:\Phi$ to indicate that the equality judgement $\vdash_{\underline{X},\underline{x}} s=t:\Phi$ can be derived as a theorem of the $2T\lambda C$-theory $\mathbf{T}$. The reader will see from the above axioms and rules that we are considering ***extensional*** theories with ***surjective pairing*** operations. The first order part of such theories and their relationship to carstesian closed categories are exposed in [LS, Part I]. Seely [Se] considers the case of similar theories in full higher order typed lambda calculus, except that he also allows a second kind of equality judgement — namely the assertion that two *types* are (extensionally) equal. Influenced by Bénabou's observations in [B] on the rôle of equality between objects (≡types) in category theory, we have specifically left out this form of judgement as inappropriate to the intuitive notion of type we have. Rather, in the system we are considering, one can assert that particular terms give an *isomorphism* between particular types.

We conclude this section with some technical remarks about the precise form of the equality judgements in *2TλC*-theories. Because the underlying language of a theory can introduce arbitrary constant types, it is perfectly possible for a type to possess no closed terms, i.e. to be uninhabited. Accordingly the logical system presented above is not equivalent to one in which the judgements are not tagged with a list $\underline{x}$ of free individual variables. On the other hand, since there are always closed types (e.g. the terminal type $\top$), it is not essential that the judgements be tagged with a list $\underline{X}$ of free type variables: using the rules (*type* ξ) and (*type* β), one has that $\mathbf{T}\vdash_{\underline{X},\underline{x}} s{=}t{:}\Phi$ iff $\mathbf{T}\vdash_{\underline{X}',\underline{x}} s{=}t{:}\Phi$, where $\underline{X}'$ comprises *exactly* those free type variables mentioned in $\underline{x}$, s, t, and Φ, and hence need not be given explicitly. However, the chosen form of judgement is convenient for giving the category theoretic semantics of *2TλC*, to which we now turn.

2 Hyperdoctrine models

In this section we review those parts of [Se] concerned with the connexion between *2TλC*-theories and a particular variety of *hyperdoctrine*. The general notion of a hyperdoctrine was introduced by Lawvere in his seminal work on the connexions between category theory and logic; see [L] for example. It has proved to be a very flexible and useful concept, not least in making precise the fundamental observation that *theories* in many different kinds of logic can be specified in a syntax-free way as *models* of various kinds of category theoretic structure, which structure can often be viewed as a particular kind of hyperdoctrine. We refer the reader to [P] and the references there for more information about hyperdoctrines in general. The particular kind of hyperdoctrine we will be concerned with here is a special case of that considered by Seely in [Se] and called a "PL category" there. He shows that there is a correspondence between these PL categories and equational theories over the higher order typed lambda calculus: this correspondence readily specializes to one between the *2TλC*-theories defined in the previous section and the kind of hyperdoctrine defined below. We begin by fixing some notation:

2.1. Cartesian closed categories. If a category **C** has a given terminal object, it will be denoted by T. For any object A in **C**, the unique morphism from A to T will be denoted $A{:}A\longrightarrow \mathrm{T}$, whereas the identity morphism on A will be denoted 1_A. If **C** has given binary products, the product of objects A and B in **C** will be denoted by

$$A \xleftarrow{\pi_1} A\times B \xrightarrow{\pi_2} B\ .$$

For any morphisms $f{:}C\longrightarrow A$, $g{:}C\longrightarrow B$ in **C**, $\langle f,g\rangle{:}C\longrightarrow A\times B$ will denote the unique morphism with $\pi_1\circ\langle f,g\rangle=f$ and $\pi_2\circ\langle f,g\rangle=g$. For such a category **C**, the exponential of objects A and B, if it exists, will be denoted $A\to B$ and the accompanying evaluation morphism by

$$ev{:}(A\to B)\times A\longrightarrow B.$$

Thus for a morphism $f{:}C\times A\longrightarrow B$, there is a unique morphism $\overline{f}{:}C\longrightarrow(A\to B)$ with $ev\circ(\overline{f}\times 1_A)=f$. As usual, a category is called a *cartesian closed category*, or more briefly a *ccc*, if it has a terminal object, binary products and exponentials.

We shall need to consider two slightly different notions of morphism between ccc's, the second a special case of the first. As usual, a functor $F{:}\mathbf{C}\longrightarrow\mathbf{D}$ is said to *preserve terminal objects* if whenever T is terminal in **C**, then $F(T)$ is terminal in **D**. Similarly F *preserves binary products* if whenever $A\xleftarrow{p}P\xrightarrow{q}B$ is a product diagram for A and B in **C**, then $F(A)\xleftarrow{Fp}F(P)\xrightarrow{Fq}F(B)$ is one for $F(A)$ and $F(B)$ in **D**. Such a functor also *preserves exponentials* if whenever E is the exponential of B by A in **C** (via an evaluation morphism $P\xrightarrow{e}B$ with product diagram $E\xleftarrow{p}P\xrightarrow{q}A$), then $F(E)$ is the exponential of $F(B)$ by $F(A)$ in **D** (via Fe, Fp and Fq). Then if **C** and **D** are ccc's, a functor $F{:}\mathbf{C}\longrightarrow\mathbf{D}$ will be called a *morphism of ccc's* if it preserves the terminal object, binary products and exponentials. F is a *strict morphism of ccc's* if furthermore it sends the given terminal object, binary products (and projections) and exponentials (and evaluation morphisms) in **C** to the given ones in **D**. Thus for a strict morphism one has for example, that $\langle F1_A,F1_B\rangle$ is the identity on $F(A\times B)=F(A)\times F(B)$, whereas for a morphism one has only that $\langle F1_A,F1_B\rangle$ is an isomorphism.

The category of small ccc's and strict morphisms of ccc's will be denoted **Ccc**. The forgetful functor to the category of sets which takes a small ccc to its underlying set of objects will be denoted $ob{:}\mathbf{Ccc}\longrightarrow\mathbf{Set}$.

2.2. Definition. A *2TλC-hyperdoctrine* **P** is specified by:

(*i*) a small category $|\mathbf{P}|$ with terminal object and binary products;

(*ii*) a distinguished object U in $|\mathbf{P}|$ which generates the other objects, I, via finite products, i.e. each I is U^n for some $n\in\mathbb{N}$ (including the case $n=0$, when $U^0=\mathrm{T}$);

(*iii*) a contravariant functor $|\mathbf{P}|^{op}\longrightarrow\mathbf{Ccc}$, such that the composition $|\mathbf{P}|^{op}\longrightarrow\mathbf{Ccc}\xrightarrow{ob}\mathbf{Set}$ is the representable functor $|\mathbf{P}|(-,U)$; the ccc assigned to an object I in $|\mathbf{P}|$ by the functor will be denoted simply by $\mathbf{P}(I,U)$, and the strict ccc morphism assigned to $\alpha{:}I\longrightarrow J$ in $|\mathbf{P}|$ will be denoted $\alpha^*{:}\mathbf{P}(J,U)\longrightarrow\mathbf{P}(I,U)$;

(iv) for each object I of $|\mathbf{P}|$, a functor $\prod_I:\mathbf{P}(I\times U,U)\longrightarrow\mathbf{P}(I,U)$ right adjoint to the functor $\pi_I^*:\mathbf{P}(I,U)\longrightarrow\mathbf{P}(I\times U,U)$, such that these functors $\prod_I$ are natural in I, i.e. for any $\alpha:I\longrightarrow J$ $\alpha^*\circ\prod_J=\prod_I\circ(\alpha\times 1_U)^*$.

$|\mathbf{P}|$ is the *base* category of the *2TλC*-hyperdoctrine $\mathbf{P}$, and the ccc $\mathbf{P}(I,U)$ is the *fibre* over an object I in the base. Note that the identity on U, regarded as on object of $\mathbf{P}(U,U)$, is a *generic* object sense that any object A in any fibre $\mathbf{P}(I,U)$ is equal to $\alpha^*(1_U)$ for some $\alpha:I\longrightarrow U$ (namely $\alpha=A$). Note also that for *any* object U^n of $|\mathbf{P}|$, $\pi_I^*:\mathbf{P}(-,U)\longrightarrow\mathbf{P}(-\times U^n,U)$ has a natural right adjoint, *viz* $\prod_{(-)}\circ\prod_{(-)\times U}\circ\cdots\circ\prod_{(-)\times U^{n-1}}$.

The cartesian closed structure of any $\mathbf{P}(I,U)$ will be denoted by $\top_I$ $\times_I$ and $\to_I$. In particular when $I=\top$ we get:

$$`\top' =_{def} \top_\top:\top\longrightarrow U; \tag{2.1}$$

and when $I=U^2$ we get:

$$`\times' =_{def} (\pi_1)\times_{U^2}(\pi_2):U\times U\longrightarrow U \quad\text{and}\quad `\to' =_{def} (\pi_1)\to_{U^2}(\pi_2):U\times U\longrightarrow U, \tag{2.2}$$

where the $\pi_i\in\mathbf{P}(U^2,U)$ are the projections. The fact that the operations $(-)^*$ are strict ccc morphisms implies that for any object I of $|\mathbf{P}|$

$$\top_I = `\top'\circ I, \tag{2.3}$$

and for any $A,B\in\mathbf{P}(I,U)$

$$A\times_I B = `\times'\circ\langle A,B\rangle \quad\text{and}\quad A\to_I B = `\to'\circ\langle A,B\rangle. \tag{2.4}$$

In $\mathbf{P}(\top,U)$ there are morphisms

$$`\mathrm{Fst}':`\top'\longrightarrow\prod_\top\prod_{\top\times U}\big((\pi_1)\times_{\top\times U^2}(\pi_2)\to_{\top\times U^2}(\pi_1)\big), \tag{2.5}$$

$$`\mathrm{Snd}':`\top'\longrightarrow\prod_\top\prod_{\top\times U}\big((\pi_1)\times_{\top\times U^2}(\pi_2)\to_{\top\times U^2}(\pi_2)\big) \tag{2.6}$$

obtained from the two projection morphisms for $(\pi_1)\times_{\top\times U^2}(\pi_2)$ by transposing across the exponential adjunction and across the adjunctions for $\prod$. Finally, note that for any $I\in|\mathbf{P}|$, $A\in\mathbf{P}(I\times U,U)$ and $B\in\mathbf{P}(I,U)$, there is an internal product projection morphism

$$\pi_B:\prod_I A\longrightarrow A\circ\langle 1_I,B\rangle \tag{2.7}$$

in $\mathbf{P}(I,U)$ given by $\pi_B=\langle 1_I,B\rangle^*(\varepsilon_A)$ where $\varepsilon_A:\pi_I^*(\prod_I A)\longrightarrow A$ is the counit of the adjunction $\pi_I^*\dashv\prod_I$ at A.

The theory of hyperdoctrines is part of the wider theory of *fibred* categories. (See [B] and the references there.) The following remarks on particular aspects of Definition 2.2 assume some familiarity with this theory:

2.3. Remark. The kind of hyperdoctrine defined above is "stricter" than usual, in the sense that all the parts of the structure which are normally asked to commute up to (canonical) isomorphisms, are here asked to commute up to equality; in particular $\mathbf{P}$ determines a fibration over $|\mathbf{P}|$ which is *split* (and has various other properties). A laxer notion still suitable for the semantics of *2TλC*-theories, would be that of a category $\mathbf{P}$, fibred over a base category $|\mathbf{P}|$ (with finite products), with stably cartesian closed fibres, containing an object G which is generic in the sense that any other object can be obtained up to isomorphism from G by change of base along a cartesian morphism, and such that $\mathbf{P}$ has U-indexed products where $U \in |\mathbf{P}|$ is the object underlying G. In fact it is the case that every such $\mathbf{P}$ is equivalent over $|\mathbf{P}|$ to a $2T\lambda C$-hyperdoctrine, and the latter notion is often more convenient to work with. However there is one aspect of Definition 2.2 which is not mere convenience, but crucial for the construction of a topos model from a $2T\lambda C$-hyperdoctrine to be given in section 4. This is the requirement that the right adjoint functor $\prod_I$ be natural in I rather than just pseudo-natural. (As noted by Coquand and Ehrhard [CE], Seely's notion of "PL category" is "strict" in all respects except this one, and in giving an equational presentation of the theory of PL categories they add in the naturality of the $\prod$-functors.)

Next we indicate how a $2T\lambda C$-hyperdoctrine $\mathbf{P}$ provides a semantics for $2T\lambda C$-theories. Roughly speaking, *the denotations of types are objects in the fibres of* $\mathbf{P}$ (which are particular morphisms in $|\mathbf{P}|$) *and the denotations of terms are morphisms in the fibres.* There is a third syntactic category, variously called "kinds" or "orders", which we have not so far mentioned since for the second order typed lambda calculus the kinds are just finite powers of the basic kind "*Type*". *The denotations of kinds are objects in* $|\mathbf{P}|$, and in particular "*Type*" *is denoted by* U. To make these statements more precise, we first have to specify an interpretation of the underlying language of the types and terms and take account of the free type and individual variables they may have:

2.4. Structures and models. Let $\mathbf{P}$ be a $2T\lambda C$-hyperdoctrine and $ar: C \longrightarrow \mathbb{N}$ some multiset of type constructor symbols. A *C-structure* M in $\mathbf{P}$ is specified by giving, for each $F \in C$, an element $MF \in \mathbf{P}(U^{ar(F)}, U)$. Then for a type Φ over C and a finite list $\underline{X} = X_0, \ldots, X_{n-1}$ of type variables containing the free type varables of Φ, define

$$[\![\Phi, \underline{X}]\!]_M \in \mathbf{P}(U^n, U)$$

by structural recursion:

- $[\![X_i, \underline{X}]\!]_M = \pi_i$, the ith projection morphism;
- $[\![F(\Phi_0, \ldots, \Phi_{m-1})]\!]_M = MF \circ \langle [\![\Phi_0, \underline{X}]\!]_M, \ldots, [\![\Phi_{m-1}, \underline{X}]\!]_M \rangle$;
- $[\![\mathsf{T}, \underline{X}]\!]_M = \mathsf{T}_{U^n}$;
- $[\![\Phi * \Psi, \underline{X}]\!]_M = [\![\Phi, \underline{X}]\!]_M *_{U^n} [\![\Psi, \underline{X}]\!]_M$, where $* = \times$ or $\to$;
- $[\![\prod X.\Phi, \underline{X}]\!]_M = \prod_{U^n} [\![\Phi, \underline{X}, X]\!]_M$.

Note that if Φ is a *closed* type, then we can take $\underline{X} = \langle\rangle$ the empty list, and get $[\![\Phi, \langle\rangle]\!]_M$ in $\mathbf{P}(\mathsf{T}, U)$: *we will write* $[\![\Phi]\!]_M$ *for* $[\![\Phi, \langle\rangle]\!]_M$. By structural induction, one can show that *substitution of types for type variables is interpreted using composition in* $|\mathbf{P}|$:

$$[\![[X := \Psi]\Phi, \underline{X}]\!]_M = [\![\Phi, \underline{X}, X]\!]_M \circ \langle 1_{U^n}, [\![\Psi, \underline{X}]\!]_M \rangle. \tag{2.8}$$

Similarly, if $X \notin FTV(\Phi)$, then

$$[\![\Phi, \underline{X}, X]\!]_M = \pi_1^* [\![\Phi, \underline{X}]\!]_M \text{, where } \pi_1 : U^n \times U \longrightarrow U^n. \tag{2.9}$$

If now $L=(C,K)$ is a language (as defined in 1.2), an *L-structure* M in $\mathbf{P}$ is specified by a C-structure together with, for each $a{:}\Phi$ in K, a global element of the object $[\![\Phi]\!]_M$, i.e. a morphism $Ma{:}\mathrm{T}_{\mathrm{T}}\longrightarrow[\![\Phi]\!]_M$ in $\mathbf{P}(\mathrm{T},U)$. Then for a $2T\lambda C$ term t over L of type Φ say, together with a finite list $\underline{x}=x_0^{\Phi_0},\dots,x_{m-1}^{\Phi_{m-1}}$ of distinct individual variables containing $FIV(t)$ and with a finite list $\underline{X}=X_0,\dots,X_{n-1}$ of distinct type variables containing $FTV(t)$ and $FTV(\Phi_0),\dots,FTV(\Phi_{m-1})$, define

$$[\![t,\underline{x},\underline{X}]\!]_M : [\![\Phi_0,\underline{X}]\!]_M\times\cdots\times[\![\Phi_{m-1},\underline{X}]\!]_M \longrightarrow [\![\Phi,\underline{X}]\!]_M$$

in $\mathbf{P}(U^n,U)$ by structural recursion:

- $[\![x_i^{\Phi_i},\underline{x},\underline{X}]\!]_M=\pi_i$, the ith projection morphism;
- $[\![\langle\rangle,\underline{x},\underline{X}]\!]_M=[\![\Phi_0,\underline{X}]\!]_M\times\cdots\times[\![\Phi_{m-1},\underline{X}]\!]_M$, the unique morphism to T_{U^n};
- $[\![\mathsf{Fst},\underline{x},\underline{X}]\!]_M=(U^n)^*(\text{'Fst'})\circ([\![\Phi_0,\underline{X}]\!]_M\times\cdots\times[\![\Phi_{m-1},\underline{X}]\!]_M)$, with 'Fst' as in (2.5);
- $[\![\mathsf{Snd},\underline{x},\underline{X}]\!]_M=(U^n)^*(\text{'Snd'})\circ([\![\Phi_0,\underline{X}]\!]_M\times\cdots\times[\![\Phi_{m-1},\underline{X}]\!]_M)$, with 'Snd' as in (2.6);
- $[\![a,\underline{x},\underline{X}]\!]_M=(U^n)^*(Ma)\circ([\![\Phi_0,\underline{X}]\!]_M\times\cdots\times[\![\Phi_{m-1},\underline{X}]\!]_M)$, for $a\in K$;
- $[\![\langle s,t\rangle,\underline{x},\underline{X}]\!]_M=\langle[\![s,\underline{x},\underline{X}]\!]_M,[\![t,\underline{x},\underline{X}]\!]_M\rangle$;
- $[\![st,\underline{x},\underline{X}]\!]_M=ev\circ\langle[\![s,\underline{x},\underline{X}]\!]_M,[\![t,\underline{x},\underline{X}]\!]_M\rangle$;
- $[\![s\Psi,\underline{x},\underline{X}]\!]_M=\pi_{[\![\Psi,\underline{X}]\!]_M}\circ[\![s,\underline{x},\underline{X}]\!]_M$, with $\pi_{[\![\Psi,\underline{X}]\!]_M}$ as in (2.7) (and using (2.8));
- $[\![\lambda x^{\Phi}s,\underline{x},\underline{X}]\!]_M=\bar{f}$, the exponential transpose of $f=[\![s,\underline{x},x,\underline{X}]\!]_M$;
- $[\![\lambda X.s,\underline{x},\underline{X}]\!]_M=\bar{g}$, the transpose across the adjunction $\pi_1^*\dashv\prod_{U^n}$ of
$$g=[\![s,\underline{x},\underline{X},X]\!]_M:\pi_1^*\Big(\prod_{i<m}[\![\Phi_i,\underline{X}]\!]_M\Big)=\prod_{i<m}[\![\Phi_i,\underline{X},X]\!]_M\longrightarrow[\![\Phi,\underline{X},X]\!]_M$$

(where the last clause makes use of (2.9)).

We shall say that the L-structure M *satisfies* an equality judgement $\vdash_{X,\underline{x}} s=t{:}\Phi$ and write

$$M\models_{X,\underline{x}} s=t{:}\Phi\ ,$$

if $[\![s,\underline{x},\underline{X}]\!]_M=[\![t,\underline{x},\underline{X}]\!]_M$ in $\mathbf{P}(U^n,U)$. Then if $\mathbf{T}=(L,A)$ is a $2T\lambda C$-theory, we will say that an L-structure M is a *model* of $\mathbf{T}$ if it satisfies all of the equality judgements which comprise the set A of axioms of $\mathbf{T}$. One easily proves:

2.5. Soundness Lemma. *The above definition of satisfaction of an equality judgement by a structure in a $2T\lambda C$-hyperdoctrine is sound for the equational logic of* 1.4. *Thus if* $\mathbf{T}$ *is a $2T\lambda C$-theory and M is a model of* $\mathbf{T}$ *in a $2T\lambda C$-hyperdoctrine* $\mathbf{P}$, *then for any equality judgement one has: if* $\mathbf{T}\vdash_{X,\underline{x}} s=t{:}\Phi$ *then* $M\models_{X,\underline{x}} s=t{:}\Phi$.

□

2.6. Classifying hyperdoctrines and generic models. If $\mathbf{T}=(L,A)$ is a $2T\lambda C$-theory, then the $2T\lambda C$ types and terms over L can be used to construct a $2T\lambda C$-hyperdoctrine, called the *classifying* hyperdoctrine of $\mathbf{T}$, denoted $\mathbf{P_T}$ and defined as follows:

The objects of $|\mathbf{P_T}|$ are in bijection with the natural numbers: the object corresponding to $n\in\mathbb{N}$ will be denoted U^n, and we will write U for U^1 and T for U^0.

Morphisms $U^n\longrightarrow U^m$ in $|\mathbf{P_T}|$ are m-tuples of morphisms $U^n\longrightarrow U$; and the latter are equivalence classes $[\Phi,\underline{X}]$ of pairs $(\Phi,\underline{X})$, where Φ is a type, $\underline{X}$ is a finite list of distinct type variables containing $FTV(\Phi)$ and the equivalence relation on such pairs is that of α-equivalence, i.e. $(\Phi,\underline{X})$ is equivalent to $([\underline{X}:=\underline{X}']\Phi,\underline{X}')$. Composition in $|\mathbf{P_T}|$ is given by substitution of types for type variables.

The objects of each fibre $\mathbf{P_T}(U^n,U)$ are necessarily the morphisms $U^n \longrightarrow U$ in $|\mathbf{P_T}|$. Given two such objects, $[\Phi,\underline{X}]$ and $[\Psi,\underline{Y}]$ say, a morphism $[\Phi,\underline{X}]\longrightarrow[\Psi,\underline{Y}]$ in $\mathbf{P_T}(U^n,U)$ is an equivalence class $[t,x^{\Phi},\underline{X},\underline{Y}]$, where $t{:}\Psi$ is a term with $FIV(t)\subseteq\{x^{\Phi}\}$, $FTV(t)\subseteq\underline{X},\underline{Y}$ (we assume $\underline{X}$ distinct from $\underline{Y}$), and where $(t_1,x_1^{\Phi},\underline{X}_1,\underline{Y}_1)$ is equivalent to $(t_2,x_2^{\Phi},\underline{X}_2,\underline{Y}_2)$ iff

$$\mathbf{T}\vdash_{\langle\rangle,\langle\rangle}\lambda\underline{X}_1\underline{Y}_1.\lambda x_1^{\Phi}.t_1=\lambda\underline{X}_2\underline{Y}_2.\lambda x_2^{\Phi}.t_2:\textstyle\prod\underline{X}\underline{Y}.\Phi\to\Psi.$$

Composition is given by substitution of terms for individual variables and the functors $(-)^*$ between the fibres are given by substitution of types for type variables in both types and terms. The terminal object is $[\top,\underline{X}]$; the product of $[\Phi,\underline{X}]$ and $[\Psi,\underline{X}]$ is $[\Phi\times\Psi,\underline{X}]$; and their exponential is $[\Phi\to\Psi,\underline{X}]$.

The right adjoint to $\pi_1^*:\mathbf{P_T}(U^n,U)\longrightarrow\mathbf{P_T}(U^n\times U,U)$ sends $[\Phi,\underline{X},X]$ to $[\prod X.\Phi,\underline{X}]$ and $[t,x^{\Phi},\underline{X},X]$ to $[\lambda X.([x^{\Phi}:=z\,X]t),z^{\prod X.\Phi},\underline{X}]$.

The verification that the above recipe does give a *2TλC*-hyperdoctrine is a straightforward exercise. (See also [Se, Proposition 4.6] for the full higher order case and [LS, I.11] for the first order case.) Almost tautologically $\mathbf{P_T}$ contains a model of $\mathbf{T}$, which we shall call the *generic* model of $\mathbf{T}$ and denote by $I_{\mathbf{T}}$. The underlying L-structure of $I_{\mathbf{T}}$ sends a type constructor F of arity n to $[F(\underline{X}),\underline{X}]:U^n\longrightarrow U$ and an individual constant a of (closed) type Φ to $[a,x^{\top},\langle\rangle]:[\top,\langle\rangle]\longrightarrow[\Phi,\langle\rangle]$. It follows from the definitions of $\mathbf{P_T}$ and $I_{\mathbf{T}}$ that:

$$\mathbf{T}\vdash_{\underline{X},\underline{x}} s=t{:}\Phi \text{ iff } I_{\mathbf{T}}\models_{\underline{X},\underline{x}} s=t{:}\Phi\ . \qquad (2.10)$$

In particular, the notion of a model in a *2TλC*-hyperdoctrine is complete for the equational logic of 1.4:

An equality judgement is a theorem of a 2TλC-theory iff it is satisfied by all models of the theory in 2TλC-hyperdoctrines (iff it is satisfied by the generic model).

However, the generic model $I_{\mathbf{T}}$ has a much stronger property than just (*2.10*), namely:

Any model of $\mathbf{T}$ *in any 2TλC-hyperdoctrine* $\mathbf{P}$ *can be obtained up to isomorphism as the image of the generic model* $I_{\mathbf{T}}$ *along an essentially unique morphism of 2TλC-hyperdoctrines* $\mathbf{P_T}\longrightarrow\mathbf{P}$.

(A 2-hyperdoctrine morphism $F{:}\mathbf{P}\longrightarrow\mathbf{Q}$ is specified by a finite product preserving functor $|F|{:}|\mathbf{P}|\longrightarrow|\mathbf{Q}|$ sending the $U\in|\mathbf{P}|$ to the $U\in|\mathbf{Q}|$, and by a natural transformation $F_{(-)}{:}\mathbf{P}(-,U)\longrightarrow\mathbf{Q}(|F|(-),U)$ whose component functors are ccc morphisms commuting with the right adjoints $\prod_I$.)

Thus $\mathbf{P_T}$ is the *2TλC*-hyperdoctrine "freely generated" by the *2TλC*-theory $\mathbf{T}$. Conversely, up to equivalence every *2TλC*-hyperdoctrine $\mathbf{P}$ can be presented as the classifying hyperdoctrine of some *2TλC*-theory $\mathbf{T}$: for the set C of type constructor symbols, for each arity n take one symbol $\ulcorner F\urcorner$ for each morphism $F{:}U^n\longrightarrow U$ in $|\mathbf{P}|$, giving an evident C-structure $M{:}\ulcorner F\urcorner\mapsto F$; for the set of individual constants K, for each closed type Φ over

C and each $a{:}\mathrm{T}_{\mathrm{T}}\longrightarrow M\Phi$ in $\mathbf{P}(\mathrm{T},U)$ take a symbol $\ulcorner a\urcorner{:}\Phi$, giving a language $L=(C,K)$ and an evident L-structure M in $\mathbf{P}$ with $M(\ulcorner a\urcorner)=a$; then let $\mathbf{T}$ be the $2T\lambda C$-theory (L,A), where A consists of all the equality judgements which are satisfied by the structure M. Evidently M is a model of $\mathbf{T}$, and one can show that the corresponding $2T\lambda C$-hyperdoctrine morphism $\mathbf{P_T}\longrightarrow\mathbf{P}$ is an equivalence.

2.7. Summary. The notion of satisfaction of an equality judgement by a structure in a $2T\lambda C$-hyperdoctrine is both sound and complete for equational logic over the second order typed lambda calculus. More importantly, the classifying hyperdoctrine construction sets up a correspondence between $2T\lambda C$-theories and $2T\lambda C$-hyperdoctrines which allows us to view the latter notion as a "presentation-free" version of the former. Something has definitely been achieved in this transfer from theories to hyperdoctrines, since $2T\lambda C$-hyperdoctrines are quite elementary kinds of structure: indeed they are models of a particular, essentially equational theory and are thus amenable to study using algebraic and category-theoretic techniques. Thus in proving the completeness and full embedding theorems of section 4, we will be working not with $2T\lambda C$-theories, but with the hyperdoctrines to which they correspond and applying category theoretic constructions to these. The equational aspect of (higher order) hyperdoctrines is emphasised by Coquand and Ehrhard [CE], and of course the first order part of this (the equational presentation of the notion of ccc) lies at the heart of recent work of Cousineau, Curien and Mauny [CCM, Cu].

3 Topos models

In this section we assume some familiarity with the theory of toposes and particularly with the use of higher order intuitionistic predicate logic to describe properties of and make constructions in a topos via its internal language: see [J, section 5.4] and [LS, Part II]. Also implicit to the material in this section is the use of fibred (and indexed) categories over a particular topos to provide an *elementary theory of categories* (both large and small) *relative to the topos.* The parts of this (not yet fully developed) theory we shall need are sufficiently simple for them to be given explicitly in terms of more "traditional" topos theory (by which we mean [J] up to, but not including its appendix). [PS] provides a lot of material on indexed category theory, and a taste of the wider aspects of the theory can be got from [B].

The reason why we have to go slightly beyond the traditional internal logic of a topos is simple: the syntax of $2T\lambda C$-theories invovles *variable types* and to model this in a topos $\mathbf{E}$ we will consider the *generalized objects* of $\mathbf{E}$. To understand what is meant by this, recall first the notion of a generalized element of an object B in **E**: for each object I of **E**, the *generalized elements of B at stage I* are the morphisms $b{:}I\longrightarrow B$ in **E**, and they are used to give the Kripke-Joyal forcing semantics of intuitionistic higher order predicate logic in toposes (see [LS, II.8]); the *global* elements of B are morphisms $\top\longrightarrow B$, but in general these are insufficient for detecting properties of the object B. Moving up a level, the collection of objects of **E** (global objects, one might call them), are generally insufficient for detecting all the properties of **E** as a constructive universe of "sets". Instead one must consider, for each object I, the *generalized objects of* **E** *at stage* I, which are by definition morphisms $p{:}E\longrightarrow I$ in **E** with codomain I. The idea behind this definition is that when **E**=**Set**, such $p{:}E\longrightarrow I$ correspond precisely to I-indexed collections of sets, $(E_i|i\in I)$ (by defining $E_i=\{e\in E\,|\,p(e)=i\}$), just as generalized elements $b{:}I\longrightarrow B$ correspond to I-indexed collections of elements of B, $(b(i)\,|\,i\in I)$.

Just as the generalized elements of B at stage I form a set $\mathbf{E}(I,B)$, so the generalized objects of **E** at stage I form a category (a topos in fact), namely the *slice* category $\mathbf{E}/I$ whose morphisms are commutative triangles over I. We begin by fixing some notation for these:

3.1. Slice categories. Let **C** be a category. For each object I of **C**, let $\Sigma_I{:}\mathbf{C}/I\longrightarrow\mathbf{C}$ denote the *forgetful functor from the slice category* $\mathbf{C}/I$: thus a typical object of $\mathbf{C}/I$ is a morphism in **C** of the form $p{:}\Sigma_I(p)\longrightarrow I$ and a typical morphism $f{:}p\longrightarrow q$ in $\mathbf{C}/I$ is given by a morphism $f=\Sigma_I(f){:}\Sigma_I(p)\longrightarrow\Sigma_I(q)$ in **C** satisfying $q\circ f=p$. The identity on I in **C** is a terminal object in $\mathbf{C}/I$. The binary product of p and q in $\mathbf{C}/I$ will be denoted $p\times_I q$ if it exists, in which case it is necessarily given by a pullback square in **C**:

$$\begin{array}{ccc}
\Sigma_I(p\times_I q) & \longrightarrow & \Sigma_I(q) \\
\downarrow & & \downarrow{\scriptstyle q} \\
\Sigma_I(p) & \xrightarrow[p]{} & I
\end{array}\quad .$$

Similarly, the exponential of p and q in $\mathbf{C}/I$ will be denoted $p\to_I q$ if it exists and called a *local exponential in* **C**. A morphism $\alpha{:}I\longrightarrow J$ in **C** will be called *squarable* if the pullback along α of any morphism with codomain J exists in **C**. For such an α, the operation of pulling back along α gives a functor between slice categories, which will be denoted

$$\alpha^*{:}\mathbf{C}/J\longrightarrow\mathbf{C}/I\ .$$

If the right adjoint to α^* exists at an object p of $\mathbf{C}/I$, it will be denoted $\Pi_\alpha(p)$: there is thus a morphism $\varepsilon{:}\alpha^*(\Pi_\alpha(p))\longrightarrow p$ in $\mathbf{C}/I$ with the universal property that for any $f{:}\alpha^*(q)\longrightarrow p$ there is a unique $\overline{f}{:}q\longrightarrow\Pi_\alpha(p)$ in $\mathbf{C}/J$ with $f=\varepsilon\circ\alpha^*(\overline{f})$. Of course when **C** is a topos it has all pullbacks, local exponentials and right adjoints to pulling back along a morphism (see [J, 1.4]); in particular, when **C**=**Set** $p{:}E\longrightarrow I$, and $q{:}F\longrightarrow I$, then for each $i\in I$

$$(\Sigma_I(p\to_I q))_i = E_i\to F_i\ ,$$ an exponential in **Set**,

and for each $j\in J$

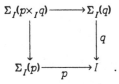

$(\Sigma_J(\Pi_\alpha p))_j = \prod\{E_i \mid \alpha(i)=j\}$, a cartesian product in **Set**.

If translated into intuitionistic higher order predicate logic, these formulas remain true for the internal logic of an arbitrary topos **E**. Thus for i a variable of type I, let E_i be an abbreviation for $\{e{:}E \mid p(e)=i\}$, etc; then **E** satisfies:

$$(\Sigma_I(p \to_I q))_i \cong \{r{:}\Omega^{E\times F} \mid \text{"}r \text{ is the graph of a function from } E_i \text{ to } F_i\text{"}\} \tag{3.1}$$

$$(\Sigma_J(\Pi_\alpha p))_j \cong \{r{:}\Omega^{I\times E} \mid \text{"}r \text{ is the graph of a partial function from } \{i{:}I \mid \alpha(i)=j\} \text{ to } E\text{" and } \forall i{:}I(\alpha(i)=j \Rightarrow r(i)\in E_i)\}\ . \tag{3.2}$$

3.2. Internal full subcategories. Consider the following construction:

> Starting with a set U and a U-indexed collection of sets $(G_u \mid u\in U)$, form the small category **U** with underlying set of objects U, with hom sets $\mathbf{U}(u,v) = \mathbf{Set}(G_u,G_v)$ and with composition and identities inherited from **Set**; by construction, the assignment $u \mapsto G_u$ extends to a diagram of type **U** in **Set** which is a full and faithful functor $G{:}\,\mathbf{U} \hookrightarrow \mathbf{Set}$.

Note that specifying the initial data for this construction is equivalent to giving a single morphism $\tau{:}G \longrightarrow U$ in **Set**, where G is the disjoint union $\bigcup\{\{u\}\times G_u \mid u\in U\}$ and $\tau{:}(u,x) \mapsto u$ (so that $G_u \cong \tau^{-1}\{u\}$). Now starting with any topos **E** and a morphism $\tau{:}G \longrightarrow U$ in **E**, it is possible to carry out the analogue of the above construction, obtaining an internal category object **U** in **E** together with a full and faithful internal diagram G of type **U** in **E**. This is achieved as follows:

Pull back τ along the two projections $\pi_i{:}U\times U \longrightarrow U$, to get $\pi_0^*(\tau) = \tau\times 1_U{:}G\times U \longrightarrow U\times U$ and $\pi_1^*(\tau) = 1_U\times\tau{:}U\times G \longrightarrow U\times U$ in $\mathbf{E}/U^2$. Now form the local exponential $\pi_0^*(\tau) \to_{U^2} \pi_1^*(\tau)$ and suppose it is given by the morphism $\langle d_0,d_1\rangle{:}U_1 \longrightarrow U\times U$ in **E**; then in the internal language of **E** one has $(G\times U)_{\langle u,v\rangle} \cong G_u$ and $(U\times G)_{\langle u,v\rangle} \cong G_v$, so that by (3.1) $(U_1)_{\langle u,v\rangle} \cong G_u \to G_v$ (where u and v are variables of type U). It follows that

$$d_0, d_1{:}U_1 \rightrightarrows U$$

is the underlying graph of a category object **U** in **E** (with composition in **U** being given by the composition morphism $((G_v \to G_w)\times(G_u \to G_v) \longrightarrow (G_u \to G_w) \mid u,v,w{:}U)$ in $\mathbf{E}/U^3$). Then $\tau{:}G \longrightarrow U$ becomes an internal diagram of type **U** via the action $((U_1)_{\langle u,v\rangle}\times G_u \longrightarrow G_v \mid u,v{:}U)$ which (using the above identifications of internal fibres) is given by the evaluation morphism $ev{:}(\pi_0^*(\tau) \to_{U^2} \pi_1^*(\tau))\times_{U^2} \pi_0^*(\tau) \longrightarrow \pi_1^*(\tau)$ in $\mathbf{E}/U^2$. The transpose $((U_1)_{\langle u,v\rangle} \longrightarrow (G_u \to G_v) \mid u,v\in U)$ of this action across the exponential adjunction, gives the effect of G on morphisms of **U**; and by definition of U_1, this is an isomorphism — and hence G is full and faithful. **U** will be called the *internal full subcategory of* **E** *determined by* $\tau{:}G \longrightarrow U$, and the diagram $G\in\mathbf{E}^{\mathbf{U}}$ the *inclusion of* **U** *into* **E**.

We are interested in such internal full subcategories which are *closed in* **E** *under certain internal products.* In order to understand precisely what is meant by this, it is important first to note that our terminology is slightly misleading, since an internal full subcategory is not a generalized collection of generalized objects of $\mathbf{E}$ (a concept which can be formalized by Bénabou's notion in [B] of "definable class of objects" of a topos), but rather a single object whose variable elements u *name* variable objects $G_u=\{g{:}G \mid \tau(g)=u\}$. Then if we assert for example, that $\mathbf{U}$ is *closed in* $\mathbf{E}$ *under binary products*, we mean that there is a morphism $\ulcorner\times\urcorner{:}U^2\longrightarrow U$ and a pullback square in $\mathbf{E}$ of the form:

$$\begin{array}{ccc} \Sigma_{U^2}(\pi_0^*(\tau)\times_{U^2}\pi_1^*(\tau)) & \dashrightarrow & G \\ {\scriptstyle \pi_0^*(\tau)\times_{U^2}\pi_1^*(\tau)}\Big\downarrow & & \Big\downarrow{\scriptstyle \tau} \\ U^2 & \xrightarrow[\ulcorner\times\urcorner]{} & U\,. \end{array} \qquad (3.3)$$

Since the pullback of τ along $\ulcorner\times\urcorner$ has internal fibres $(G_{\ulcorner\times\urcorner(u,v)} \mid u,v{:}U)$, the above pullback square furnishes an internal family of isomorphisms $G_{\ulcorner\times\urcorner(u,v)}\cong G_u\times G_v$ $(u,v{:}U)$, so that $\ulcorner\times\urcorner(u,v)$ names the product of the objects named by u and v. Similarly, $\mathbf{U}$ *is closed in* $\mathbf{E}$ *under exponentiation* if there is a morphism $\ulcorner\rightarrow\urcorner{:}U^2\longrightarrow U$ and a pullback square of the form:

$$\begin{array}{ccc} \Sigma_{U^2}(\pi_0(\tau)\rightarrow_{U^2}\pi_1(\tau)) & \dashrightarrow & G \\ {\scriptstyle \pi_0^*(\tau)\rightarrow_{U^2}\pi_1^*(\tau)}\Big\downarrow & & \Big\downarrow{\scriptstyle \tau} \\ U^2 & \xrightarrow[\ulcorner\rightarrow\urcorner]{} & U\,. \end{array} \qquad (3.4)$$

$\mathbf{U}$ *contains the terminal object* of $\mathbf{E}$ if there is a morphism $\ulcorner\mathrm{T}\urcorner{:}\mathrm{T}\longrightarrow U$ and a pullback square of the form:

$$\begin{array}{ccc} \mathrm{T} & \dashrightarrow & G \\ {\scriptstyle 1_{\mathrm{T}}}\Big\downarrow & & \Big\downarrow{\scriptstyle \tau} \\ \mathrm{T} & \xrightarrow[\ulcorner\mathrm{T}\urcorner]{} & U\,. \end{array} \qquad (3.5)$$

Finally, $\mathbf{U}$ *is closed in* $\mathbf{E}$ *under* U*-indexed products* if there is a morphism $\ulcorner\Pi\urcorner{:}(U\rightarrow U)\longrightarrow U$ and a pullback of the form:

$$\begin{array}{ccc} \Sigma_{U\rightarrow U}(\Pi_p\, ev^*(\tau)) & \dashrightarrow & G \\ {\scriptstyle \Pi_p\, ev^*(\tau)}\Big\downarrow & & \Big\downarrow{\scriptstyle \tau} \\ U\rightarrow U & \xrightarrow[\ulcorner\Pi\urcorner]{} & U \end{array} \qquad (3.6)$$

where $p{:}(U\rightarrow U)\times U\longrightarrow(U\rightarrow U)$ is the first projection and $ev{:}(U\rightarrow U)\times U\longrightarrow U$ is the evaluation morphism. Using (3.2), one sees that the effect of this last pullback square is to provide an internal family of isomorphisms $G_{\ulcorner\Pi\urcorner(f)}\cong\prod(G_{f(u)} \mid u{:}U)$ indexed by f of type $U\rightarrow U$: thus $\ulcorner\Pi\urcorner(f)$ names the product of the U-indexed family of objects named by $f(u)$ for $u{:}U$. Collecting together the data required to specify such an internal full subcategory, we arrive at the following:

3.3. Definition. A *topos model* of the second order typed lambda calculus is given by an elementary topos $\mathbf{E}$ together with morphisms $\tau{:}G\longrightarrow U$, $\ulcorner\mathrm{T}\urcorner{:}\mathrm{T}\longrightarrow U$, $\ulcorner\times\urcorner,\ulcorner\rightarrow\urcorner{:}U^2\longrightarrow U$ and $\ulcorner\Pi\urcorner{:}(U\rightarrow U)\longrightarrow U$ in $\mathbf{E}$ and pullback squares of the form (3.3),(3.4),(3.5) and (3.6). $\mathbf{E}$ will be called the *ambient topos* of the model; $\mathbf{U}$ will denote the internal full subcategory of $\mathbf{E}$ determined by τ.

Topos models *are* models of the second order typed lambda calculus because:

3.4. Lemma. *Every topos model of the second order typed lambda calculus gives rise to a $2T\lambda C$-hyperdoctrine $\mathbf{P}$ via the category valued hom functor $\mathbf{E}(-,\mathbf{U})$.*

Proof. Since the internal full subcategory $\mathbf{U}$ is closed in $\mathbf{E}$ under finite products and exponentials, $\mathbf{U}$ is automatically a ccc object in $\mathbf{E}$. Consequently its generalized elements at any stage I form a ccc (with set of objects $\mathbf{E}(I,U)$ and set of morphisms $\mathbf{E}(I,U_1)$); and for any $\alpha:I\longrightarrow J$, the operation α^* gives a strict morphism of ccc's. Thus $\mathbf{E}(-,\mathbf{U})$ is a functor $\mathbf{E}^{op}\longrightarrow\mathbf{Ccc}$. Letting $|\mathbf{P}|$ be the full subcategory of $\mathbf{E}$ whose objects are the finite powers of U, we can restrict this functor to get $\mathbf{P}=\mathbf{E}(-,\mathbf{U}):|\mathbf{P}|^{op}\longrightarrow\mathbf{Ccc}$ satisfying parts (*i*), (*ii*) and (*iii*) of Definition 2.2. For part (*iv*) of the definition, we use the fact that $\mathbf{U}$ is closed in $\mathbf{E}$ under U-indexed products. Thus $\prod_I:\mathbf{P}(I\times U,U)\longrightarrow\mathbf{P}(I,U)$ is defined on objects by sending $u:I\times U\longrightarrow U$ to $\prod_I(u)=\ulcorner\Pi\urcorner\circ\overline{u}$ where $\overline{u}:I\longrightarrow(U\to U)$ is the exponential transpose of u. Similarly, $\prod_I$ is defined on morphisms by sending $m:I\times U\longrightarrow U_1$ to $\prod_I(m)=\ulcorner\Pi_1\urcorner\circ\overline{m}$, where $\ulcorner\Pi_1\urcorner:(U\to U_1)\longrightarrow U_1$ sends U-indexed collections of morphisms in $\mathbf{U}$ to their product in $\mathbf{E}$, (which again lies in $\mathbf{U}$ since it is a *full* sucategory and closed under such products): in other words, $\ulcorner\Pi_1\urcorner$ is uniquely defined by requiring $d_i\circ\ulcorner\Pi_1\urcorner=\ulcorner\Pi\urcorner\circ(1_U\to d_i)$ for $i=0,1$ and (using the internal language of $\mathbf{E}$) further requiring that for $m:U\to U_1$

$$\ulcorner\Pi_1\urcorner(m):G_{\ulcorner\Pi\urcorner(d_0m)}\to G_{\ulcorner\Pi\urcorner(d_1m)}$$

correspond under the isomorphisms

$$G_{\ulcorner\Pi\urcorner(d_im)}\cong\prod\{G_{d_im(u)}\mid u:U\}\qquad(i=0,1)$$

to $\prod\{m(u)\mid u:U\}$. Evidently this definition of $\prod_I:\mathbf{P}(I\times U,U)\longrightarrow\mathbf{P}(I,U)$ is natural in I. That it gives a right adjoint to π_I^* follows from the fact that $\pi_I^*:\mathbf{E}(-,\mathbf{U})\longrightarrow\mathbf{E}(-\times U,\mathbf{U})\cong\mathbf{E}(-,\mathbf{U}^U)$ is the representable functor induced by the diagonal internal functor $\overline{\pi}_I:\mathbf{U}\longrightarrow\mathbf{U}^U$ — and taking U-indexed products is right adjoint to this.

□

Since each topos model $\mathbf{E},\mathbf{U}$ determines a $2T\lambda C$-hyperdoctrine, it provides a semantics for the types and terms of the second order typed lambda calculus, with the types being denoted by particular morphisms in $\mathbf{E}$ with codomain U and the terms by morphisms with codomain U_1. In particular the definitions in 2.4 specialize to give us the notion of a *topos model of a $2T\lambda C$-theory.*

As we explained in the previous section, arbitrary $2T\lambda C$-hyperdoctrines provide a semantics which is completely general. (Perhaps neutral is a better word.) In contrast, topos models embody a very particular and apparently naive idea of the meaning of the various symbols of the calculus: A topos is itself a model of higher order intuitionistic predicate logic; in a topos model of *2TλC*, the closed types get interpreted as elements of a family U of (names of) "sets" in this logic, and more generally the polymorphic (i.e. non-closed) types are given by *arbitrary* functions from this family to itself; furthermore, $\times$ is interpreted by taking actual *cartesian products*, $\rightarrow$ by taking *full function exponentials* and $\prod$ by taking products of "sets" in the family indexed by the elements of U. Thus in this notion of model we allow ourselves to work in a "non-standard" universe of sets but make up for this by insisting that the type forming operations be interpreted in a completely standard way. This is in contrast to more traditional formulations of the notion of model of polymorphic lambda calculus (such as in [BM] or [R2]) which are couched in classical set theory, but allow some of the operations ($\prod$ in particular) to be non-standard. The intersection of these two approaches is trivial: when $\mathbf{E}=\mathbf{Set}$, a topos model is given by a set U of sets closed under finite products, exponentiation and U-indexed products; and a simple cardinality argument (using the principles of classical logic !) shows that any such U must have as elements only sets with at most one element. There are therefore no non-trivial topos models of *2TλC* when the ambient topos is the category of sets.

However the more liberal nature of the internal logic of toposes in general means that non-trivial examples of Definition 3.3 do exist. The first such is due to Hyland and Moggi, with $\mathbf{E}$ being Hyland's effective topos [H], U the object of $\neg\neg$-closed partial equivalence relations on the natural number object N and $\tau{:}G\longrightarrow U$ having internal fibres $G_u=Eu/u|_{Eu}$, where $Eu=\{n{:}N\,|\,u(n,n)\}$ and $u|_{Eu}$ is the equivalence relation obtained by restriction. Dana Scott has dubbed the (generalized) elements of U the *modest sets*: they have far greater closure properties than those required by Definition 3.3 — as Hyland, Robinson and Rosolini show in [HRR], the limit or colimt of *any* internal diagram of modest sets is again modest. (The object of *all* partial equivalence relations on N in the effective topos enjoys similar properties.) There are in all likelyhood very many interesting topos models of *2TλC* and related type theories. In the next section we will show how to manufacture topos models from *2TλC*-hyperdoctrines, so that in particular there are *enough* such models to distinguish the theorems of a *2TλC*-theory from the non-theorems.

4 Full embedding and completeness

In this section we prove the main result of the paper by showing how each *2TλC*-hyperdoctrine can be expanded to a topos model. To do so we will employ two standard category theoretic tools, namely the *Grothendieck construction* of a split fibration from a category-valued functor and the *Yoneda embedding* of a category into a topos of presheaves.

4.1. The Grothendieck construction. Suppose that **B** is a category and that $\mathbf{P}:\mathbf{B}^{op}\longrightarrow\mathbf{Cat}$ is a contravariant functor from **B** into the category of small categories. Construct a new category $Gr(\mathbf{P})$ and a functor $P:Gr(\mathbf{P})\longrightarrow\mathbf{B}$ from **P** as follows:

> The objects of $Gr(\mathbf{P})$ are pairs (I,A), where I is an object of **B** and A an object of $\mathbf{P}(I)$. Given two such objects $(I,A),(J,B)$, the morphisms $(I,A)\longrightarrow(J,B)$ in $Gr(\mathbf{P})$ are given by pairs (α,f), where $\alpha:I\longrightarrow J$ in **B** and $f:A\longrightarrow\mathbf{P}(\alpha)B$ in $\mathbf{P}(I)$. The composition of $(\alpha,f):(I,A)\longrightarrow(J,B)$ and $(\beta,g):(J,B)\longrightarrow(K,C)$ in $Gr(\mathbf{P})$ is $(\beta\circ\alpha,\mathbf{P}(\alpha)(g)\circ f)$. The identity on (I,A) is $(1_I,1_A)$. The functor $P:Gr(\mathbf{P})\longrightarrow\mathbf{B}$ sends an object (I,A) to I and a morphism (α,f) to α.

The kind of functors into **B** which arise in this way can be characterized by category theoretic properties: they are the *split fibrations* over **B**. More generally, the construction can be applied to *pseudofunctors* $\mathbf{B}^{op}\longrightarrow\mathbf{Cat}$, setting up a correspondence between these and *cloven fibrations* over **B**. We shall not need to use these concepts explicitly here, and refer the interested reader to [Gr].

Now suppose that **P** is a $2T\lambda C$-hyperdoctrine. Recalling the notation of Definition 2.2, we can apply the Grothendieck construction to the composition of $\mathbf{P}(-,U):|\mathbf{P}|^{op}\longrightarrow\mathbf{Ccc}$ with the forgetful functor $\mathbf{Ccc}\longrightarrow\mathbf{Cat}$; we will denote the resulting category and functor simply by:

$$P:Gr(\mathbf{P})\longrightarrow|\mathbf{P}|\,.$$

Because of the special nature of **P**, note that the objects of $Gr(\mathbf{P})$ are given just by morphisms $A:I\longrightarrow U$ in $|\mathbf{P}|$ with codomain U, and that a morphism $(A:I\longrightarrow U)\longrightarrow(B:J\longrightarrow U)$ in $Gr(\mathbf{P})$ is given by a pair (α,f), where $\alpha:I\longrightarrow J$ in $|\mathbf{P}|$ and $f:A\longrightarrow\alpha^*(B)=(B\circ\alpha)$ in $\mathbf{P}(I,U)$. The following properties of $Gr(\mathbf{P})$ and P are easily verified:

4.2. Lemma. *Let* **P** *be a* $2T\lambda C$*-hyperdoctrine. Then:*

(i) $Gr(\mathbf{P})$ *has and* P *preserves finite products. The terminal object in* $Gr(\mathbf{P})$ *is* $(\top_\top:\top\longrightarrow U)$ *and the binary product of* $(A:I\longrightarrow U)$ *and* $(B:J\longrightarrow U)$ *is* $(\pi_1^*(A)\times_{I\times J}\pi_2^*(B):I\times J\longrightarrow U)$ *(where the* π_i *are the product projections for* $I\times J$ *in* $|\mathbf{P}|$*).*

(ii) P *has a full and faithful right adjoint* $T:|\mathbf{P}|\longrightarrow Gr(\mathbf{P})$*, given on objects by sending* I *to the terminal object in the fibre over* I*, i.e.* $T(I)=(\top_I:I\longrightarrow U)$.

□

We now focus attention on a particular morphism in $Gr(\mathbf{P})$, namely:

$$t=_{def}(1_U,1_U):(1_U:U\longrightarrow U)\longrightarrow(\top_U:U\longrightarrow U)=T(U). \qquad (4.1)$$

(Our convention of using the same letter to denote both an object in a category and its associated morphism to the terminal object has become rather confusing at this point: in *(4.1)* the first "1_U" denotes the identity on U in $|\mathbf{P}|$, whereas the second denotes the unique morphism in the ccc $\mathbf{P}(U,U)$ from the object 1_U to the terminal object.) Although $Gr(\mathbf{P})$ does not have all pullbacks, the morphism t is squarable, i.e. the pullback of it along any morphism with codomain $T(U)$ exists. This is because such a morphism

$(\alpha,f):(A:I\longrightarrow U)\longrightarrow T(U)$ in $Gr(\mathbf{P})$ necessarily has $f=A:A\longrightarrow T_I$, and then

$$\begin{array}{ccc}(A\times_I\alpha:I\longrightarrow U) & \xrightarrow{(\alpha,\pi_2)} & (1_U:U\longrightarrow U)\\ {\scriptstyle (1_I,\pi_1)}\downarrow & & \downarrow{\scriptstyle t}\\ (A:I\longrightarrow U) & \xrightarrow[(\alpha,A)]{} & T(U)\end{array}$$

is easily verified to be a pullback square in $Gr(\mathbf{P})$. Let Σ denote the collection of morphisms in $Gr(\mathbf{P})$ which can be obtained by pullback from t: thus (α,f) is in Σ iff α is an isomorphism and f a product projection. This class Σ inherits good properties from $\mathbf{P}$ with respect to local exponentiation and right adjoints to pulling back (*cf.* 3.1):

4.3. Lemma. *If* $\mathbf{P}$ *is a* $2T\lambda C$*-hyperdoctrine and* $\Sigma\subseteq mor(Gr(\mathbf{P}))$ *defined as above, then:*

(*i*) *Any morphism in* Σ *is squarable.*

(*ii*) *For any* $p:Y\longrightarrow X$ *and* $q:Z\longrightarrow X$ *in* Σ, *their local exponential* $p\rightarrow_X q$ *exists in* $Gr(\mathbf{P})$ *and is an element of* Σ.

(*iii*) *If* $r:W\longrightarrow X\times T(U)$ *is in* Σ, *then* $\Pi_{\pi_1}(r)$, *the right adjoint at* r *to the pullback functor* $\pi_1^*:Gr(\mathbf{P})/X\rightarrow Gr(\mathbf{P})/X\times T(U)$, *exists and is an element of* Σ.

Proof. Since all the morphisms in Σ are obtained by pullback, (*i*) follows immediately from the fact that a pullback of a pullback is a pullback. Thus for any $(1_J,\pi_1):(B\times_J C:J\longrightarrow U)$ in Σ and any $(\alpha,f):(A:I\longrightarrow U)\longrightarrow(B:J\longrightarrow U)$ in $Gr(\mathbf{P})$, the pullback $(\alpha,f)^*(1_J,\pi_1)$ is

$$(1_I,\pi_1):(A\times_I\alpha^*C:I\longrightarrow U)\longrightarrow(A:I\longrightarrow U) \tag{4.2}$$

For (*ii*), suppose that $X=(A:I\longrightarrow U)$, $Y=(A\times_I B:I\longrightarrow U)$, $Z=(A\times_I C:I\longrightarrow U)$, $p=(1_I,\pi_1:A\times_I B\longrightarrow A)$ and $q=(1_I,\pi_1:A\times_I C\longrightarrow A)$. Then for any morphism in $Gr(\mathbf{P})$ with codomain X, $(\alpha,f):(D:J\longrightarrow U)\longrightarrow X$ say, it is easy to see that $(\alpha,f)\times_X p$ is

$$(\alpha,f\circ\pi_1):(D\times_J B\alpha:J\longrightarrow U)\longrightarrow X.$$

Hence specifying a morphism $(\alpha,f)\times_X p\longrightarrow q$ in $Gr(\mathbf{P})/X$ amounts to giving a morphism $D\times_J B\alpha\longrightarrow C\alpha$ in $\mathbf{P}(J,U)$; and transposing across the exponential adjunction, this amounts to giving a morphism $D\longrightarrow(B\alpha\rightarrow_J C\alpha)=\alpha^*(B\rightarrow_I C)$. It follows from this that we can take $p\rightarrow_X q$ to be

$$(1_I,\pi_1):(A\times_I(B\rightarrow_I C):I\longrightarrow U)\longrightarrow X, \tag{4.3}$$

since morphisms in $Gr(\mathbf{P})/X$ into this from (α,f) are also specified by morphisms $D\longrightarrow\alpha^*(B\rightarrow_I C)$. Note that as required, (*4.3*) is in Σ.

Similar arguments show that for (*iii*) we can take $\Pi_{\pi_1}(r)$ to be

$$(1_I,\pi_1):(A\times_I\textstyle\prod_I(B):I\longrightarrow U)\longrightarrow X, \tag{4.4}$$

when $X=(A:I\longrightarrow U)$, $W=(\pi_1^*(A)\times_{I\times U}B:I\times U\longrightarrow U)$ and $r=(1_{I\times U},\pi_1):W\longrightarrow(\pi_1^*(A):I\times U\longrightarrow U)\cong X\times T(U)$.

□

Although $Gr(\mathbf{P})$ is far from being a topos (or even locally cartesian closed), Lemma 4.3 means that the construction given in 3.2 (which uses pullbacks and local exponentials) can be carried out starting with the morphism t in $Gr(\mathbf{P})$, to yield an internal full subcategory there with good closure properties. To actually get an internal full subcategory in a topos, we will use:

4.4. The Yoneda embedding. If $\mathbf{C}$ is a small category, $[\mathbf{C}^{op},\mathbf{Set}]$ will denote the *topos of presheaves* on $\mathbf{C}$, i.e the category of contravariant, set-valued functors on $\mathbf{C}$ and natural transformations between such. $H:\mathbf{C}\hookrightarrow[\mathbf{C}^{op},\mathbf{Set}]$ will denote the *Yoneda embedding* — the full and faithful functor sending an object I in $\mathbf{C}$ to the hom functor $H(I)=\mathbf{C}(-,I)$, and sending a morphism $\alpha:I\longrightarrow J$ to the natural transformation $H(\alpha)=\alpha^*:\mathbf{C}(-,J)\longrightarrow\mathbf{C}(-,I)$ whose components are given by precomposition with α. We will need some properties of the Yoneda embedding under taking slice categories, and for this it is convenient to work with an equivalent version of the topos $[\mathbf{C}^{op},\mathbf{Set}]$ in terms of "discrete fibrations":

Since each set is a discrete category, any functor $X:\mathbf{C}^{op}\longrightarrow\mathbf{Set}$ can be regarded as category-valued and hence one can apply the Grothendieck construction of 4.1 to it to obtain $P:Gr(X)\longrightarrow\mathbf{C}$. $Gr(X)$ is often called the *category of elements of* X, since its objects are pairs (I,x) with $I\in ob\mathbf{C}$ and $x\in X(I)$ (and its morphisms $(I,x)\longrightarrow(J,y)$ are just those $\alpha:I\longrightarrow J$ in $\mathbf{C}$ with $X(\alpha)(y)=x$). The functors $P:\mathbf{X}\longrightarrow\mathbf{C}$ that arise in this way are the *discrete fibrations*, which by definition are those with

$$\begin{array}{ccc} mor(\mathbf{X}) & \xrightarrow{cod} & ob(\mathbf{X}) \\ {\scriptstyle mor(P)}\downarrow & & \downarrow{\scriptstyle ob(P)} \\ mor(\mathbf{C}) & \xrightarrow[cod]{} & ob(\mathbf{C}) \end{array}$$

a pullback square in **Set**. More precisely, the Grothendieck construction (together with a similar construction on natural transformations) gives a functor $Gr:[\mathbf{C}^{op},\mathbf{Set}]\longrightarrow\mathbf{Cat}/\mathbf{C}$ which is an equivalence of categories between $[\mathbf{C}^{op},\mathbf{Set}]$ and $\mathbf{Dfib}(\mathbf{C})$, the full subcategory of $\mathbf{Cat}/\mathbf{C}$ whose objects are discrete fibrations. Under this equivalence the Yoneda embedding $H:\mathbf{C}\hookrightarrow[\mathbf{C}^{op},\mathbf{Set}]$ is identified with the functor $\mathbf{C}\longrightarrow\mathbf{Dfib}(\mathbf{C})$ which sends an object I of $\mathbf{C}$ to the functor $\Sigma_I:\mathbf{C}/I\longrightarrow\mathbf{C}$ of 3.1. Now simple properties of pullbacks imply that for functors $P:\mathbf{X}\longrightarrow\mathbf{C}$ and $Q:\mathbf{Y}\longrightarrow\mathbf{X}$, if P is a discrete fibration, then $Q\circ P$ is one iff Q is; consequently $\mathbf{Dfib}(\mathbf{C})/P\cong\mathbf{Dfib}(\mathbf{X})$. Hence for any $X:\mathbf{C}^{op}\longrightarrow\mathbf{Set}$, there is an equivalence of categories:

$$[\mathbf{C}^{op},\mathbf{Set}]/X\simeq[(GrX)^{op},\mathbf{Set}]\,. \tag{4.5}$$

In particular, when X is a hom functor $H(I)$, (4.5) becomes:

$$[\mathbf{C}^{op},\mathbf{Set}]/H(I)\simeq[(\mathbf{C}/I)^{op},\mathbf{Set}]\,, \tag{4.6}$$

and under this equivalence, the Yoneda embedding for $\mathbf{C}/I$ is identified with the functor $\mathbf{C}/I\longrightarrow[\mathbf{C}^{op},\mathbf{Set}]/H(I)$ which is "apply H to morphisms".

It is well known that H preserves any limits which exist in $\mathbf{C}$. It is also the case that H preserves any existing exponentials — a fact put to good use by Dana Scott in [Sc1]. In fact something more general is true:

4.5. Lemma. *The Yoneda embedding $H:\mathbf{C}\hookrightarrow[\mathbf{C}^{op},\mathbf{Set}]$ preserves any local exponentials and instances of right adjoints to pulling back (Π-functors) that happen to exist in* **C**.

Proof. Local exponentials are definable in terms of Π-functors: if (p is squarable and) $p\to_I q$ exists, then so does $\Pi_p(p^*(q))$ and they are canonically isomorphic. So it suffices to show that H preserves any instances of Π-functors. Thus given $\alpha:I\longrightarrow J$ and $p:E\longrightarrow I$ in **C**, if $\Pi_\alpha(p)$ exists, with counit morphism $\epsilon:\alpha^*(\Pi_\alpha(p))\longrightarrow p$ say, then we must show that transposing the morphism

$$(H\alpha)^*(H(\Pi_\alpha p))\cong H(\alpha^*(\Pi_\alpha p))\xrightarrow{H\epsilon}H(p)$$

across the adjunction $(H\alpha)^*\dashv\Pi_{H\alpha}$ gives an isomorphism $H(\Pi_\alpha p)\cong\Pi_{H\alpha}(Hp)$ in $[\mathbf{C}^{op},\mathbf{Set}]/H(J)$. Transferring the problem to $[(\mathbf{C}/J)^{op},\mathbf{Set}]$ via the equivalence (4.6), we can calculate that for any q in $\mathbf{C}/J$:

$$\begin{aligned}
H(\Pi_\alpha p)(q) &\cong [(\mathbf{C}/J)^{op},\mathbf{Set}](H(q),H(\Pi_\alpha p)) && \text{(Yoneda lemma)}\\
&\cong (\mathbf{C}/J)(q,\Pi_\alpha p) && \text{(H full and faithful)}\\
&\cong (\mathbf{C}/I)(\alpha^*(q),p) && \text{(α^* left adjoint to Π_α)}\\
&\cong [(\mathbf{C}/I)^{op},\mathbf{Set}]((H\alpha)^*(Hq),Hp) && \text{(H full, faithful and pullback preserving)}\\
&\cong [(\mathbf{C}/J)^{op},\mathbf{Set}](Hq,\Pi_{H\alpha}(Hp)) && \text{($(H\alpha)^*$ left adjoint to $\Pi_{H\alpha}$)}\\
&\cong (\Pi_{H\alpha}(Hp))(q)\,. && \text{(Yoneda lemma)}
\end{aligned}$$

These isomorphisms are natural in q and give the required isomorphism $H(\Pi_\alpha p)\cong\Pi_{H\alpha}(Hp)$.

□

We can now state and prove our main result:

4.6. Theorem. (Full embedding in topos models.) *Let* **P** *be a 2TλC-hyperdoctrine. Apply the Grothendieck construction to* **P** *to obtain the category $Gr(\mathbf{P})$, containing the morphism $t:(1_U:U\to U)\longrightarrow T(U)$ of (4.1). Then the internal full subcategory* **U** *of the topos of presheaves $[(Gr\mathbf{P})^{op},\mathbf{Set}]$ determined by $H(t)$ is a topos model of the second order typed lambda calculus. Moreover, the 2TλC-hyperdoctrine* **P** *is fully embedded in the 2TλC-hyperdoctrine determined by this topos model, in the sense that there is a full, faithful and finite product preserving functor*

$$|\mathbf{P}|^{op}\xhookrightarrow{T}Gr(\mathbf{P})^{op}\xhookrightarrow{H}[(Gr\mathbf{P})^{op},\mathbf{Set}]$$

and a natural isomorphism $\mathbf{P}(-,U)\cong[(Gr\mathbf{P})^{op},\mathbf{Set}](HT(-),\mathbf{U})$.

Proof. Let us write G for the object $(1_U:U\to U)$ in $Gr(\mathbf{P})$. Referring to Definition 3.3, to see that $H(t):H(G)\longrightarrow HT(U)$ determines a topos model, we have to produce morphisms $\ulcorner\mathrm{T}\urcorner:\mathrm{T}\longrightarrow HT(U)$, $\ulcorner\times\urcorner,\ulcorner\to\urcorner:HT(U)^2\longrightarrow HT(U)$ and $\ulcorner\Pi\urcorner:(HT(U)\to HT(U))\longrightarrow HT(U)$ in $[(Gr\mathbf{P})^{op},\mathbf{Set}]$, together with corresponding pullback squares of the form (3.3), (3.4), (3.5) and (3.6). The first three morphisms are rather easy to produce, since they already exist at the level of $|\mathbf{P}|$:

Let $\ulcorner\mathrm{T}\urcorner:\mathrm{T}\longrightarrow U$ and $\ulcorner\times\urcorner,\ulcorner\to\urcorner:U^2\longrightarrow U$ in $|\mathbf{P}|$ be as in (2.1) and (2.2). Now in $Gr(\mathbf{P})$, by (4.2) we have for the two product projections $\pi_i:TU^2\longrightarrow TU$ (i=0,1) that:

$$\pi_i^*(t) = (1_{U^2},\pi_i):(\pi_i : U^2 \longrightarrow U) \longrightarrow TU^2 .$$

Hence by (4.2) again

$$\pi_0^*(t) \times_{TU^2} \pi_1^*(t) = (1_{U^2}, \pi_0 \times_{U^2} \pi_1) = (1_{U^2}, `\times') = T(`\times')^*(t) \tag{4.7}$$

in $Gr(\mathbf{P})/TU^2$, giving a pullback square in $Gr(\mathbf{P})$ of the form (3.3). Similarly, using (4.3) we have:

$$\pi_0^*(t) \to_{TU^2} \pi_1^*(t) = (1_{U^2}, \pi_0 \to_{U^2} \pi_1) = (1_{U^2}, `\to') = T(`\to')^*(t) \tag{4.8}$$

giving the required pullback of the form (3.4); and by (4.2) again,

$$1_{T(\top)} = (1_\top, \top_\top) = (1_\top, `\top') = T(`\top')^*(t) \tag{4.9}$$

giving the pullback of the form (3.5) in $Gr(\mathbf{P})$. Then applying H to these pullback squares and using the result of Lemma 4.5 that H preserves the (pullbacks and) local exponentials involved in them, we get the required pullbacks in $[(Gr\mathbf{P})^{op},\mathbf{Set}]$ for the morphisms

$$\top \cong HT(\top) \xrightarrow{HT(`\top')} HT(U) \text{ and } HT(U)^2 \cong HT(U^2) \xrightarrow{HT(`*')} HT(U) \quad (\text{where } * = \times \text{ or } \to).$$

To produce a morphism $`\Pi':(HT(U) \to HT(U)) \longrightarrow HT(U)$ and a pullback of the form (3.6) is more complicated, since $|\mathbf{P}|$ is not a ccc and the exponential $HT(U) \to HT(U)$ is not a representable presheaf. However, consider the following calculation:

$$\begin{aligned}
(HT(U) \to HT(U))(-) &\cong [(Gr\mathbf{P})^{op},\mathbf{Set}](H(-), HT(U) \to HT(U)) && \text{(Yoneda lemma)}\\
&\cong [(Gr\mathbf{P})^{op},\mathbf{Set}](H(-) \times HT(U), HT(U)) && \text{(definition of exponential)}\\
&\cong [(Gr\mathbf{P})^{op},\mathbf{Set}](H(- \times TU), H(TU)) && (H \text{ preserves products})\\
&\cong Gr(\mathbf{P})(- \times TU, TU) && (H \text{ full and faithful})\\
&\cong |\mathbf{P}|(P(- \times TU), U) && (P \text{ left adjoint to } T)\\
&\cong |\mathbf{P}|(P(-) \times U, U). && (P \text{ preserves products and } PT = 1)
\end{aligned}$$

We can therefore identify the exponential $HT(U) \to HT(U)$ with the functor $|\mathbf{P}|(P(-) \times U, U):Gr(\mathbf{P})^{op} \longrightarrow \mathbf{Set}$. Similarly, the hom functor $HT(U)$ can itself be identified with $|\mathbf{P}|(P(-),U):Gr(\mathbf{P})^{op} \longrightarrow \mathbf{Set}$. But then part (*iv*) of Definition 2.2 gives a natural transformation $\prod_{P(-)}$: $|\mathbf{P}|(P(-) \times U, U) \longrightarrow |\mathbf{P}|(P(-),U)$ and we define $`\to':(HT(U) \to HT(U)) \longrightarrow HT(U)$ to be this. To get a pullback square of the form (3.6) as well, we have to show that $(`\to')^*(Ht) \cong \Pi_p ev^*(Ht)$ in $[(Gr\mathbf{P})^{op},\mathbf{Set}]/(HT(U) \to HT(U))$. Since by (4.5) the latter is equivalent to the topos of presheaves on $Gr(|\mathbf{P}|(P(-) \times U, U))$, it is sufficient to exhibit bijections

$$[(Gr\mathbf{P})^{op},\mathbf{Set}]/(HT(U) \to HT(U))(\alpha, (`\to')^*(Ht)) \cong [(Gr\mathbf{P})^{op},\mathbf{Set}]/(HT(U) \to HT(U))(\alpha, \Pi_p ev^*(Ht))$$

natural in $\alpha:H(X) \longrightarrow (HT(U) \to HT(U))$ and then apply the Yoneda lemma. But if $\hat{\alpha}:P(X) \times U \longrightarrow U$ corresponds to α under the identifications made above, then the definition of $`\to'$ gives:

$$`\to' \circ \alpha = H\left(\overline{\textstyle\prod_{PX} \hat{\alpha}}\right), \tag{4.10}$$

where $\overline{\prod_{PX} \hat{\alpha}}$ is the transpose of $\prod_{PX} \hat{\alpha}:P(X) \longrightarrow U$ across the adjunction $P \dashv T$. Now $\hat{\alpha}$ also corresponds to a morphism $X \times T(U) \longrightarrow T(U)$ in $Gr(\mathbf{P})$ — call it a, say: then by (4.4)

$$\left(\overline{\textstyle\prod_{PX} \hat{\alpha}}\right)^*(t) = \Pi_{\pi_1}(a^*(t)) \tag{4.11}$$

with π_1 the first projection $X \times T(U) \longrightarrow X$. Now apply H to (4.11) and use (4.10), plus the fact that H preseves finite limits and Π-functors (Lemma 4.5) and sends a to $\bar{\alpha}$, the exponential transpose of α; we get:

$$(`\!\to\!`\circ\alpha)^*(Ht) \cong \Pi_p \bar{\alpha}^*(Ht) \tag{4.12}$$

in $[(Gr\mathbf{P})^{op},\mathbf{Set}]/H(X)$, with p the first projection $H(X)\times HT(U) \longrightarrow H(X)$. Since $\bar{\alpha} = ev\circ(\alpha\times 1_{HX})$ and (by standard properties of Π-functors) $\Pi_p\circ(\alpha\times 1_{HX})^* \cong \alpha^*\circ\Pi_p$, (4.12) gives:

$$\alpha^*((`\!\to\!`)^*(Ht)) \cong \alpha^*(\Pi_p ev^*(Ht)).$$

Using this isomorphism we get a bijection between morphisms $\alpha \longrightarrow (`\!\to\!`)^*(Ht)$ and morphisms $\alpha \longrightarrow \Pi_p ev^*(Ht)$ in the slice category $[(Gr\mathbf{P})^{op},\mathbf{Set}]/(HT(U)\to HT(U))$. The constructions performed to get this bijection are evidently natural in α: so as remarked above, we can infer that they are induced by an isomorphism $(`\!\to\!`)^*(Ht) \cong \Pi_p ev^*(Ht)$, as required.

This establishes that we have a topos model $\mathbf{U}$ in $[(Gr\mathbf{P})^{op},\mathbf{Set}]$. Since by 4.2(*ii*) and standard properties of the Yoneda embedding, HT *is* full, faithful and finite product preserving, to complete the proof of Theorem 4.6 we just have to exhibit the natural isomorphism:

$$\mathbf{P}(-,U) \cong [(Gr\mathbf{P})^{op},\mathbf{Set}](HT(-),\mathbf{U}). \tag{4.13}$$

Our calculations in the earlier part of this proof imply that the category object $\mathbf{U}$ is the image under H of the category object $\mathbf{V}$ in $Gr(\mathbf{P})$ with underlying graph:

$$(U^2, `\!\to\!`) \underset{d_1}{\overset{d_0}{\rightrightarrows}} (U, \top_U) = T(U),$$

where $d_i = (\pi_i, `\!\to\!`)$ $(i=0,1)$. Thus for (4.13) it suffices to give a natural isomorphism

$$v : \mathbf{P}(-,U) \cong Gr(\mathbf{P})(T(-),\mathbf{V}).$$

For each $I \in |\mathbf{P}|$, define v on objects $A \in |\mathbf{P}|(I,U)$ by $v_I(A) = T(A)$; since T is full and faithful, this gives a bijection. To define v on morphisms, given $A,B : I \longrightarrow U$ in $|\mathbf{P}|$, note that morphisms $\langle TA,TB\rangle \longrightarrow \langle d_0,d_1\rangle$ in $Gr(\mathbf{P})/TU^2$ are specified by morphisms $\top_I \longrightarrow \langle A,B\rangle^*(`\!\to\!`) = A \to_I B$ in $\mathbf{P}(I,U)$, which correspond under the exponential adjunction to morphisms $A \longrightarrow B$ in $\mathbf{P}(I,U)$: then define v_I on morphisms by sending $f : A \longrightarrow B$ to the morphism $\langle TA,TB\rangle \longrightarrow \langle d_0,d_1\rangle$ over TU^2 specified by the exponential transpose of f. Routine calculations show that this recipe makes v_I into a functor $\mathbf{P}(I,U) \longrightarrow Gr(\mathbf{P})(TI,\mathbf{V})$, and its construction is evidently natural in I. It is a bijection on objects and on hom sets, and hence is an isomorphism of categories (and also of $2T\lambda C$-hyperdoctrines therefore). This completes the proof of the theorem.

□

We can apply Theorem 4.6 to the classifying hyperdoctrine $\mathbf{P_T}$ of a $2T\lambda C$-theory $\mathbf{T}$ (*cf.* 2.6). $\mathbf{P_T}$ contains the generic model $I_{\mathbf{T}}$ and from (2.10) we have that:

$$\mathbf{T} \vdash_{X,\mathbf{x}} s=t:\Phi \text{ iff } I_{\mathbf{T}} \models_{X,\mathbf{x}} s=t:\Phi.$$

Transporting $I_{\mathbf{T}}$ along the isomorphism $\mathbf{P_T}(-,U) \cong [(Gr\mathbf{P_T})^{op},\mathbf{Set}](HT(-),\mathbf{U})$ of Theorem 4.6, gives a topos model of $\mathbf{T}$ with the same property. So we deduce:

4.7. Corollary. (Completeness of topos models.) *Let* **T** *be a $2T\lambda C$-theory. Then the theorems of* **T** *are precisely those equality judgements in the language of* **T** *which are satisfied by all topos models of* **T**; *in fact there is a single topos model whose true equality judgements are exactly the theorems of* **T**.

□

The logical significance of Theorem 4.6 is greater than just the above corollary. Recall the corresponence, reviewed in section 2, between theories in the second order typed lambda calculus and hyperdoctrines; recall also the correspondence, mentioned at the beginning of section 3, between theories in *higher order intuitionistic predicate logic* (or *HOL*, for short) and elementary toposes. In the light of these correspondences, we may rephrase the full embedding theorem as follows:

Each $2T\lambda C$-theory can be interpreted in a theory in HOL so that:

(*i*) *the types of the $2T\lambda C$-theory have a* standard *interpretation in the HOL-theory* (*i.e.* $\top$ *is terminal,* $\times, \prod$ *are products and* $\rightarrow$ *is exponentiation*);

(*ii*) *any two closed terms of the $2T\lambda C$-theory which can be proved to be equal in the HOL-theory are already provably equal in the $2T\lambda C$-theory*;

(*iii*) *any closed term of a type coming from the $2T\lambda C$-theory which can be proved to exist using HOL is provably equal to a $2T\lambda C$-term.*

4.8. Extensions. We conclude by mentioning that the method of proving the full embedding result also suffices to prove similar results for related type theories. In particular theories in the full *higher order typed lambda calculus* (Girard's system F_ω augmented with finite product types) correspond to the kind of hyperdoctrine where $|\mathbf{P}|$ is required to be cartesian closed (and for any object V of $|\mathbf{P}|$, $\pi_I^*:\mathbf{P}(-,U)\longrightarrow\mathbf{P}(-\times V,U)$ is required to have a natural right adjoint). Since the Yoneda embedding preserves exponentials, the internal full subcategory **U** constructed in $[(Gr\mathbf{P})^{op},\mathbf{Set}]$ will now also be closed in the topos under internal products indexed by any object in the sub-ccc of the topos generated by the object of objects of **U**: indeed, as remarked in the proof of Theorem 4.6, the proof that **U** has the required closure properties is easier when $|\mathbf{P}|$ is a ccc. One could also augment the type theory with polymorphic sums, arriving at Seely's notion [Se] of a "PL-theory" and the corresponding kind of hyperdoctrine: the extra stucture is modelled by left adjoints to $(-)^*$ functors, and these are carried through into the topos model.

References

[B] J.Bénabou, *Fibred categories and the foundations of naive category theory*, J. Symbolic Logic 50(1985) 10-37.

[BM] K.B.Bruce and A.R.Meyer, *The semantics of second order polymorphic lambda calculus.* In: G.Kahn *et al* (eds), *Semantics of Data Types,* Lecture Notes in Computer Science No.173 (Springer-Verlag, Berlin - Heidelberg - New York - Tokyo, 1984) pp 131-144.

[CCM] G.Cousineau, P.L.Curien and M.Mauny, *The categorical abstract machine.* In: J.P.Jouannaud (ed.), *Functional Programming Languages and Computer Architecture,* Lecture Notes in Computer Science No.201 (Springer-Verlag, Berlin - Heidelberg - New York - Tokyo, 1985) pp 50-64.

[CE] T.Coquand and T.Ehrhard, *An equational presentation of higher order logic.* In this volume.

[CFS] A.Carboni, P.J.Freyd and A.Scedrov, in preparation.

[Cu] P.L.Curien, *Categorical combinators, sequential algorithms and functional programming,* Research Notes in Theoretical Computer Science (Pitman, London, 1986).

[F] M.Fourman, *Sheaf models for set theory,* J. Pure Appl. Algebra 19(1980) 91-101.

[FS] P.J.Freyd and A.Scedrov, *Some semantic aspects of polymorphic lambda calculus.* In: *Proc. 2nd Annual Symp. on Logic in Computer Science* (IEEE, 1987) pp 315-319.

[G1] J.-Y.Girard, *Interprétation fonctionelle et elimination des coupures dans l'arithmétique d'ordre supérieur,* Thèse de Doctorat d'Etat (Paris, 1972).

[G2] J.-Y.Girard, *The system F of variable types, fifteen years later,* Theoretical Computer Science 45(1986) 159-192.

[Gr] J.W.Gray, *Fibred and cofibred categories.* In: S.Eilenberg *et al* (eds), *Proc. LaJolla Conference on Categorical Algebra* (Springer-Verlag, Berlin - Heidelberg, 1966) pp 21-83.

[H] J.M.E.Hyland, *The effective topos.* In: A.S.Troelstra and D.vanDalen (eds), *The L.E.J.Brouwer Centenary Symposium* (North-Holland, Amsterdam, 1982) pp 165-216.

[HRR] J.M.E.Hyland, E.Robinson and G.Rosolini, *Discrete objects in the effective topos.* To appear.

[J] P.T.Johnstone, *Topos Theory,* L.M.S. Monographs No.10 (Academic Press, London, 1977).

[LS] J.Lambek and P.J.Scott, *Introduction to Higher Order Categorical Logic,* Cambridge Studies in Advanced Mathematics No.7 (Cambridge University Press, 1986).

[L] F.W.Lawvere, *Equality in hyperdoctrines and comprehension schema as an adjoint functor.* In: A.Heller (ed.), *Proc. New York Symp. on Applications of Categorical Algebra* (Amer. Math. Soc., Providence, 1970) pp 1-14.

[McC] D.C.McCarty, *Realizability and recursive set theory,* Annals Pure Appl. Logic 32(1986) 153-183.

[P] A.Poigné, *Tutorial on category theory and logic.* In: D.Pitt *et al* (eds), *Category theory and computer programming,* Lecture Notes in Computer Science No.240 (Springer-Verlag, Berlin - Heidelberg - New York, 1986) pp 103-142.

[PS] R.Paré and D.Schumacher, *Abstract families and the adjoint functor theorems.* In: P.T.Johnstone and R.Paré (eds), *Indexed Categories and their Applications,* Lecture Notes in Mathematics No.661 (Springer-Verlag, Berlin - Heidelberg - New York, 1978) pp 1-125.

[R1] J.C.Reynolds, *Types, abstraction and parametric polymorphism,* Information Processing 83(1983) 513-523.

[R2] J.C.Reynolds, *Polymorphism is not set-theoretic.* In: G.Kahn *et al* (eds), *Semantics of Data Types,* Lecture Notes in Computer Science Vol.173 (Springer-Verlag, Berlin - Heidelberg - New York - Tokyo, 1984) pp 145-156.

[Sc1] D.S.Scott, *Relating theories of the lambda calculus.* In: J.R.Hindley and J.P.Seldin (eds), *To H.B.Curry: essays on combinatory logic, lambda calculus and formalism* (Academic Press, London, 1980) pp 403-450.

[Sc2] D.S.Scott, *The presheaf model for set theory*. University of Oxford, February 1980. Manuscript.

[Se] R.A.G.Seely, *Categorical semantics for higher order polymorphic lambda calculus*, J. Symbolic Logic, to appear.

An equational presentation of higher order logic

Thierry Coquand and Thomas Ehrhard
INRIA, ENS and Cambridge

Introduction

We propose an equational version, based on a paper by Seely [Seely 86], of the higher-order type system of J.Y. Girard [Girard72]. The situation is here similar to the one in cartesian closed category w.r.t. simple typed calculus: there exists now a lot of equational presentations of this theory [Curien 86,Lambek-Scott,Scott,BreMey] [1]. One of the most obvious advantages of this kind of presentation is to avoid the problem of α-conversion (that appears crucially for an implementation). This is particularly important in higher-order calculi since the traditional restriction for the binding of type variables seems difficult to solve [2].

The main idea of Seely's formalism is to see quantifications as adjunctions. It is an old paradigm, first discovered by Lawvere ([Lawvere]), and developped by Seely in two papers ([Seely 82] and [Seely 86]). What we present here is just a syntactic and equational formalization of Seely's categorical presentation. We shall illustrate our formalism by showing that it furnishes a nice way for checking that the constructions of [Girard85] and [CoGuWi] are models of $F\omega$. We shall adopt the non-standard notation $f\,;g$ for the composition of $f : A \to B$ and $g : B \to C$.

1 Some motivations

One of the main goals of this paper is to present an "algebraic" treatment (i.e. without mention of the notion of bound variables) of the proof-theory of higher-order logic. We shall first recall what is higher-order logic, and how it is usually formalized.

The formalization of Church [Church] is the following. We define first the notion of *orders*. These are defined inductively by BNF syntax

$$A := \Omega \mid A \Rightarrow B \mid A \times B.$$

Intuitively, Ω is the order of truth values (viewed as a boolean algebra in classical logic, and a Heyting algebra in intuitionistic logic). $A \times B$ is the product, and $A \Rightarrow B$ the exponentiation of the orders A and B. An order can be approximatively thought of as a set (see [Lambek-Scott] for the precise connection between orders and sets). We can add also basic orders (in [Church], an order for representing the set of all integers is added).

Next we consider the typed λ-calculus (with orders as type) built on the constant $\Rightarrow$, of order $(\Omega \times \Omega) \Rightarrow \Omega$, and the polymorphic constant Π, of order $(A \Rightarrow \Omega) \Rightarrow \Omega$.

Now we have defined our language. Intuitively, λ-terms of type Ω are our formulae. As we have λ-abstraction, we can form arbitrary high-level concepts. For instance, the concept of

[1] An abstract machine has even been developped and implemented on the basis of one of those [CAM].

[2] Actually, this problem appears already in the formalization of first-order logic.

inclusion is a polymorphic one, of type $(A \Rightarrow \Omega) \Rightarrow (A \Rightarrow \Omega) \Rightarrow \Omega$ and is represented by the term $\lambda P^{A\Rightarrow\Omega}.\lambda Q^{A\Rightarrow\Omega}.\Pi(\lambda x^A.((\Rightarrow (Px))(Qx)))$. As the notation is somewhat cumbersome, we shall write $\sigma \times \tau$ (resp. $\sigma \Rightarrow \tau$) for $((\times\sigma)\tau)$ (resp. $((\Rightarrow \sigma)\tau)$), and $\forall x^A.\sigma$ for $\Pi(\lambda x^A.\sigma)$.

If we want a logical calculus, we must now define what is a provable formula. There are two possibilities, depending on whether we want to do classical or intuitionistic logic. It is well known how to translate classical into intuitionistic logic using Gödel translation, so that we shall restrict ourselves to intuitionistic logic (in any way, one can look at [Church] which is in a classical framework).

We shall present this calculus in a non-usual form, but that will make easier the connections with what follows. Our calculus will be a sequent calculus. The sequents are of the form $\sigma \to \tau$, where σ and τ are formulae. We shall introduce a basic formula $\mathbf{T}$, which represents the absolute true formula, so that a formula σ will be said *provable* if, and only if, $\mathbf{T} \to \sigma$ will be provable. Note that the terms and formulae are λ-terms, and we'll always consider them modulo β, η-reduction.

The axioms of our system are

$$(\varphi \Rightarrow \psi) \times \varphi \to \psi \quad \varphi \times \psi \to \varphi \quad \varphi \times \psi \to \psi \quad \varphi \to \mathbf{T}$$

and

$$\forall x^A.\varphi \to [t/x]\varphi$$

where t is any term of order A (rule of $\forall$-elimination). We have furthermore the following inference rules

$$\frac{\varphi \to \tau \quad \tau \to \psi}{\varphi \to \psi} \quad \frac{\varphi \to \psi \quad \varphi \to \tau}{\varphi \to \psi \times \tau} \quad \frac{\varphi \times \psi \to \tau}{\varphi \to \psi \Rightarrow \tau}$$

and, if x (of order A) *does not* appear free in φ

$$\frac{\varphi \to \psi}{\varphi \to \forall x^A.\psi.}$$

The system is now the intuitionistic version of higher-order logic of Church [Church]. This presentation is very near the presentation of Spector [Girard72]. Note that this version is purely *intensional* (the only equality between formulae is the β, η-conversion). The reader can compare with the extensional version of [Lambek-Scott] (we shall see in the last section how to add a natural number object to this intensional version).

1.1 The system $F\omega$

There exists also a natural deduction presentation of the previous system. We can then study the natural deduction proof of this system, seen as λ-terms [Howard]. We obtain the system $F\omega$, which is the natural framework for the study of proof in higher-order logic [Girard72]. This calculus corresponds to the simple type calculus. The system of equations we shall present is the equivalent of the equational formalism of [Lambek-Scott] relatively to higher-order logic (instead of intuitionistic propositional calculus).

So, what follows can be thought of as a system of notation for proofs in higher-order logic, with furthermore a notion of equality between these proofs. We shall see that we can interpret all the known notions of equality (β and η conversions).

Our system is intermediate between the lambda-calculus version of [Girard72] and the categorical version of [Seely 86]. The choice of an equational presentation of Seely's formalism combines two advantages: we do not have the syntactic problem of bound variables of the lambda-calculus and we do not have the semantic problem of equality "up to isomorphism" of category theory. Furthermore, we don't seem to lose something w.r.t. the model theory of higher type system, since we'll see that the domain models of [Girard85] and [CoGuWi] are indeed models of this equational theory[3].

2 A short presentation of Seely's categorical framework.

The main object of [Seely 86] is to give a semantical interpretation to the $F\omega$ system in indexed categories. Our paper gives an equational presentation of the corresponding formal theory. In order to give a more intuitive view of that formalism, we briefly present a simplified version of Seely's semantics.

Let S be a cartesian closed category (CCC for short), in which we shall distinguish an object called Ω. Next, we consider an *explicit* indexed category on S, that is a functor $G : S^{\mathrm{op}} \to \mathrm{Cat}$. We are so in a very special case of indexed category [Pare-Schu], which works in general with pseudo-functor, that is where the functoriality conditions hold "up to isomorphism". Surprisingly, it seems that this restriction is not important w.r.t. model theory, as we shall see later. For each $A \in S$, the category $G(A)$ must be a CCC, and if σ is a morphism from A to B in S, the functor $G(\sigma)$ must preserve the structure of CCC.

We assume a bijective correspondance between $\mathrm{Obj}\,(G(A))$ and $\mathrm{Hom}_S\,(A\ ,\Omega)$ for every $A \in S$, in such a way that we shall identify objects of $G(A)$ and arrows $A \to \Omega$. For all morphism $\sigma : A \to B$ in S, we note σ^* the functor $G(\sigma) : G(B) \to G(A)$, and we assume that that functor acts on objects of $G(B)$ by right composition with σ, that is $\sigma^*(\tau) = \sigma\ ;\tau$.

Now, for all $B \in S$, let G^B be the functor defined by $G^B(A) = G(A \times B)$ and $G^B(f) = G(f \times \mathrm{Id}_B)$. We may define a natural transformation Fst^*_B from G to G^B: if A is an object of S, $\mathrm{Fst}^*_B(A)$ is $G(\mathrm{Fst}_{A,B})$ (where $\mathrm{Fst}_{A,B}$ is the first projection from $A \times B$). We demand first that each $\mathrm{Fst}^*_B(A)$ has a right adjoint $\forall_B(A)$. Let $\Lambda_B(A)$ be the bijection of hom-sets of the adjunction:

$$\Lambda_B(A) : \mathrm{Hom}_{G^B(A)}\,((\mathrm{Fst}\ ;\sigma)\ ,\tau) \cong \mathrm{Hom}_{G(A)}\,(\sigma\ ,\forall_B(A)(\tau))\,.$$

We demand furthermore that $\forall_B$ form a natural transformation, and that $\Lambda_B(A)$ be natural in A. On the objects, the first request is equivalent to the fact that there exists $\Pi : (B \Rightarrow \Omega) \to \Omega$, and that, if σ is an object in $G^B(A)$,

$$\forall_B(A)(\sigma) = \mathrm{Cur}(\sigma)\ ;\Pi$$

(precisely, Π is $\forall_B(B \Rightarrow \Omega)(\mathrm{App})$) On the morphisms, this will add one new equation that we shall make precise later. For the writing of the equational version of this categorical presentation, we shall define the adjunction $\forall_B(A)$ by the bijection $\Lambda_B(A)$, and its co-unit morphism, that we shall note Proj (so that Proj is the morphism in $G^B(A)$ from $(\mathrm{Fst}\ ;\ \forall(\tau))$ to τ defined by $\mathrm{Proj} = \Lambda^{-1}\left(\mathrm{Id}_{\forall(\tau)}\right)$).

3 The equational theory

We shall now give a completely formal equational presentation. The reader must refer to the previous section for the intuition behind the typing rules and the equations.

3.1 The signature

We present a two-sorted equational theory, as for vector spaces. There will be *operators* (in analogy with scalars) and *terms* (in analogy with vectors). The operator meta-variables will be written in greek letters and the term meta-variables in roman letters. As in vector spaces, there will be one mixed binary operator (that we call *action*) between operators and terms.

---

[3] which is *more* than simply models in Seely's sense The difference between our notion of models and the categorical one is that we give explicitely, by skolemisation, the limits and colimits, and not only "up to isomorphism".

The signature of operators – This is the same as the one for Combinatory Categorical Logic (see [Curien 86], this will be abbreviated in CCL) with three additional constants.

The binary operators are: ";", "$\langle \_ , \_ \rangle$". There is one unary operator: Cur(_). The usual constants are: Fst, Snd, App, Id. All this constitutes the signature of CCL. We add $\Rightarrow$, Π, and $\times$ as constants.

The signature of terms – It includes the signature of CCL, that we shall overload between the two sorts. We add one unary operator: $\Lambda(\_)$, and one constant: Proj.

The mixed operator – It takes respectively one operator and one term and produces one term. We shall write it $*$.

3.2 Typing

Now we present a way to type those categorical operators and terms. We must first add *orders* [Girard72,Seely 86]. These are defined inductively by BNF syntax

$$A := \Omega \mid A \Rightarrow B \mid A \times B.$$

A typing statement has the following form

$$x : y \rightarrow z,$$

where, either x is an operator and y, z are orders, or x is a term and y, z are operators.

Typing rules for operators – They are the ones of the cartesian closed categories, with the addition of typing rules for the supplementary constants.

$$\frac{\sigma : A \rightarrow B \quad \tau : B \rightarrow C}{\sigma ; \tau : A \rightarrow C} \quad \frac{\sigma : A \rightarrow B \quad \tau : A \rightarrow C}{\langle \sigma , \tau \rangle : A \rightarrow B \times C} \quad \frac{\sigma : A \times B \rightarrow C}{\mathrm{Cur}(\sigma) : A \rightarrow B \Rightarrow C}$$

$$\mathrm{Fst} : A \times B \rightarrow A \quad \mathrm{Snd} : A \times B \rightarrow B \quad \mathrm{App} : (A \Rightarrow B) \times A \rightarrow B \quad \mathrm{Id} : A \rightarrow A$$

For the supplementary constants, we add

$$\times : (\Omega \times \Omega) \rightarrow \Omega \quad \Rightarrow : (\Omega \times \Omega) \rightarrow \Omega \quad \Pi : (A \Rightarrow \Omega) \rightarrow \Omega.$$

Equations on operators – These are the equations of CCL [Curien 86].

$$\begin{aligned}
\sigma ; (\tau ; \theta) &= (\sigma ; \tau) ; \theta \\
\sigma ; \mathrm{Id} &= \sigma \\
\mathrm{Id} ; \sigma &= \sigma \\
\langle \sigma , \tau \rangle ; \mathrm{Fst} &= \sigma \\
\langle \sigma , \tau \rangle ; \mathrm{Snd} &= \tau \\
\langle \mathrm{Fst} , \mathrm{Snd} \rangle &= \mathrm{Id} \\
\theta ; \langle \sigma , \tau \rangle &= \langle \theta ; \sigma , \theta ; \tau \rangle \\
\langle \mathrm{Cur}(\sigma) , \tau \rangle ; \mathrm{App} &= \langle \mathrm{Id} , \tau \rangle ; \sigma \\
\tau ; \mathrm{Cur}(\sigma) &= \mathrm{Cur}(\langle \mathrm{Fst} ; \tau , \mathrm{Snd} \rangle ; \sigma) \\
\mathrm{Cur}(\mathrm{App}) &= \mathrm{Id}
\end{aligned}$$

Typing the terms – We shall write $\sigma \times \tau$ (resp. $\sigma \Rightarrow \tau$) for $\langle \sigma , \tau \rangle ; \times$ (resp. $\langle \sigma , \tau \rangle ; \Rightarrow$), and $\forall(\sigma)$ for $\mathrm{Cur}(\sigma) ; \Pi$.

We first have something anologous to the operators typing rules:

$$\frac{t : \sigma \to \tau \quad u : \tau \to \rho}{t ; u : \sigma \to \rho} \quad \frac{t : \sigma \to \tau \quad u : \sigma \to \rho}{\langle t , u \rangle : \sigma \to \tau \times \rho} \quad \frac{t : \sigma \times \tau \to \rho}{\mathrm{Cur}(t) : \sigma \to \tau \Rightarrow \rho}$$

$$\frac{\sigma : A \to \Omega \quad \tau : A \to \Omega}{\mathrm{Fst} : \sigma \times \tau \to \sigma} \quad \frac{\sigma : A \to \Omega \quad \tau : A \to \Omega}{\mathrm{Snd} : \sigma \times \tau \to \tau}$$

$$\frac{\sigma : A \to \Omega \quad \tau : A \to \Omega}{\mathrm{App} : (\sigma \Rightarrow \tau) \times \sigma \to \tau} \quad \frac{\sigma : A \to \Omega}{\mathrm{Id} : \sigma \to \sigma}$$

For the supplementary terms, we add

$$\frac{t : (\mathrm{Fst} ; \sigma) \to \tau}{\Lambda(t) : \sigma \to \forall(\tau)}$$

$$\mathrm{Proj} : (\mathrm{Fst} ; \forall(\sigma)) \to \sigma.$$

Finally we must add the type equality rule

$$\frac{t : \sigma \to \tau \quad \sigma = \sigma_1 \quad \tau = \tau_1}{t : \sigma_1 \to \tau_1}$$

Here the equality is to be understood as the equality defined by the equational theory on operators given previously. This is an original feature. The definition of typing only makes sense modulo an equational theory. The general framework for this kind of theories is the notion of graphical algebras of [Burroni].

Typing rule for the action –

$$\frac{\sigma : A \to B \quad \tau, \rho : B \to \Omega \quad t : \tau \to \rho}{\sigma * t : (\sigma ; \tau) \to (\sigma ; \rho)} \quad .$$

Remark – Relatively to the Curry-Howard paradigm, we note the following logical interpretation of the additional terms (for the terms of CCL, see [Curien 86]): $\Lambda(\_)$ corresponds to the $\forall$-introduction inference rule, Proj represents the $\forall$-elimination axiom schema, and $*$ is the notion of instantiation of a schema (that is usually expressed only at the meta-level). The reader can compare these rules with the ones of section 1.

In what follows, the operators are considered modulo CCL-equality, as defined above. The axioms of the system become

$$(\varphi \Rightarrow \psi) \times \varphi \to \psi \quad \varphi \times \psi \to \varphi \quad \varphi \times \psi \to \psi$$

and

$$\mathrm{Fst} ; \forall(\varphi) \to \varphi,$$

but we need furthermore the instantiation schema

$$\frac{\varphi \to \psi}{(\tau ; \varphi) \to (\tau ; \psi)}.$$

The inference rules are now

$$\frac{\varphi \to \tau \quad \tau \to \psi}{\varphi \to \psi} \quad \frac{\varphi \to \psi \quad \varphi \to \tau}{\varphi \to \psi \times \tau} \quad \frac{\varphi \times \psi \to \tau}{\varphi \to \psi \Rightarrow \tau}$$

and

$$\frac{(\mathrm{Fst} \; ; \varphi) \to \psi}{\varphi \to \forall(\psi)} \; .$$

We can thus read the typing rules as the formulation of a Gentzen-like system for higher-order logic, but *without* the boring problem of variable names, cf. [Takeuti] for a classical presentation of higher-order logic. Note the way we express the $\forall$-introduction: the condition that a variable does not appear free becomes the fact that the left formula is of the form Fst ; φ. We get the "ordinary" $\forall$-elimination by a combination of an axiom and the instantiation schema. It reads:

$$\forall(\varphi) \to \langle \mathrm{Id} \, , \tau \rangle \; ; \varphi,$$

and $\langle \mathrm{Id} \, , \tau \rangle \; ; \varphi$ represents the "substitution" of τ in φ.

3.3 The equations for terms

The first equations are the one of CCL [Curien 86].

$$\begin{aligned}
s \; ; (t \; ; u) &= (s \; ; t) \; ; u \\
s \; ; \mathrm{Id} &= s \\
\mathrm{Id} \; ; s &= s \\
\langle s \, , t \rangle \; ; \mathrm{Fst} &= s \\
\langle s \, , t \rangle \; ; \mathrm{Snd} &= t \\
\langle \mathrm{Fst} \, , \mathrm{Snd} \rangle &= \mathrm{Id} \\
u \; ; \langle s \, , t \rangle &= \langle u \; ; s \, , u \; ; t \rangle \\
\langle \mathrm{Cur}(s) \, , t \rangle \; ; \mathrm{App} &= \langle \mathrm{Id} \, , t \rangle \; ; s \\
t \; ; \mathrm{Cur}(s) &= \mathrm{Cur}(\langle \mathrm{Fst} \; ; t \, , \mathrm{Snd} \rangle \; ; s) \\
\mathrm{Cur}(\mathrm{App}) &= \mathrm{Id}
\end{aligned}$$

Then we express that $*$ is a contravariant functor:

$$\begin{aligned}
(\sigma \; ; \tau) * t &= \sigma * (\tau * t) \\
\mathrm{Id} * t &= t
\end{aligned}$$

The next equations express the fact that operators act on this structure of CCL for terms by $*$ (more precisely, the first two equations express the covariant functoriality of each σ^*, and the other ones say that they preserve CCC structure on the nose).

$$\begin{aligned}
\sigma * (t \; ; u) &= (\sigma * t) \; ; (\sigma * u) \\
\sigma * \mathrm{Id} &= \mathrm{Id} \\
\sigma * \langle t \, , u \rangle &= \langle \sigma * t \, , \sigma * u \rangle \\
\sigma * \mathrm{Cur}(t) &= \mathrm{Cur}(\sigma * t) \\
\sigma * \mathrm{Fst} &= \mathrm{Fst} \\
\sigma * \mathrm{Snd} &= \mathrm{Snd} \\
\sigma * \mathrm{App} &= \mathrm{App}
\end{aligned}$$

Next, we give the equations which express the fact that $\Lambda(\_)$ and Proj define the right adjunction of Fst $*$ _. From now on, in order to avoid parentheses, we shall make the convention that the precedence of "$*$" ist greater than the one of ";".

$$\begin{aligned} \mathrm{Fst} * \Lambda(s) \,;\, \mathrm{Proj} &= s \\ s \,;\, \Lambda(t) &= \Lambda(\mathrm{Fst} * s \,;\, t) \\ \Lambda(\mathrm{Proj}) &= \mathrm{Id}. \end{aligned}$$

Furthermore, as seen before, we express that these adjunctions are natural. More precisely, we write that $\Lambda(\_)$ and Proj are natural, that is:

$$\begin{aligned} \Lambda(\langle \mathrm{Fst} \,;\, \sigma \,,\, \mathrm{Snd}\rangle * s) &= \sigma * \Lambda(s) \\ \langle \mathrm{Fst} \,;\, \sigma \,,\, \mathrm{Snd}\rangle * \mathrm{Proj} &= \mathrm{Proj} \end{aligned}$$

P.L. Curien pointed out the important consequence of that last equation and of $\Lambda(\mathrm{Proj}) = \mathrm{Id}$, which expresses, as we shall see, extensionality for the application of a term to an operator: $\Lambda(\langle \mathrm{Id} \,,\, \mathrm{Snd}\rangle * \mathrm{Proj}) = \mathrm{Id}$.

Some remarks – First, we can notice the following lemma (proved by structural induction):

Lemma 1 *If $t : \sigma \to \tau$ is a derived proposition, then there exists an order A such that $\sigma : A \to \Omega$ and $\tau : A \to \Omega$ are derived.*

We can furthermore see that all our equations are typable relatively to this system. For example the equation

$$(\mathrm{Fst} * \Lambda(s)) \,;\, \mathrm{Proj} = s$$

is typable with $s : (\mathrm{Fst} \,;\, \sigma) \to \tau$, $\sigma : A \to \Omega$, $\tau : A \times B \to \Omega$, $\mathrm{Fst} : A \times B \to A$, and $\mathrm{Proj} : (\mathrm{Fst} \,;\, \Pi(\tau)) \to \tau$.

Finally, we make the following conjecture:

Conjecture 1 *It is decidable whether or not a given term is typable, and furthermore, if the term is typable, it has a most general type (in the appropriate sense).*

From higher order to first order logic – As previously said, the main idea of this formalism is to code quantifications into adjunctions. We did it for universal quantification, but it is possible for existential too as R. Seely did in his papers. P.L. Curien has developped the corresponding equational formalism, see [Curien 87]. Of course, this idea is not specific to higher order logic, and R. Seely applied it to first order in his paper [Seely 82]. It is easy to write the corresponding equational theory, in the same spirit as for higher order.

In fact, the obtained theory is very similar to the one for higher order logic where we introduce the categorical equivalent of atomic predicate symbols. The base category in this case needs only to have products. The objects of this base category represent the types of individuals, and the arrows the terms. The objects of the fiber of an object A represent first-order propositions about individuals of type A. The logical formulae so represented are the first-order ones built on $\Rightarrow$ and $\forall$. We are obliged to write explicitly the object parts of the naturality of adjunctions since we don't assume any more bijections between the object parts of the slide categories and hom-sets in the base category.

4 Translation of $F\omega$ in categorical terms.

4.1 A formalism "à la de Bruijn" for the $F\omega$ system.

Let us consider the minimal system defined by the following syntax:
Orders:

$$A := \Omega \mid A \Rightarrow B$$

Operators:

$$\sigma := k \mid \sigma \Rightarrow \tau \mid \forall A\,\sigma \mid [A]\,\sigma \mid \sigma\,\tau$$

Terms:

$$t := n \mid s\,t \mid [\sigma]\,t \mid t\,\{\sigma\} \mid \Lambda A\,t$$

In those BNF definitions, k and n stand for de Bruijn numbers.

For an operator σ, we shall note FV (σ) the number of free variables of the corresponding operator in the usual notation. In the same way, we define FV (t) (number of free variables of term) and $\mathrm{FV_O}\,(t)$ (number of free variables of operator) for any term t.

The typing and the equality of operators. – A context of orders is a finite list $\alpha = (\alpha_0, \cdots, \alpha_{k-1})$ of orders. $|\alpha|$ is the length k of α and $A \cdot \alpha$ is the context $\alpha = (A, \alpha_0, \cdots, \alpha_{k-1})$. More generally, for $i \leq k$, $\alpha[i \leftarrow A]$ is the context $(\alpha_0, \cdots, \alpha_{i-1}, A, \alpha_i, \cdots, \alpha_{k-1})$. If α is a context of orders, we define the predicate $\alpha \vdash \sigma : A$, where σ is a an operator such that FV $(\sigma) \leq |\alpha|$ and A is an order, inductively on order structure by

$$\alpha \vdash k : \alpha_k \qquad \frac{\alpha \vdash \sigma : \Omega \quad \alpha \vdash \tau : \Omega}{\alpha \vdash \sigma \Rightarrow \tau : \Omega} \qquad \frac{A \cdot \alpha \vdash \sigma : \Omega}{\alpha \vdash \forall A\,\sigma : \Omega}$$

$$\frac{A \cdot \alpha \vdash \sigma : B}{\alpha \vdash [A]\,\sigma : A \Rightarrow B} \qquad \frac{\alpha \vdash \sigma : A \Rightarrow B \quad \alpha \vdash \tau : A}{\alpha \vdash \sigma\tau : B}$$

On operators, we shall consider a $\beta\eta$ theory, and we shall note $=$ the equality modulo $\beta\eta$ conversion. The usual de Bruijn's operators of shifting and substitution will be noted respectively $\_[j \uparrow k \rightarrow]$ and $\_[k \leftarrow \_]$ for the shifting of j from depth k and the substitution at depth k (see [Bruijn]). We shall abbreviate $\_[1 \uparrow 0 \rightarrow]$ in $\_\uparrow$. They are defined inductvely on operator structure.

The typing of terms – For a given context of orders α, a valid context of types is a finite list a of types, ie. of operators a_i such that $\alpha \vdash a_i : \Omega$ for all i. If a satisfies this condition w.r.t. α, we define the predicate $\alpha, a \vdash t : \sigma$, where t is a term such that FV $(t) \leq |a|$ and $\mathrm{FV_O}\,(t) \leq |\alpha|$ and σ is an operator, inductively on term structure by

$$\alpha, a \vdash n : a_n \qquad \frac{\alpha, a \vdash t : \sigma \Rightarrow \tau \quad \alpha, a \vdash u : \sigma}{\alpha, a \vdash t\,u : \tau} \qquad \frac{\alpha, \sigma \cdot a \vdash t : \tau}{\alpha, a \vdash [\sigma]\,t : \sigma \Rightarrow \tau}$$

$$\frac{\alpha, a \vdash t : \forall A\sigma \quad \alpha \vdash \tau : A}{\alpha, a \vdash t\{\tau\} : \sigma\,[0 \leftarrow \tau]} \qquad \frac{A \cdot \alpha, a \Uparrow \vdash t : \sigma}{\alpha, a \vdash \Lambda A t : \forall A\,\sigma} \qquad \frac{\alpha, a \vdash t : \sigma \quad \sigma = \sigma'}{\alpha, a \vdash t : \sigma'}$$

where $a\,[j \uparrow k \rightarrow] = (a_1\,[j \uparrow k \rightarrow], a_2\,[j \uparrow k \rightarrow], \cdots)$.

It is easy to define de Bruijn operators of shifting and substitution for the variables of operator and for the variables of term of a term, and that is done inductively on term structure. The first ones will be noted $\_[\mathbf{j} \uparrow \mathbf{k} \rightarrow]$ and $\_[\mathbf{k} \leftarrow \_]$ and the second ones $\_[m \uparrow n \rightarrow]$ and $\_[n \leftarrow \_]$. We shall abbreviate $\_[\mathbf{1} \uparrow \mathbf{0} \rightarrow]$ in $\_\Uparrow$.

On terms, we shall consider a $\beta\eta$ theory for the application to terms and to operators. In our formalism, the corresponding equations will be written

$$
\begin{aligned}
([\sigma]\ s)\ t &= s\,[0 \leftarrow t] \\
[\sigma]\ (s \uparrow 0) &= s \\
(\Lambda A\, s)\,\{\sigma\} &= s\,[0 \leftarrow \sigma] \\
\Lambda A\ (s \Uparrow \{0\}) &= s
\end{aligned}
$$

4.2 The translation

Essentially, the operators are interpreted as the terms of the simply typed λ-calculus in CCL [Curien 86].

The translation of operators – Every order of the $F\omega$ system is naturally interpreted as the corresponding order in the categorical formalism. We note $[A]$ this interpretation for an order A. If $\alpha = (\alpha_n, \cdots, \alpha_1)$ is a context of orders, we define $\beta = [\alpha]$ as the finite product $(\cdots(T \times [\alpha_1])\cdots) \times [\alpha_n]$ where T is any order (it will never be acceded)[4]. Finally, for all operator σ such that $\alpha \vdash \sigma : A$, we define $[\![\sigma]\!]_\beta$, operator of our categorical formalism such that $[\![\sigma]\!]_\beta : \beta \rightarrow [A]$, inductively as follows (it is the same translation as the one given in [Curien 86,Huet] for λ-calculus in de Bruijn notation):

$$
\begin{aligned}
[\![k]\!]_\beta &= \mathrm{Fst}^k\ ;\ \mathrm{Snd} \\
[\![\sigma\,\tau]\!]_\beta &= \langle [\![\sigma]\!]_\beta\ ,\ [\![\tau]\!]_\beta \rangle\ ;\ \mathrm{App} \\
[\![[A]\ \sigma]\!]_\beta &= \mathrm{Cur}([\![\sigma]\!]_{\beta\times[A]}) \\
[\![\sigma \Rightarrow \tau]\!]_\beta &= [\![\sigma]\!]_\beta \Rightarrow [\![\tau]\!]_\beta \\
[\![\forall A\,\sigma]\!]_\beta &= [\![[A]\ \sigma]\!]_\beta\ ;\ \Pi
\end{aligned}
$$

That last equation means that the role of the morphism Π is to transform a propositional schema into a universally quantified proposition.

The translation of terms – Let α be a context of orders and $a = (a_n, \cdots, a_1)$ a context of types in that context of orders. If $\beta = [\alpha]$, we define $b = [\![a]\!]_\beta$ as the product $(\cdots(\theta \times [\![a_1]\!]_\beta)\cdots) \times [\![a_n]\!]_\beta$ where θ is the operator described below (again, it never will be acceded). We choose an object θ_T in the category $G(T)$. It will represent the empty context of types in the empty context of orders. Now for each context α of orders we can consider the projection on T $\mathrm{Nil}_\alpha : [\alpha] \rightarrow T$ wich is a composition of first projections. We take in each category $G([\alpha])$ as starting object for building contexts the object $\mathrm{Nil}_\alpha\ ;\ \theta$. Once more, if we have a terminal object in each fiber (strictly preserved by each transition functor), we can choose it. Now, for any term t of the system $F\omega$ such that $\alpha, a \vdash t : \sigma$, we define $[\![t]\!]^\beta_b$, categorical term satisfying $[\![t]\!]^\beta_b : b \rightarrow [\![\sigma]\!]_\beta$, inductively on term structure by

$$
\begin{aligned}
[\![n]\!]^\beta_b &= \mathrm{Fst}^n\ ;\ \mathrm{Snd} \\
[\![t\,u]\!]^\beta_b &= \langle [\![t]\!]^\beta_b\ ,\ [\![u]\!]^\beta_b \rangle\ ;\ \mathrm{App} \\
[\![[\sigma]\ t]\!]^\beta_b &= \mathrm{Cur}([\![t]\!]^\beta_{b\times[\![\sigma]\!]_\beta}) \\
[\![t\,\{\tau\}]\!]^\beta_b &= [\![t]\!]^\beta_b\ ;\ (\langle \mathrm{Id}\ ,\ [\![\tau]\!]_\beta \rangle * \mathrm{Proj}) \\
[\![\Lambda A\,t]\!]^\beta_b &= \Lambda([\![t]\!]^{\beta\times[A]}_{\mathrm{Fst};b})
\end{aligned}
$$

[4]We can take for T a terminal object if we have one, but it is important to note that we actually can interpret the empty context by an arbitrary object. The same remark will apply for the context of types.

Let us give a small example. The traduction of the polymorphic identity $\Lambda A:\Omega.\lambda x:A.x$ of type $\Pi A:\Omega.A\Rightarrow A$ will be the morphism $\Lambda(\mathrm{Cur}(\mathrm{Snd}))$ of type $\theta\to\mathrm{Cur}(\mathrm{Snd}\Rightarrow\mathrm{Snd})$; Π in the fiber $G(T)$.

In order to prove a soundness theorem, we need some substitution lemmas which make the connection between the categorical and de Bruijn formalisms. We don't give all of them, but just the ones which concern the action of operators on terms:

Lemma 2 *If $\alpha, a\vdash t:\sigma$ then $A\cdot\alpha, a\uparrow\vdash t\Uparrow:\sigma\uparrow$ and*

$$\mathrm{Fst}*[\![t]\!]^{\beta}_{b}=[\![t\Uparrow]\!]^{\beta\times[A]}_{\mathrm{Fst};b}$$

where $\beta=[\alpha]$ and $b=[\![a]\!]_{\beta}$.

Lemma 3 *If $A\cdot\alpha, a\Uparrow\vdash t:\tau$ and $\alpha\vdash\sigma:A$ then*

$$\langle\mathrm{Id}\,,[\![\sigma]\!]_{\beta}\rangle*[\![t]\!]^{\beta\times[A]}_{\mathrm{Fst};b}=[\![t\,[0\leftarrow\sigma]]\!]^{\beta}_{b}$$

where $\beta=[\alpha]$ and $b=[\![a]\!]_{\beta}$.

Soundness – With those lemmas, it is possible to prove the following

Theorem 1

$$s=t\Rightarrow[\![s]\!]^{b}_{\beta}=[\![t]\!]^{b}_{\beta}$$

where the left equality is the smallest congruence on $F\omega$ terms containing the stated β- and η-rules, and the right equality is the one that we have defined in the preceding section for categorical terms.

We shall just show what happens for application to operators.

For the β-rule, we have:

$$[\![(\Lambda A\,t)\,\{\sigma\}]\!]^{\beta}_{b}=\Lambda([\![t]\!]^{\beta\times[A]}_{\mathrm{Fst};b})\;;\;(\langle\mathrm{Id}\,,[\![\sigma]\!]_{\beta}\rangle*\mathrm{Proj})$$

by applying the definition of term interpretation. Now we get by an easy categorical calculus, using the fact that "$*$" is an action:

$$[\![(\Lambda A\,t)\,\{\sigma\}]\!]^{\beta}_{b}=\langle\mathrm{Id}\,,[\![\sigma]\!]_{\beta}\rangle*\Big(\big(\mathrm{Fst}*\Lambda([\![t]\!]^{\beta\times[A]}_{\mathrm{Fst};b})\big)\;;\;\mathrm{Proj}\Big)$$

and we can use the first rule of adjunction which yields:

$$[\![(\Lambda A\,t)\,\{\sigma\}]\!]^{\beta}_{b}=\langle\mathrm{Id}\,,[\![\sigma]\!]_{\beta}\rangle*[\![t]\!]^{\beta\times[A]}_{\mathrm{Fst};b}$$

and we conclude by using the substitution lemma.

For the η-rule:

$$[\![\Lambda A\,(t\uparrow\{0\})]\!]^{\beta}_{b}=\Lambda([\![t]\!]^{\beta\times[A]}_{\mathrm{Fst};b}\;;\;(\langle\mathrm{Id}\,,\mathrm{Snd}\rangle*\mathrm{Proj}))$$

but we know by substitution lemma that $[\![t]\!]^{\beta\times[A]}_{\mathrm{Fst};b}=\mathrm{Fst}*[\![t]\!]^{\beta}_{b}$ and we conclude by using the second equation of adjunction, and the categorical equation of extensionality.

Remark – It is possible to extend $F\omega$ by adding a product at the level of the types and at the levels of the orders, and to extend the previous translation to this calculus. Then the inverse translation doesn't raise any problem and is a direct extension of the one of [Curien 86].

5 Application to the semantics of $F\omega$

What follows is an application of the previous work. It is shown how to use the categorical combinators to prove that something is really a model of $F\omega$, simply by checking some equations.

The most naive idea in building a Seely's model for $F\omega$ would be to take for S the category of all categories and for Ω the category of sets. Doing that, we define the required adjunctions in terms of limits, and our attempt obviously fails since there are not all (large) limits in Ω. But J.Y. Girard showed in [Girard85] how to get a model for F using much subtler structures (Qualitative Domains). In [CoGuWi] a more general framework has been given to those constructions, that show that these ideas may apply as well with other notions of domains. Actually the abstract version is simpler in the case of complete algebraic lattices than in the case of dI-domains (where the richer structure of domains causes strange phenomena). We shall for this reason restrict ourselves to the simpler case of complete algebraic lattices.

The main idea is to define variable types as some functors (just like in the naive case, but we don't take all of them) and to define then the notion of "object of variable type". Then the semantical interpretation of the universal quantification of a variable type will be the collection of all objects of this variable type. Of course all the difficulty is to keep this collection small enough, and to give it the required structure; it was the object of the two papers cited above.

5.1 O-categories

O-categories are categories where hom-sets are ordered and we shall require filtered families of morphisms to have a l.u.b. . Composition must be monotonic and continuous. When $\mathbf{C}$ is a O-category, we can (cf [Plotkin-Smyth]) define a category $\mathbf{C}^+$ of which objects are those of $\mathbf{C}$ and each $f \in \mathrm{Hom}_{\mathbf{C}^+}(A, B)$ is a pair $\langle \varphi, \psi \rangle$ where $\varphi \in \mathrm{Hom}_{\mathbf{C}}(A, B)$ and $\psi \in \mathrm{Hom}_{\mathbf{C}}(B, A)$ satisfy $\varphi ; \psi \geq \mathrm{Id}$ and $\psi ; \varphi \leq \mathrm{Id}$[5]. We shall note $\psi = f^-$ and $\varphi = f^+$. The category structure of C^+ is defined by $\mathrm{Id} = \langle \mathrm{Id}, \mathrm{Id} \rangle$ and $f ; g = \langle f^+ ; g^+, g^- ; f^- \rangle$.

A O-category $\mathbf{C}$ will be said to be a O-CCC (cartesian closed category) if it is a CCC in the usual way and if furthermore product and exponentiation of arrows are continuous and monotonic in *both* components.

Let $\mathbf{A}$ be a category, $\mathbf{C}$ be an O-CCC. We shall see that it is possible to put on the collection of continuous functors from $\mathbf{A}$ to $\mathbf{C}^+$ a structure of O-CCC (where continuous means to preserve filtered colimits). Let F and G be two functors $\mathbf{A} \to \mathbf{C}^+$. It is possible to define two functors $F \times G$ and $F \Rightarrow G$ in the following way:
Object parts of those functors are given for $X \in \mathbf{A}$ by $(F \times G)(X) = F(X) \times G(X)$ and $(F \Rightarrow G)(X) = F(X) \Rightarrow G(X)$.
For the arrow part, we set if $f \in \mathrm{Hom}_{\mathbf{A}}(X, Y)$

$$\begin{aligned}
(F \times G)(f)^+ &= F(f)^+ \times G(f)^- \\
(F \times G)(f)^- &= F(f)^- \times G(f)^- \\
(F \Rightarrow G)(f)^+ &= F(f)^- \Rightarrow G(f)^+ \\
(F \Rightarrow G)(f)^- &= F(f)^+ \Rightarrow G(f)^-
\end{aligned}$$

(Actually, $\_\times\_$ and $\_\Rightarrow\_$ may be seen as functors $\mathbf{C}^+ \times \mathbf{C}^+ \to \mathbf{C}^+$ covariant in both components.)

Definition 1 *The category $\mathcal{F}(A, C)$ is defined as having objects the continuous functors from $\mathbf{A}$ to $\mathbf{C}^+$ and arrows $F \to G$ the families $(\Phi(X))_{X \in \mathbf{A}}$ such that*

- $\Phi(X) \in \mathrm{Hom}_{\mathbf{C}}(F(X), G(X))$

[5] This definition is not the original one of [Plotkin-Smyth], however, the reader can check that all the ordinary properties are still valid with this definition, cf. [Gunter85] for the construction of the inverse limit in this framework.

- *for all* $X, Y \in \mathbf{A}$ *and* $f \in \mathrm{Hom}_{\mathbf{A}}(X, Y)$*, we have* $F(f)^- ; \Phi(X) ; G(f)^+ \leq \Phi(Y)$ *(Φ is then said to be monotonic)*
- *whenever* X *is the filtered limit of* (X_i, f_i)*, that is* id_X *is the directed sup of the family of projectors* $f_i^- ; f_i^+$*, then* $\Phi(X)$ *is the sup of the family* $F(f_i)^- ; \Phi(X_i) ; G(f_i)^+$ *(Φ is then said to be continuous).*

With the notations of this definition we have the following

Theorem 2 *$\mathcal{F}(A, C)$ is a* **O**-*CCC taking* $\langle \Phi , \Psi \rangle(X) = \langle \Phi(X) , \Psi(X) \rangle$, $\mathrm{Cur}(\Phi)(X) = \mathrm{Cur}(\Phi(X))$ *and* $\Phi \leq \Psi$ *if, and only if, for all* $X \in \mathbf{A}$ *we have* $\Phi(X) \leq \Psi(X)$ *in* $\mathrm{Hom}_{\mathbf{C}}(F(X), G(X))$.

This very general theorem sets up the framework in which we shall develop the domain model of our equational theory. The base category will be a category of categories, with continuous functors as morphisms. The object Ω will be the category $\mathbf{Dom}^+$, where **Dom** is **O**-CCC of complete algebraic lattices. The fiber of an object A is then the ccc associated to the **O**-CCC $\mathcal{F}(A, \mathbf{Dom}^+)$.

5.2 Higher-order product

Let **Dom** be the usual **O**-CCC of complete algebraic lattices. We first recall the basic result of [CoGuWi].

Definition 2 *We say that an object* x *in a category* A *is finitely presentable if and only if the functor* $\mathrm{Hom}_A(x , \_)$ *commutes with filtered colimits.*

Let A *be a category. We say that* A *is* algebroidal *if, and only if, there exists a set* S *of objects of* A *such that each object in* A *is the filtered limit of objects in* S*, and that each object in* S *is finitely presentable. We say then that this set* generates *the category* A.

Definition 3 *Let* A *be an algebroidal category, and* F *a continuous functor from* A *to* $\mathbf{Dom}^+$. *A continuous section of* F *is a family* (t_X) *indexed over objects of* $\mathbf{Dom}^+$ *such that* t_X *is an element of* $F(X)$ *and if* X *is the filtered limit of* (X_i, f_i)*, then* t_X *is the sup of the directed family* $(F(f_i)^+(t_{X_i}))$.

The class of such sections, for the pointwise ordering is nothing else but the category of continuous sections of the Grothendiek fibration of the functor F. The surprising point is that this category is actually a complete algebraic lattice, as soon as the category A is algebroidal.

Theorem 3 *Let* A *be an algebroidal category, and* F *a continuous functor from* A *to* $\mathbf{Dom}^+$. *Then the class of continuous section of* F *is a complete algebraic lattice for the pointwise ordering. We shall write* $\Pi(F)$ *this complete algebraic lattice.*

The point is already the fact that $\Pi(F)$ is a *set* if A is algebroidal. Indeed, a continuous section is completely determined by its values on a set which generates the category A. The completeness of the poset $\Pi(F)$ is straightforward. Verifying that it is a complete algebraic lattice (for the pointwise ordering) is done by building explicitely the finite elements. These elements are the continuous sections $(u_X^{A,a})$, determined by a pair (A, a), where A is a finite complete algebraic lattice and a a finite element of $F(A)$, defined by

$$u_X^{A,a} = \bigvee \{F(f)^+(a) / f : A \to X\}.$$

The fact that they form a basis of finite objects is then a formal verification left to the reader. This remark, together with the fact that, in $\mathbf{Dom}^+$, all object is a directed limit of finite domains, is the essence of why it is possible to give an interpretation of impredicative products.

The fundamental remark is about our previous definition of a morphism $t : \sigma \to \tau$ in the category $\mathcal{F}(A, \mathbf{Dom}^+)$ between two functors $\sigma : A \to \Omega$ and $\tau : A \to \Omega$. It is nothing else but a continuous section of the functor $\sigma \Rightarrow \tau$. This remark will be essential in the construction of Λ.

For the definition of the base category, we shall use a concept well-known in category theory. This is the concept of l.f.p. categories [GabUlm], i.e. algebroidal categories with (small) limits and filtered colimits. What we shall use is that the category of l.f.p. categories, with as morphisms the continuous functors (i.e. functors which commute with filtered colimits) is a (very large) ccc. Furthermore, the category of left adjoints $\mathbf{Dom}^+$, of the O-CCC of complete algebraic lattices **Dom** is itself a l.f.p. category. It will play the role of the object Ω in this model. We can then take for the base category S this ccc of l.f.p. categories. What is above shows that we can consider $\Pi(F)$ for every object A in S and every continuous functor $F : A \to \Omega$.

5.3 Le model

We shall now build a model of our equational theory using this category S. Since S is a ccc, we get the interpretation of Cur(), Fst, Snd, Id, App, $\langle \sigma , \tau \rangle$, $\sigma ; \tau$ at the level of operators.

Next, let us examine the level of terms. We define a functor $G : S^{op} \to$ Cat. $G(A)$ is the ccc associated to the O-CCC $\mathcal{F}(A, \mathbf{Dom}^+)$ (cf. theorem 2). So the objects of $G(A)$ are the elements of $\mathrm{Hom}_S (A , \Omega)$ for every $A \in S$. Since $G(A)$ is a ccc, we have an interpretation of Cur(), Fst, Snd, Id, App, $\langle t , u \rangle$, $t ; u$ at the level of terms.

If $\rho \in \mathrm{Hom}_S (A , B)$, we define $G(\rho) \in \mathrm{Hom}_{Cat} (G(B) , G(A))$ by composition with ρ as follows. Let σ be an object of $G(B)$. So σ is a continuous functor from B to $\mathbf{Dom}^+$. Then $G(\rho)(\sigma)$ is the continuous functor $\rho ; \sigma$ from A to Ω. If $t \in \mathrm{Hom}_{G(B)} (\sigma , \tau)$, then $G(\rho)(t)$ is defined by $G(\rho)(t)_X = t_{\rho(X)}$ for $X \in A$. This is the interpretation of the action $\sigma * t$.

Let us now define $\Pi : (A \Rightarrow \Omega) \to \Omega$, Λ and Proj.

The definition of Π has already been given on the objects (domain of continuous sections of the Grothendiek fibration). The definition on the arrows is the following: let $s \in \mathrm{Hom}_{A \Rightarrow \Omega} (\sigma , \tau)$, and (t_X) be a continuous section of the functor σ. Then, $\Pi(s)(t_X)$ is the family $(s_X(t_X))$ (notice that we have used that s is a continuous section of the functor $\sigma \Rightarrow \tau$). We can check that this is indeed a continuous section of the functor τ.

The definition of Λ results again from the remark that the morphisms in each fiber are continuous sections. Hence, if $\sigma : A \to \Omega$, $\tau : A \times B \to \Omega$, and $t : (\mathrm{Fst} ; \sigma) \to \tau$, t is a continuous section of the functor $(\mathrm{Fst} ; \sigma) \Rightarrow \tau$. Then $\Lambda(t) : \sigma \to \forall(\tau)$ is the continuous section of the functor $\sigma \Rightarrow \forall(\tau)$ defined by $(\Lambda(t)_X)_Y = t_{(X,Y)}$ (so that Λ is like a curryfication, but mixing two levels).

Finally, we define Proj : $(\mathrm{Fst} ; \forall(\sigma)) \to \sigma$. For $X \in A$, $Y \in B$, $\mathrm{Proj}_{(X,Y)}$ is the projection function on the domain Y, $\forall(\sigma(X, \_)) \to \sigma(X, Y)$. More precisely, $\mathrm{Proj}_{(X,Y)}(t)$ is t_Y, if $t = (t_Z)$ is a continuous section of the functor $\sigma(X, \_)$. We can then check that Proj itself is a continuous section of the functor $(\mathrm{Fst} ; \forall(\sigma)) \Rightarrow \sigma$.

We can see now one of the advantages of equational presentations. To check that indeed we have a model of $F\omega$, all we have to check is that the equations of the section 3.2 are satisfied.

remark – We have chosen the complete algebraic lattice semantics since it is the simplest one and has been used in denotational semantics [Stoy]. But the same construction can be done with Scott Domains, and with embeddings instead of left adjoints. We can give a similar construction also in the framework of stable maps and stable ordering [Girard85], and then get a categorical (and equational) version of [Girard85], following [CoGuWi]. The reader can check that the description of the model in this section is completely "polymorphic" in the precise notion of domains.

6 How to add recursive types

We want to show how to extend recursive types to our formalism. We shall limit ourselves to one simple example; the type of integers, but our treatment extends directly to other recursive types used in computer science: lists, trees, ...

6.1 How to add a type of integers

We can think of the objects of the base category S as types of a programming language, and the objects of $G(A)$ as propositions about programs of type A. It is known how to add a type of integers to the cartesian closed category S (natural number object), but if we want an equational presentation of it, we are forced to consider only a weak notion of natural number object (only the existence part is verified). The fact that we consider proofs explicitely will allow ourselves to express equationally the full initiality.

First, we add a terminal object **T** to the base category S, with the polymorphic (in ML sense) morphism **Nil** : $A \rightarrow$ **T**, and the equation $f =$ **Nil**, if f is of type $A \rightarrow$ **T**. We shall need furthermore the arrow **true** of type **T** $\rightarrow \Omega$.

For the expression of the weak initiality, we add a new constant type **Nat** to the category S, and two morphisms, **Zero** : **T** $\rightarrow$ **Nat** and **Succ** : **Nat** $\rightarrow$ **Nat**. It is important to notice that furthermore, we add a (polymorphic) recursion operator **Rec**, such that, if $a : \mathbf{T} \rightarrow A$ and $f : A \rightarrow A$, then $\mathbf{Rec}(a,f) : \mathbf{Nat} \rightarrow A$, with the equations

$$\mathbf{Zero} \ ; \mathbf{Rec}(a,f) = a \quad \mathbf{Succ} \ ; \mathbf{Rec}(a,f) = \mathbf{Rec}(a,f) \ ; f.$$

But these equations fail to express totally the fact that **Nat** is a natural number object: indeed, we have not expressed the fact that $\mathbf{Rec}(a,f)$ is the unique arrow such that the two previous equations are verified (and it does not seem possible to express this fact equationally, we need Horn clauses).

To express this unicity, we shall add another constant, but at the level of the *terms*. We add the constant **rec**, such that, if P is an operator of type $\mathbf{Nat} \rightarrow \Omega$, if t is a term of type $(\mathbf{Zero} \ ; P)$ (in $G(\mathbf{T})$), u of type $P \rightarrow P$ (in $G(\mathbf{Nat})$), then $\mathbf{rec}(t,u)$ is of type $(\mathbf{Nil} \ ; \mathbf{true}) \rightarrow P$. Note that $\mathbf{rec}(t,u)$ represents the proof that P is universally valid. Indeed, if n is an arrow of type $\mathbf{T} \rightarrow \mathbf{Nat}$, then $n * (\mathbf{rec}(t,u))$ is of type $\mathbf{true} \rightarrow (n \ ; P)$ in $G(\mathbf{T})$, and so expresses the fact that $n \ ; P$ is provable. We add the equations

$$\mathbf{Zero} * (\mathbf{rec}(t,u)) = t \quad \mathbf{Succ} * (\mathbf{rec}(t,u)) = \mathbf{rec}(t,u) \ ; u.$$

This operator **rec** represents the usual Peano recurrence axioms, and is the representation in our formalism of the unicity.

Note that such a construction is actually possible for any recursive type. That is we have both a *recursion* operator, and an *induction* principle, that can be expressed in our formalism as two recursions oprators, but at two distinct levels.

6.2 How to get an extensional theory

We have now (the notion corresponding to) a natural number object. We'll sketch how to interpret the type system of [Lambek-Scott] in our logic. The main difference besides the fact that their logic is *extensional*, is that they don't have a recursion operator, but they postulate, besides Peano induction axiom, the other Peano axioms, that $S(n) \neq 0$, and that $S(n) = S(m) \Rightarrow n = m$. We'll see that these other axioms are *derivable* in our system (that is, we have expressed the full initiality "no junk, no confusion" with two recursion operators).

We'll define an interpretation of this extensional logic in our system. First, we interpret intensional type theory. For this, we use the equivalence between this system and $F\omega$, and we know how to interpret intensional higher-order logic in $F\omega$ (we forget the terms): we interpret a type of higher-order logic by an order, a term (of higher-order logic) by an operator, so that a proposition is interpreted as an operator of type Ω, and finally, we say that a proposition is true if, and only if, it corresponds to a non-empty operator of type Ω (at the categorical level, this amounts to build the "tripos" associated to the given fibration [HyJoPi], [Seely 86]).

Now, it is well known (cf. for instance [Gand]) how to interpret extensional higher-order logic in intensional higher-order logic. We define, by induction over A, a syntactic term E_A of type $A \to A \to \Omega$, which will interpret the extensional equality of the system of Lambek-Scott [Lambek-Scott]. The equality E_Ω is the equivalence, $E_{\mathbf{Nat}}$ is "Leibniz"-equality (i.e. intensional equality, that is $E_{\mathbf{Nat}}$ is $(\lambda p)(\lambda q)(\forall P : \mathbf{Nat} \to \Omega)(P\ p) \Rightarrow (P\ q))$, and $E_{\mathbf{T}}(p,q)$ is the fact that both p and q are intensionaly equal to the only constant of type $\mathbf{T}$. Then, $E_{A\to B}$ is $(\lambda f)(\lambda g)(\forall x : A)(E_A\ x\ x) \Rightarrow (E_B\ (f\ x)\ (g\ x))$.

All this is well-known. What is new here, is that we have one operator **Rec**, which is of type $\sigma, (\mathbf{Nat}, \sigma \to \sigma) \to (\mathbf{Nat} \to \sigma)$, and which satisfies the equations $\mathbf{Rec}(a,f,0) = a$ and $\mathbf{Rec}(a,f,S(n)) = f(n, \mathbf{Rec}(a,f,n))$ (so that the functional part contains Gödel system T), and we have (via **rec**) the ordinary Peano induction axioms. The remarkable fact is then that the interaction of the induction axiom and the recursion operator produces the validity of the other Peano axioms. For instance, we have that $(S(n) = 0) \Rightarrow \perp$ since we can define by recursion a predicate P of type $\mathbf{Nat} \to \Omega$ such that $P(0) = \mathit{True}$ and $P(S(n)) = \mathit{False}$ (this is similar to the proof of the same fact in Martin-Löf type theory with one universe [Martin-Löf]).

We thus get an extensional logic, at least as powerful as the system of [Lambek-Scott], where we can express and prove the full initiality of our type of natural numbers. The difference with a type system like the one of [Lambek-Scott] is that we have as a sub-system a functional system with recursion operators (here the system T). Intuitively, this system combines together computation (Gödel system T) and deduction (higher-order logic).

Conclusion

We have presented a purely equational system, in which we can interpret higher-order logic. It is possible to see this system as a three levels equational theory, in the tradition of the "Algèbres Graphiques" of Burroni [Burroni,Lambek-Scott]. We can think of this system intuitively in the following way: the level of orders corresponds to the ordinary types of a programming language, at the level of operators we get the programs (and some programs are specifications of programs), and the terms represent the proofs of the specifications on programs. This shows that our formalism is a serious candidate for a language of proofs upon a functional language like ML. It is likely that the notion of models of higher-order typed λ-calculi should be developped in this formalism. This notion becomes then a particular case of the clear and simple notion of model of equational theory where all we have to do is to check that certain equations hold. Finally, an important point is the introduction of a two-level equational theory for describing certain indexed categories. This kind of formalism, more syntactical than and not strictly equivalent to the ordinary categorical presentation ([Seely 86], [Pare-Schu])[6] seems to have an interest on its own.

---

[6] Indeed, a syntactic presentation of a categorical concept involves a uniform explicit choice of objects in a class of isomorphic ones. A simple example is the difference between a category with products and a category with explicit products.

References

[Berry] Berry, G. "Stable models of the typed λ-calculi" ICALP, Springer-Verlag, LNCS 62, 1978 pp. 72-89

[BreMey] V. Breazu-Tannen and A. Meyer. "Lambda Calculus with Constrained Types" Abstract (1985)

[Burroni] A. Burroni. "Algèbres graphiques" Cahiers de Topologie et Geometrie Differentielle Volume XXII-3 1981

[CAM] G. Cousineau, P.L. Curien and M. Mauny. "The Categorical Abstract Machine." In Functional Programming Languages and Computer Architecture, Ed. J. P. Jouannaud, Springer-Verlag LNCS 201 (1985) 50–64.

[CoGuWi] T. Coquand, C. Gunter, G. Winskel. "Polymorphism and domain equations" Technical report Cambridge n°107.

[Curien 86] P. L. Curien. "Categorical Combinators, Sequential Algorithms and Functional Programming." Research Notes in Theoretical computer Science, Pitman (1986).

[Curien 87] P. L. Curien "Categorical combinators for (existential) quantification" unpublished notes.

[Bruijn] N.G. de Bruijn. "Lambda-Calculus Notation with Nameless Dummies, a Tool for Automatic Formula Manipulation, with Application to the Church-Rosser Theorem." Indag. Math. **34,5** (1972), 381–392.

[Church] A. Church. "A formulation of the simple theory of types." Journal of Symbolic Logic **5,1** (1940) 56–68.

[GabUlm] P. Gabriel and F. Ulmer. "Lokal Präsentierbare Kategorien" Lecture Notes in Mathematics, Vol. 221 (Springer, Berlin).

[Gand] R.O. Gandy. "On the axiom of extensionality-Part I." J.S.L. 21 (1956).

[Girard72] J.Y. Girard. "Interprétation fonctionnelle et élimination des coupures dans l'arithmétique d'ordre supérieure." Thèse d'Etat, Université Paris VII (1972).

[Girard85] J.Y. Girard. "The system F of variable types, fifteen years later" in TCS 86.

[Gunter85] C. Gunter. "Profinite Solutions for Recursive Domain Equations" PhD thesis, CMU, (1985).

[Howard] W. A. Howard. "The formulæ-as-types notion of construction." Unpublished manuscript (1969). Reprinted in to H. B. Curry: Essays on Combinatory Logic, Lambda Calculus and Formalism, Eds J. P. Seldin and J. R. Hindley, Academic Press (1980).

[Huet] G. Huet. "Cartesian closed categories and Lambda-Calculus" in Combinators and functional programming languages Eds P.L. Curien, G. Cousineau et B. Robinet. LNCS 242

[HyJoPi] J.M.E. Hyland, P.T. Johnstone, A.M. Pitts. "Tripos theory." Math. Proc. Camb. Phil. Soc. 88 (1980).

[Lambek-Scott] J. Lambek and P. J. Scott. "Introduction to higher order categorical logic" Cambridge studies in advanced mathematics, Cambridge University Press, 1986.

[Lawvere] Lawvere. "Adjointness in foundations" Dialectica 23 (1969) 281-296.

[Martin-Löf] P. Martin-Löf. "Intuitionistic Type Theory." Bibliopolis (1982).

[McCracken] N. McCracken. "An investigation of a programming language with a polymorphic type structure." Ph.D. Dissertation, Syracuse University (1979).

[Pare-Schu] R. Paré, D. Schumacher. "Indexed Categories and Their Applications." Springer-Verlag, Lecture Notes in Mathematics, 661.

[Plotkin-Smyth] M.B. Plotkin and G.D. Smyth. "The category-theoretic solution of recursive domain equations" SIAM J. COMPUT. Vol. 11, No 4, November 1982

[Scott] D. Scott. "Relating Theories of the Lambda-Calculus." in To H. B. Curry: Essays on Combinatory Logic, Lambda-calculus and Formalism, Eds. J. P. Seldin and J. R. Hindley, Academic Press (1980).

[Seely 82] R.A.G Seely. "Hyperdoctrines, natural deduction and the Beck condition" Zeitschrift für Mat. Logik und Grundlagen d. Math. 29 505-542.

[Seely 84] R.A.G Seely. "Locally Cartesian Closed Categories and Type Theory" Math. Proc. Camb. Phil. Soc. 95, 1984.

[Seely 86] R.A.G. Seely. "Categorical Semantics for Higher Order Polymorphic Lambda Calculus" Draft.

[Stoy] J.E. Stoy. "Denotational Semantics: the Scott-Strache Approach to the Programming Language Theory" The M.I.T. Press, Cambridge, Massachussetts, and London, England.

[Takeuti] G. Takeuti. "Proof theory." Studies in Logic **81** Amsterdam (1975).

ENRICHED CATEGORIES FOR LOCAL AND INTERACTION CALCULI

Stefano Kasangian Dipartimento di Matematica, Università di Milano, 20133 Milano (Italy)
Anna Labella Dipartimento di Matematica, Università di Roma "La Sapienza", 00185 Roma (Italy)
Alberto Pettorossi IASI-CNR, Viale Manzoni 30, 00185 Roma (Italy)

ABSTRACT

The construction of models for distributed computations plays a very important role in designing and developing parallel computing systems. Various algebraic approaches have been proposed in the past as, for instance, the communicating computing agents of [Mil80], [BeK85], and [BHR84].

In our work we propose a general method for defining the categorical models for classes of algebras of distributed computing agents. If the *static* and *dynamic* operations [Mil80] of the algebras enjoy suitable properties, we can construct *enriched categories* which are models of distributed computations, including also the case of concurrent finite automata which cooperate via protocols. The construction is uniform with respect to the particular algebra one may wish to consider.

1. INTRODUCTION

We assume that computations are performed by a collection of agents, which have local computing capabilities and cooperate towards the achievement of a global goal (for instance, the solution of a common problem) by sharing and exchanging information via an interaction calculus of some kind.

As a particular instance, we may think of CCS processes [Mil80] which cooperate using "handshaking" communications. In [Mil80] it is clearly demonstrated through many examples, the power and the flexibility of those calculi, based on the idea of "interactions" among "local computations".

The case in which the interactions take place through shared memories may also be considered within the framework of handshaking communications. Indeed, the shared memories can be viewed as *passive* agents which are involved in the computations and allow for reading and writing operations only.

Many calculi for parallel and distributed computations have been suggested in the literature (see for instance: [AuB84], [BeK85], [BHR84], [Car85], [DeM86], [HAB78], [Hoa78], [Maz78], [Mil79], [Pet62], [TaW82]), and many research topics are under investigation in this area. In our study we abstract away from all details and particular hypotheses relative to the various approaches which have been proposed. As already indicated, we only assume that there are two kinds of calculi: i) the *local calculi* describing the computations performed within each computing agent, and ii) the *interaction calculi* describing the ways in which computations of different agents are synchronized together.

We will study the relationships among those kinds of calculi, and we will give the conditions to be satisfied for their harmonious merging in order to obtain *global calculi* (also called *distributed calculi*).

We will propose a general framework for the construction of the categorical semantics for a large class of algebras for distributed computations. Our constructions borrow simple tools from Category Theory, and provide a somewhat novel way of understanding the nature of concurrent computing.

Related work may be found in [Win84, LaP86]. In [Win84] various categories for modelling concurrency were presented and their relationships were investigated.

In [LaP86] we constructed the categorical models for a class of local and interaction calculi for synchronized computations. We gave a categorical interpretation of Synchronization Algebras [Mil83] as a monoid of Actions equipped with a partial synchronization operation. We were able to define *good categorical semantics* for global calculi when the local ones are the nondeterministic processes of [Mil80, BeK85, BHR84] and the interaction calculi are indeed Synchronization Algebras.

Here we extended the results presented in [LaP86] to the case where the sequential composition of local calculi within processes can be modelled as a tensor [:] on a category **P** of computations, which is monoidal closed. The tensor [:] is the generalization of the concatenation operation : of the monoid of Actions. We also generalize the synchronization operation $*$ by introducing a bifunctor [$*$] on **P**×**P**.

2. INTRODUCTORY EXAMPLES OF DISTRIBUTED COMPUTATIONS

The basic idea of the local and interaction calculi among processes can be shown in the theory of **Binary Strings**. In that theory we may think of the elements of the free monoid $P=\{0,1\}^*$ as processes, defined as follows: $P ::= \varepsilon \mid 0 \mid 1 \mid P:P$.

The operator : denotes the concatenation operation. (Instead of : we often use juxtaposition only.) In P there is an "empty process" which is denoted by the empty string ε. It behaves as the identity, that is, $p:\varepsilon = \varepsilon:p = p$ for any $p \in P$.

Informally speaking, processes in P are agents which perform actions in a sequential way. Those actions are either 0 or 1. Each action is performed in one unit of time, and time flows in a linearly ordered way.

The *local calculus* for those processes is determined by the concatenation operation : in the following way. A given process $p \equiv p1:p2$ (where ≡ denotes syntactical identity) may become the process p2 after performing the actions occurring in p1, and we write $p \to p2$. The relation $p \to p2$ is the left cancellation of some actions of p, and it is an order relation. Therefore we may say that for any process in P time flows "from left to right". For instance, looking at the actions of $p \equiv 0:0:1:0:1$ we have that: $00101 \to 0101 \to 101 \to 01 \to 1 \to \varepsilon$, and also: $00101 \to 01$.

Any given process determines the relation "→" without the interaction of other processes. For that reason we say that "→" denotes the local calculus within each agent.

Notice that two processes related by "→" are also related by an "intermediate process". In fact, with reference to the example given above, we have that: 00101 is related to 01 by the process 001, and we write: "00101 —(001)→ 01".

Therefore, the "→" relation defines a labelled transition system [Plo82], whose labels are the connecting processes. This means that an instance of "→" is itself a process. That idea is crucial for the construction of the models we will present in the next Section using enriched categories.

Notice also that the local calculus is supposed to be *complete* in the sense that, for any process p2 in P and any relation → labelled by p1, there exists in P a process p such that p —(p1)→ p2.

In P there are two atomic processes which correspond to the atomic actions 0 and 1. A similar correspondence between atomic actions and processes can be found in [BeK85].

We may now introduce an *interaction calculus* among the processes in P. Let us assume that it is defined, for example, by the exclusive-or operator, denoted by ∗, which acts in a bitwise way. We have, for instance: 00101 ∗ 10111 = 10010.

Often it is important that the interaction calculus *distribute* over the local calculus, so that the following holds: (p1:p2) ∗ (q1:q2) = (p1∗q1) : (p2∗q2) (Interchange Law).
In that case, in fact, the categorical models for the local and interaction calculi can be constructed in a simple way, as we will see later.

Those categorical constructions are essentially based on the fact that for computing the interaction (or synchronization) of two processes we need only to compute the synchronization of their corresponding actions. That means that we need only to know how ∗ behaves on 0's and 1's. Indeed, in our example we defined ∗ as follows: x∗y ≡ if x=y then 0 else 1, for any x and y in {0,1}.

In the sequel we will also study the problem of defining the interaction between two processes of different "lengths", as for instance, 00101 and 101. One may want the result of 00101∗101 to be either 100 (by synchronizing the two given strings from the left) or undefined (simply because they have different lengths). Which choice should be made? The models we will construct show the categorical properties of both choices and their relationship. In particular we will show that the first choice provides what we will call a *good semantics*. ■

The second example we want to describe are the **Nondeterministic Computations**, where the local calculus is nondeterministic. (The calculus that we will actually present is a variant of Synchronous CCS [Mil83]. We will develop the general case in Section 5.) We consider that processes may perform at any instant in time one action in a given set of possible actions. We assume that the *local calculus* within each agent is defined by a term (or process) of the form:

P ::= NIL | λ:P | P+P | x | rec x. P,

where: i) λ belongs to a given set of (atomic) Actions={1,α,$\underline{\alpha}$,β,$\underline{\beta}$,...} where the complement operation (denoted here by the underlining) is an involution, ii) + is associative and commutative, iii) x is a formal variable, and iv) rec x. P implicitly defines an infinite term as specified below.

Notice that the process formation rule λ:P can be viewed as a particular instance of the rule P:P of the Binary Strings example, by assuming that the left process in P:P is atomic and it corresponds to the action λ. Notice also that in general, the operation : does not distributes over +.

The operational semantics of the local calculus is defined by the following derivation rules:

(Local 1) λ:P —(λ)→ P

(Local 2.1) if P —(λ)→ P' then P+Q —(λ)→ P'

(Local 2.2) if P —(λ)→ P' then Q+P —(λ)→ P'

As usual, we may also consider infinite computations via the recursion operator "rec x. P", such that:

(Local 3) if P[rec x. P / x] —(λ)→ P' then rec x. P —(λ)→ P',

where P[a / b] denotes substitution of a for b in P. For more details see [Mil83].

The *interaction calculus* is determined by a synchronization algebra on Actions where the operator ∗ (denoted by | in [Mil80]) realizes the interaction between two local calculi as specified by the following derivation rule for any λ, $\mu \in$ Actions:

(Interaction) if P —(λ)→ P' and Q —(μ)→ Q' then P∗Q —(λ∗μ)→ P'∗Q'.

In our example we have that: i) ∗ is commutative; ii) for any $\lambda \in$ Actions $\lambda * 1 = 1 * \lambda = \lambda$, and $\lambda * \underline{\lambda} = 1$; iii) in all other cases ∗ is undefined. The Interaction rule realizes the synchronization between the computations λ and μ of the agents P and Q, respectively. Notice that for action synchronization we used the same operator ∗ used for process synchronization. This notation is consistent with the fact that we consider atomic actions as processes.

The synchronization ∗ is assumed to be distributive with respect to +. ■

In the sequel we will provide using the results of our main Proposition 1, a categorical model for calculi where the operator ∗ is more generally a partial operation defined over an arbitrary alphabet. In that way we will encompass a whole class of distributed calculi (for instance: [Mil80, BeK85, BHR84]).

Moreover, Definition 1 will tell us the general conditions under which the local and the interaction calculi fit together. In the case of CCS [Mil80] those conditions are essentially expressed by the Expansion Theorem, whereby the operator ∗ of the interaction calculus can be recursively reduced to sums (+) and action-prefixings (:), which are operations of the local calculi. Distributed calculi should enjoy this reduction property, because it allows us to consider the result of an interaction calculus as an instance of a local calculus, and therefore, to build "larger" computing agents from "smaller" ones.

Our distinction between local calculi and interaction calculi is basically taken from [Mil80], where the author classifies the operations of CCS into *static* and *dynamic* ones. However, we assume here a slightly different standpoint, because we separate the two sets of operations into distinct calculi. Similar ideas have been considered also in [AsR86].

In our treatment we will also touch on the problem of the *equivalence between agents*. This problem arises from the need of generating the computing agents in different ways, and considering those agents to be equivalent with respect to a given observation function [DeM85].

3. LOCAL, INTERACTION, AND GLOBAL CALCULI: THE MAIN RESULTS

In general, we will consider processes as black boxes, and the only thing we know about them is that they are able to perform computations. Therefore, we assume that processes are abstract entities of a "world of processes" which are related via computations. The basic idea is that, in general, processes perform computations and they become new processes. Indeed, the examples we have given in the previous Section illustrate that point.

We will first study the structure of those computation relations which are assumed to be objects of an abstract entity, called "global calculus". We assume that we are given the following two operations:
i) the concatenation operation (:) denoting the action of putting one computation "after" another, and
ii) the synchronization operation (∗) denoting the action of merging two distinct computations, so that during the i-th unit of time their corresponding i-th atomic actions are performed "together".
As we will see, a given global calculus determines a monoidal closed category **P** of computations.

Obviously, various definitions of concatenation and synchronization can be given. We do not restrict ourselves to any particular definition here, and the models we will construct are valid for any choice satisfying the required properties indicated below.

Notice also that the case when many different notions of synchronization are considered together within the same algebra for computations [Hen83], can be analyzed by generalizing our constructions.

Once we have defined the category **P**, we can consider a category **W** corresponding to a "world of processes" as a category enriched over **P** [Kel82].
The following picture visualizes our categorical constructions.

Category **W** of processes (p1, p2,...) with computations as arrows among them

Category **P** of computations (c1, c2,...) which is a global calculus closed under concatenation [:] (local calculus) and synchronization [∗] (interaction calculus)

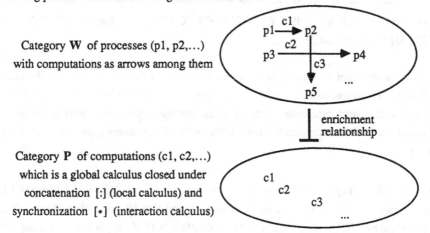

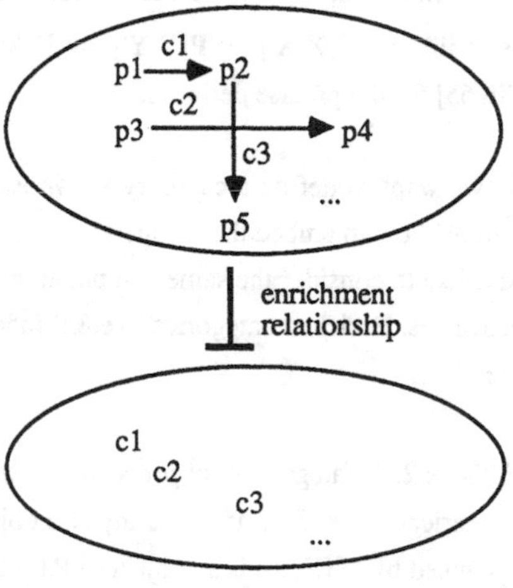

Let us first define the category of the global calculus.

DEFINITION 1. A *global calculus* is a category **P**, also denoted by (**P**,[:],[∗]), with two bifunctors [:] and [∗] such that:

i) **P** is monoidal closed, the tensor being [:]: $\mathbf{P}\times\mathbf{P}\to\mathbf{P}$ (it defines the *local calculus*), and

ii) the bifunctor [∗]: $\mathbf{P}\times\mathbf{P}\to\mathbf{P}$, defining the *interaction calculus* (also called *synchronization calculus*), is a monoidal closed functor. ■

Let us now explain the above basic definition.

Informally speaking, a category **P** is said to be *monoidal* with tensor [:] if it is possible to consider **P** as a generalized monoid, in the sense that [:] is associative and it has a unit element [Kel82].

Besides its own morphisms, the category **P** has an additional structure of morphisms, which is due to the fact that **P** is *monoidal closed*. In fact, in analogy to what happens in the category of sets, we have that for any X in **P** there is a functor **P**[X, -]: **P** → **P**, right adjoint to -[:]X.

Therefore we get an *internal hom* structure in **P**, that is, we can associate with any two objects X and Y of **P** the object **P**[X,Y], which is an object of **P**, and we will consider **P**[X,Y] as the internal hom between X and Y (for details see [Kel82]).

Notice also that the monoidal closed category **P** is the categorical analogous of the monoid {0,1}* we have considered in the Binary String example.

Recall that for the functor [∗] to be *monoidal* means that it preserves the monoidal structure of **P**. It can be shown that monoidality is roughly equivalent to the existence of morphisms η and ϕ in **P** such that:

i) η: I → I [∗] I where I is the unit element of [:],

ii) ϕ: (X[∗]Y) [:] (X'[∗]Y') → (X[:]X') [∗] (Y[:]Y') for any X,X',Y,Y' in **P**.

In that case we can say that a *Generalized Interchange Law* between [:] and [∗] holds in (**P**,[:],[∗]). That law is the natural condition one may require if one wants to interpret [:] as sequential composition of computations and [∗] as synchronization among computations.

The above Interchange Law is also a natural property for the harmonious merging of the local calculus with the interaction one.

Recall further that [∗] is *closed* if the above condition i) on I holds, and we have the following morphism: iii) γ: **P**[X,X'] [∗] **P**[Y,Y'] → **P**[X[∗]Y, X'[∗]Y'] for any X,X',Y,Y' in **P**. (See [EiK65] for the precise definition).

Now we want to define a category **W** whose objects are processes. Using the Definition 1, we already know how to embed the computations performed by processes into a global calculus, and now we would like to consider the same computations as morphisms between objects of the new category **W** of processes. Enriched categories [Kel82] indeed provide the natural approach to the definition of **W** we need.

DEFINITION 2. A category **W** of *processes computing according to the global calculus* (**P**,[:],[∗]) is a category enriched over **P**, that is, for any two objects X and Y in **W**, the "object of morphisms" from X to Y denoted by **W**[X,Y], is an object of **P** (not simply a "set" of morphisms). The composition law of the object of morphisms is defined via the tensor [:], and an identity element is defined for each object **W**[X,X]. ■

The reader may notice that one can consider the category **P** as a category which is enriched over **P** itself.

As an immediate consequence of the above definitions we have that the computations in **W** are closed under composition via the concatenation operation [:], but they fail to be closed under the interaction operation [∗]. In fact, in general, we do not know how to specify in a suitable way (with respect to the definition of **W**) the domain and the codomain of the "object of morphims" **W**[X,X'][∗]**W**[Y,Y'] which denotes the synchronization between the computations **W**[X,X'] and **W**[Y,Y']. The rest of the Section is indeed devoted to the solution of this problem.

For that purpose it is necessary to lift in a suitable way the functor [∗] from the level of the base category **P** to the level of the enriched category **W**, so that the category **W** turns out to be closed under [∗]. We proceed as follows.

First, it is easy to show that **P**×**P** is a monoidal closed category with the following tensor (induced componentwise by [:]): [:]×[:] : (**P**×**P**)×(**P**×**P**) → (**P**×**P**).

Then we can consider the product **W**×**W** as a category enriched over **P**×**P** again componentwise. If we apply the bifunctor [∗] on **P**×**P**, we get as *effect* [EiK65] of the monoidal functor the category $\mathbf{W}\times_*\mathbf{W}$ which is the enrichment of **W**×**W** over **P**, as the following diagram shows:

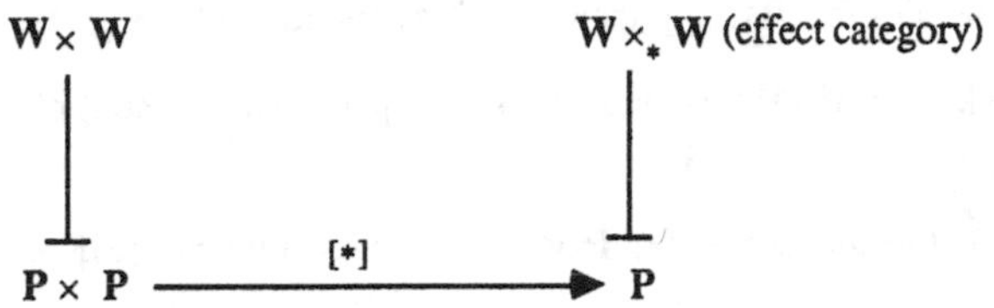

where the vertical lines denote the enrichment relationship.

The objects of $\mathbf{W}\times_*\mathbf{W}$ are the objects of **W** × **W** and the morphisms are obtained as follows:

$$\mathbf{W}\times_*\mathbf{W}[(X,Y),(X',Y')] = \mathbf{W}[X,X']\,[*]\,\mathbf{W}[Y,Y'].$$

The object of morphisms **W**[X,X'] from X to X' is interpreted as the fragment of the local calculus by which the agent X becomes the agent X'. Similarly for **W**[Y,Y']. Therefore, when two agents X and Y are put together, their interaction calculus determined by [∗] produces the pair (X',Y') after the "synchronized computation" given by **W**[X,X'] [∗] **W**[Y,Y'].

The problem of constructing the categorical models for distributed computations consists basically in representing a "synchronized computation" performed by two objects (or processes) of **W** (that is, an object of $\mathbf{W}\times_*\mathbf{W}$) as a local calculus performed by a single object (or process) in **W**.

DEFINITION 3. A *semantics* for a global calculus (**P**,[:],[∗]) is a pair (**W**,[✱$_\mathbf{W}$]) where **W** is a category enriched over **P**, and [✱$_\mathbf{W}$]: $\mathbf{W}\times_*\mathbf{W} \to \mathbf{W}$ is a **P**-functor [Kel82].

We say that a semantics is *good* if [✱$_\mathbf{W}$] is hereditarily full. ■

Let us explain the categorical notions we used in the above Definition 3.

A **P**-functor F from the category **C** to the category **D** (both enriched over **P**) is a function between the objects, such that for any X and Y, there is a morphism of **P** from **C**[X,Y] to **D**[F(X),F(Y)] satisfying the commutativity of suitable diagrams, which basically specifies the preservation of the enrichment structure [Kel82, pp. 24-25].

The **P**-functor F is said to be *hereditarily full* iff for any X in **C** and Y' in **D** s.t. **D**[Y',F(X)] is not the initial object of **P**, there exists a family of objects $\{Y_i \mid i \in I\}$ in **C** such that: i) $\forall i \in I \;\; F(Y_i)=Y'$, and ii) the family of induced morphisms $\{\mathbf{C}[Y_i,X] \to \mathbf{D}[Y',F(X)] \mid i \in I\}$ is epimorphic in **P**.

(Recall that a family $\{f_i \mid i \in I\}$ is epimorphic iff $\forall h,g$ if $\forall i\; h \bullet f_i = g \bullet f_i$ then $h=g$, where $\bullet$ denotes function composition.)

The hereditarily fullness condition intuitively expresses the fact that F is "non-redundant", in the sense that, if we interpret the objects of **W** as processes, the actions performed by a process which is in the image of F, can also be performed in an essentially the same way, by any process of its inverse-image.

The following proposition provides us with a class of models for any given global calculus.

PROPOSITION 1. Given a global calculus (**P**,[:],[∗]), we have that: i) (**P**,[$\underline{*}$]), is a semantics for (**P**,[:],[∗]), where the functor [$\underline{*}$] induced by [∗], has the domain $\mathbf{P}\times_*\mathbf{P}$ (instead of $\mathbf{P}\times\mathbf{P}$); and ii) if ($\mathbf{W}$,[$\ast_\mathbf{W}$]) is a semantics for (**P**,[:],[∗]) then there is a semantics ($\mathbf{A}$,[$\ast_\mathbf{A}$]) for any subcategory **A** of **W** which is the codomain of a **P**-functor $\underline{a}$: $\mathbf{W} \to \mathbf{A}$. In particular, point ii) of this Proposition 1 holds by replacing **W** by **P**. ■

PROOF. i) [$\underline{*}$] is a **P**-functor because [∗]: $\mathbf{P}\times\mathbf{P} \to \mathbf{P}$ is a closed functor [EiK65]. ii) The functor [$\ast_\mathbf{A}$]: $\mathbf{A}\times_*\mathbf{A} \to \mathbf{A}$ is defined as follows: $[\ast_\mathbf{A}] = \underline{a} \bullet [\ast_\mathbf{W}] \bullet (i \times i)$ where i is the inclusion from **A** to **W**. ■

The notion of good semantics will be used later to characterize the properties of some categorical models which we will construct in the examples below.

The **P**-functor $\underline{a}$ can also be considered as an observability function [DeM85] on the objects of **P**, which denote computations of the local calculus. In fact $\underline{a}$ induces an equivalence relation on those objects and it determines their representatives [KaL87].

By using Proposition 1 we will construct in the following Sections the categorical models for synchronization calculi like ACP, SCCS, CCS, and CSP [BeK85, Mil83, BHR84]. This approach based on enriched categories seems to be simpler and more general than the ones presented in [LaP86, Win84].

4. THE BINARY STRINGS EXAMPLE

Let us consider the monoid $P=\{0,1\}^*$ mentioned in Section 2. The corresponding global calculus is the category **P** which has the sets of words of P as objects and the inclusions as morphisms [Bet80]. (The reader may easily check that there is no loss of generality in considering P instead of any other free monoid.)

The tensor [:] is the generalized concatenation between sets of words (that is, the Frobenius product) and for any object S and T in **P**, $\mathbf{P}[S,T]=\{r \mid r \in P$ such that for any $s \in S \;\; r{:}s \in T\}$.

The interaction calculus [∗], specifying a "tight synchronization" between words, is defined as follows:

$$S\,[*]\,T = \{s_1{*}t_1: \ldots : s_n{*}t_n \mid s \equiv s_1: \ldots :s_n \in S,\ t \equiv t_1: \ldots :t_n \in T \text{ where for } i=1,\ldots,n\ s_i \text{ and } t_i \in \{0,1\}, \text{ and } s_i{*}t_i \text{ is defined}\}.$$

Notice that it is possible to define the functor [∗] because among the objects of **P** we consider also the empty set. Indeed, we take as objects of **P** sets of words, and not just words.

Notice that the choice of defining ∗ only for words of the same length is consistent with the validity of the Interchange Law between [:] and [∗]. In this example the Interchange Law holds only in the generalized sense. In fact, we allow the left hand side of the following inclusion to be empty:

$$(S[*]T)\,[:]\,(S'[*]T') \subseteq (S[:]S')\,[*]\,(T[:]T').$$

As stated by Proposition 1, (**P**,[$\underline{*}$]) is a semantics for (**P**,[:],[∗]). It is *not* a good semantics because [$\underline{*}$] is not a hereditarily full functor. In fact, it is impossible, in general, to get an epimorphic family of morphisms $\{\mathbf{P}[S_i,T][*]\mathbf{P}[S'_i,T'] \subseteq \mathbf{P}[S_i[*]S'_i,T[*]T'] \mid i \in I\}$, as one can see by considering the case where $S_i[*]S'_i=T[*]T'=\emptyset$ and ∗ is never defined. Therefore [$\underline{*}$] does not satisfy the condition ii) of the definition of a hereditarily full functor.

One can easily verify that a good semantics for (**P**,[:],[∗]) is given by (**P'**,[✱]), where:

$Ob(\mathbf{P'}) = Ob(\mathbf{P}) - \emptyset$, and the morphisms of **P'** are defined as follows:

for any S and T $\mathbf{P'}[S,T]=\{r \mid r \in \mathbf{P}[S,T] \text{ such that } \forall t \in T \text{ if for some } x\ \ t \equiv r{:}x \text{ then } x \in S\}$.

The "loose synchronization" functor [✱] is defined as follows:

$$S\,[✱]\,T = \{s_1{*}t_1: \ldots : s_k{*}t_k \mid s \equiv s_1: \ldots :s_{ns} \in S,\ t \equiv t_1: \ldots :t_{nt} \in T \text{ where for any } i\ s_i \text{ and } t_i \in \{0,1\}, \text{ and for } i=1,\ldots,k\ \ s_i{*}t_i \text{ is defined, and } s_{k+1}{*}t_{k+1} \text{ is not defined}\}.$$

Notice the relationship between the two functors [∗] and [✱]. The first is a monoidal closed functor, the second is not. Since [∗] is monoidal, it can be used for defining an interaction calculus which satisfies the Interchange Law with [:]. Both [$\underline{*}$] (induced by [∗] as stated by Proposition 1) and [✱] provide a semantics for (**P**,[:],[∗]) and they correspond to the two different ways of extending to words the bitwise definition of ∗ which we suggested in Section 2. The functor [✱] is hereditarily full. To sum up we have the following:

PROPOSITION 2. Using the definitions given in this Section (**P**,[:],[∗]) is a global calculus for Binary Strings, and (**P'**,[✱]) is a good semantics for it. ■

5. THE NONDETERMINISTIC COMPUTATIONS EXAMPLE

In this example we show that the categorical models of Synchronous CCS (and similarly defined calculi) introduced in [LaP86] , can also be constructed using Enriched Categories.

In [LaP86] we introduced the notion of *pretrees*. They are (possibly infinite) terms constructed from NIL, σ:- and -+-, where σ belongs to a given free monoid A=(Actions)*, where Actions is a given set of atomic actions. NIL stands for ε, i.e. the empty sequence in A.

No equivalences are assumed among the terms defined above, besides the ones which hold in A and the ones indicated below.

The category **P**, which is the basis for the models we will construct, is the category of pretrees which are associative and commutative with respect to +. We include among the objects of **P** also the *deadlock pretree* δ, such that: $\delta + T = T + \delta = T$. Notice that in general, in **P** we have: $T + NIL \neq T$.

In **P** there is a morphism from the pretree S to the pretree T iff T can be obtained from S by a (possibly empty) sequence of subtree rewritings of the form:

rule 1: $S1 + S1 \Rightarrow S1$,

followed by a second (possibly empty) sequence of rewritings of the form:

rule 2: $S2 \Rightarrow S2 + S3$ for some pretree S3.

(For different rewritings S1, S2, and S3 may be instantiated to different pretrees).

Informally speaking, the morphisms in **P** are action-preserving maps of *root-to-leaf paths* of pretrees. For instance, we have that: i) for any pretree T, there is only one morphism: $\delta \rightarrow T$; ii) there are 4 morphisms from $\alpha:\beta:(\gamma:NIL+\gamma:NIL)$ to $\alpha:(\beta:(\gamma:NIL+\gamma:NIL)+\zeta:NIL)$ corresponding to the 4 ways of mapping $\gamma:NIL+\gamma:NIL$ into itself; and iii) there is no morphism from $\alpha:\beta:NIL$ to $\alpha:(\beta:\gamma:NIL+\zeta:NIL)$.

The functor [:] is the concatenation of pretrees (which is a generalized version of the usual concatenation of words) and it is defined as follows:

i) $\delta[:]T = T[:]\delta = \delta$; ii) $NIL[:]T = T[:]NIL = T$; iii) $(\Sigma\alpha_i:S_i)[:]T = \Sigma\alpha_i:(S_i[:]T)$ where $\alpha_i \in A$.

Notice that $\varepsilon:NIL = NIL$ and in iii) α_i may be ε. For instance, we have that:

$\alpha:((\beta:\gamma:NIL) + \zeta:NIL)\ [:]\ (NIL + \zeta:NIL) = \alpha:((\beta:\gamma:(NIL + \zeta:NIL)) + \zeta:(NIL + \zeta:NIL))$.

For any X the right adjoint to -[:]X is **P**[X,-] defined as follows:

- for any Y, **P**[δ,Y]=U where U is the smallest infinite pretree t which satisfies the following equation: $t = \Sigma\alpha_i:t$, where for any i and j $\alpha_i \neq \alpha_j$ and $\alpha_i \in Actions \cup \{\varepsilon\}$;
- if $X \neq \delta$ then **P**[X,Y] is the pretree Z such that:
  i) there is a morphism $Z[:]X \rightarrow Y$ such that rule 2 only is applied when the morphism is computed on Z,
  ii) $\forall$ Z' satisfying condition i), there is a monomorphisms from Z' to Z (that is, rule 2 only is applied when computing $Z' \rightarrow Z$).

Informally speaking, if $X \neq \delta$ then **P**[X,Y] is the largest initial subpretree of Y such that if we add X to all its leaves, we get a pretree which is the domain of a morphism into Y.

The "tight synchronization" [*] between pretrees in **P** is a functor induced by the given synchronization * over the monoid A. As in Section 4 we assume that for any x and y in A:

$x*y=z$ iff ($x \equiv x_1 \ldots x_n$ and $y \equiv y_1 \ldots y_n$ and $z \equiv z_1 \ldots z_n$ and for any i $x_i, y_i, z_i \in Actions$ and $x_i * y_i = z_i$).

In particular, if either x and y have different lengths or $x_i * y_i$ is not defined then $x*y$ is not defined.

The functor [∗] can be inductively defined as follows:

i) $\delta[*]T = T[*]\delta = \delta$;

ii) for any T which is *not* a sum of pretrees:

$NIL[*]T = T[*]NIL = NIL$ if $T \equiv NIL$, otherwise it is equal to δ;

iii) $(\Sigma\alpha_i:S_i)[*](\Sigma\beta_j:T_j) = \Sigma(\alpha_i*\beta_j):(S_i[*]T_j)$ where α_i and $\beta_j \in Actions \cup \{\varepsilon\}$.

Notice that by definition [∗] distributes over +. If we assume, for instance, that $\forall x \in Actions$ $x*x=x$ then we have: $\alpha:(\beta:\gamma.NIL + \zeta:NIL)\ [*]\ \alpha:(\beta:NIL+\zeta:NIL) = \alpha:(\delta+\zeta:NIL) = \alpha:\zeta:NIL$.

By Proposition 1 (**P**,[∗̲]) is a semantics for (**P**,[:],[∗]), but as in the previous example (see Section 4) it is *not* a good one.

Let **P'** be a category whose objects are those of **P** except δ.

The morphisms of **P'** are defined as follows: for any X and Y, **P'**[X,Y] is the pretree Z such that:

i) there is morphism from Z[:]X to Y such that rule 2 only is applied when the morphism is computed on Z, and it is the identity when it is computed on X; and

ii) ∀ Z' satisfying condition i) there is a monomorphisms from Z' to Z.

As a consequence, we have that **P'** is a subcategory of **P**.

Let [✱] be a functor on **P'** defined as follows:

i) NIL [✱] T =T [✱] NIL = NIL;

ii) $(\Sigma\alpha_i:S_i)[✱](\Sigma\beta_j:T_j) = \Sigma(\alpha_i*\beta_j):(S_i[✱]T_j)$ where α_i and $\beta_j \in Actions \cup \{\varepsilon\}$.

[✱] specifies a "loose synchronization" between pretrees.

PROPOSITION 3. Using the definitions given in this Section (**P**,[:],[∗]) is a global calculus for Nondeterministic Computations, and (**P'**,[✱]) is a good semantics for it. ■

DEFINITION 4. A pretree S is said to be *nice* if it has a unique normal form with respect to the rewriting rules: i) S+S ⇒ S, and ii) S+NIL ⇒ S. ■

Let **N** be a subcategory of **P'** such that its objects are nice objects of **P'** and it is closed w.r.t. [✱].

Let **A** be the subcategory of **N** generated by the image of the functor a̲ which maps any pretree into its normal form.

PROPOSITION 4. The functor a̲: **N**→**A** defined above is **P**-functor. It is a left inverse of the inclusion functor i: **A**→**N**. The semantics (**A**,[$✱_A$]) induced from (**N**,[✱]) by a̲ is a good semantics (see Proposition 1), and it is terminal among all good semantics (**W**,[$✱_W$]) for the global calculus (**P**,[:],[∗]), such that the internal homs of **W** are objects in **N** or δ. ■

6. THE EXAMPLE OF PROTOCOLS AMONG FINITE AUTOMATA

Now we show that the situation where two finite state automata communicate on their input alphabets via protocols, fits in the general framework described in Section 3.

We assume here the abstract point of view by which protocols are instances of the interaction calculus between automata, which operate according to their local calculi.

Let us recall that, by a quite recent approach in terms of *Generalized Logic* in the sense of Lawvere [Law 73], the dynamics of automata over a (non necessarily free) monoid M can be seen as enriched (small) categories on the monoidal biclosed category $\mathcal{P}$(M), also denoted by **M**, where $\mathcal{P}$ is the usual powerset operator.

As in the Binary Strings example, let us simply remark that [Bet80, KKR83]:
i) objects of **M** are languages, ii) morphisms are inclusions, iii) the tensor product [:] is the Frobenius product of languages, and iv) for any A and B the internal hom **M**[A,B] is defined as in Section 4. Hence, **M** with [:] is our *local calculus* in this case.

The reader should again notice the connection between **M** and the category **P** of pretrees we mentioned in the previous Section. In fact, **M** can be obtained from **P** by considering, given any pretree, the set of action sequences of its paths.

An automaton **R** is a **M**-category whose objects are the *states* $\{r_i \mid i \in I\}$ of **R** and the object of morphisms **R**$[r_1,r_2]$ is the set of action sequences connecting r_1 and r_2.

In this framework we can introduce the *interaction calculus* in a way which is an instance of the one indicated in Section 3. In fact, let M be Σ^*, where Σ is a given finite alphabet. It is then possible to model communication protocols between automata as a Synchronization Algebra, by introducing a partial operation $*$: M×M → M satisfying the usual conditions we stated in the Binary Strings example, and hence, a functor [$*$] from **M**×**M** to **M**, which defines an interaction calculus.

In order to find a semantics for (**M**,[:],[$*$]) we can apply the general theory presented in Section 3, and define a *universal automaton* **U** such that any automaton **R** with just one initial state is at the same time, an object of **U** and an enriched subcategory of **U**. (Indeed we can uniquely associate with any object in **U** the subcategory of all its reachable states.) For any other automaton **S**, we are then able to get the synchronized automaton **R[✱]S**, which is again a subcategory of **U**. Then, we can use the general machinery described in the previous examples to get a good semantics on **U**, by assuming **P'=U** (see Propositions 2 and 3).

On the other hand notice that if we consider the automata **R** and **S** as **M**-enriched categories we are let to a notion of semantics which is "two-sided" in a sense that will be made clear below.

Given the automata **R** and **S** as **M**-categories, we can consider the (**M**×**M**)-category **R**×**S**, with the obvious hom structure in **M**×**M**. As before, we take the *effect* of the monoidal functor [$*$], which assigns to **R**×**S** the **M**-category **Q** = **R**$\times_*$**S** whose objects are pairs of states (r,s), and whose morphisms are the following ones: **Q**[(r,s),(r',s')] = **R**[r,r'] [$*$] **S**[s,s']. Therefore, we may produce as semantics the automaton **Q** which is the analogous of a network in the sense of [GMY84].

Notice that in order to define a semantics of the synchronization between automata we now use a bifunctor from two *different* **M**-subcategories of **U** (e.g., **R** and **S**) to a third one (e.g. **Q**). Observe further that by construction **Q** embodies the non-redundancy property associated with the definition of a good semantics, because **Q** itself is the effect category.

An analogous approach could have been developed for the examples of the Sections 4 and 5.

In this example we also get a new insight on the synchronization operation. In fact, if the communications between **R** and **S** occur through a partial operation $*$ defined as in [BHR84] (see for instance [GMY84]) we can define a *pair* of **M**-functors: $\pi_{\mathbf{R}}:\mathbf{Q}\to\mathbf{R}$ and $\pi_{\mathbf{S}}:\mathbf{Q}\to\mathbf{S}$, because the following two inclusions hold: $\mathbf{Q}[(r,s),(r',s')] \subseteq \mathbf{R}[r,r']$, and $\mathbf{Q}[(r,s),(r',s')] \subseteq \mathbf{S}[s,s']$.

This means that the image of (for instance) $\pi_{\mathbf{R}}$ is the set of action sequences in **R** which are "synchronizable" with some other action sequence in **S**.

A more detailed study of the model for distributed computations among automata we have given above, can be made by taking into account the particular properties of the protocols one wishes to consider.

7. CONCLUSIONS

We have constructed the categorical semantics of distributed calculi for concurrency in terms of categories enriched over a suitable monoidal category **P** [Kel82] equipped with two bifunctors: the tensor product [:] and the synchronization [$*$]. This construction is not complicated and it naturally generalizes the categories over the monoid of Actions we described in [LaP86]. We have shown that:

i) for any local calculus which can be modelled within the category **P**, and any interaction calculus satisfying a given Interchange Law, there exists a class of (possibly) good semantics for the corresponding global calculus (see Proposition 1);

ii) the standard case of Synchronization Algebras of Milner and Hoare (and similarly defined ones) can be modelled (using Proposition 1) by choosing as the base category for the enrichment the category **P** of pretrees [LaP86];

iii) the synchronized cooperation between finite state automata which interact on their input alphabets via protocols, enjoys good categorical semantics.

8. ACKNOWLEDGMENTS

The IASI Institute of the Italian National Research Council in Rome (Italy) gave financial support.

9. REFERENCES

[AsR86] Astesiano, E. and Reggio, G.: "A Syntax-Directed Approach to the Semantics of Concurrent Languages" Proc. Information Processing 86. (H-J. Kugler, ed.) Elsevier Science Publ. (North Holland) (1986), 571-576.

[AuB84] Austry, D. and Boudol, G.: "Algèbre de Processus et Synchronisation" Theoretical Computer Science 30 (1984), 91-131.

[BeK85] Bergstra, J.A. and Klop, J.W.: "Algebra of Communicating Processes with Abstraction". Theoretical Computer Science Vol.37, No.1 (1985), 77-121.

[Bet80] Betti, R.: "Automi e categorie chiuse" Boll. Unione Matem. Ital. (5) 17-13 (1980), 44-58.

[BHR84] Brookes, S.D., Hoare, C.A.R., and Roscoe, A.W.: "A Theory of Communicating Sequential Processes" J.A.C.M. 31, 3 (1984), 560-599.

[Car85] Cardelli, L.: "AMBER" Proc. Treizième Ecole de Printemps d'Informatique Theorique, La Val D'Ajol, Vosges (France) (May 1985).

[DeM85] Degano, P. and Montanari, U.: "Specification Languages for Distributed Systems". Proc. Mathematical Foundations of Software Development LNCS n.185 Springer-Verlag, Berlin (1985), 29-51.

[EiK65] Eilenberg, S. and Kelly, G.M.: "Closed Categories" Proc. of the Conference on Categorical Algebra. La Jolla 1965, Springer-Verlag (1966), 421-562.

[GMY84] Gouda, M.G., Manning, E.G., and Yu, Y.T.: "On Progress of Communication between Two Finite State Machines" Information and Control 63 (1984), 200-216.

[GoB84] Goguen, J., and Burstall, R.M.: "Some Fundamental Algebraic Tools for the Semantics of Computation, Part 1: Comma Categories, Colimits, Signature and Theories". Theoretical Computer Science 31(2), (1984), 175-209.

[HAB78] Hewitt, C., Atkinson, R., and Baker, H.: "Semantics of Communicating Parallel Processes" Proc. Summer School on Foundations of Artificial Intelligence, ISI, Università di Pisa, Pisa (June 1978).

[Hen83] Hennessy, M.: "Synchronous and Asynchronous Experiments on Processes" Information and Control 59 (1983), 36-83.

[Hoa78] Hoare, C.A.R.: "Communicating Sequential Processes" Communications A.C.M. Vol.21, n.8, (1978), 666-677.

[KaL87] Kasangian, S. and Labella, A.: "Enriched Categorical Semantics for Distributed Calculi" (forthcoming paper) (1987).

[Kel82] Kelly, G.M.: "Basic Concepts of Enriched Category Theory", Cambridge University Press, Cambridge (1982).

[KKR83] Kasangian, S., Kelly, G.M., and Rossi, F.: "Cofibrations and the Realization of Nondeterministic Automata" Cahiers de Topologie et Géométrie Différentielle. XXIV 1 (1983), 23-46.

[LaP86] Labella, A. and Pettorossi, A.: "Categorical Models of Process Cooperation" Proc. Category Theory and Computer Programming, Guildford, U.K., September 1985, Lecture Notes in Computer Science n.240, Springer-Verlag, Berlin (1986), 282-298.

[Law73] Lawvere, F.W.: "Metric Spaces, Generalized Logic, and Closed Categories" Rendiconti del Seminario Matematico e Fisico, Milano 43 (1973), 135-166.

[Mac71] Mac Lane, S.: "Categories for the Working Mathematician". Springer Verlag Berlin (1971).

[Maz78] Mazurkiewicz, A.: "Concurrent Program Schemes and their Interpretations" DAIMI PB 78 Aarhus University Publ., Denmark, (1978).

[Mil79] Milne, G.: "Synchronized Behaviour Algebras: A Model for Interacting Systems" Department of Computer Science, University of Southern California, Los Angeles, USA, (1979).

[Mil80] Milner, R.: "A Calculus of Communicating Systems" LNCS n.92 Springer-Verlag, Berlin (1980).

[Mil83] Milner, R.: "Calculi for Synchrony and Asynchrony" Theoretical Computer Science 25 (1983), 267-310.

[Pet62] Petri, C.A.: "Fundamentals of a Theory of Asynchronous Information Flow" In: Proceedings of IFIP Congress 62, North-Holland, Amsterdam (1962).

[Plo81] Plotkin, G.: "Lectures Notes on Domain Theory". Computer Science Department, Edinburgh University, Edinburgh (Scotland) (1981).

[Plo82] Plotkin, G.: "An Operational Semantics for CSP". Proc. Formal Description of Programming Concepts II (ed. D. Bjørner), IFIP TC-2, Garmisch-Partenkirchen, Germany (1982), 199-223.

[TaW82] Taylor, R. and Wilson, P.: "OCCAM: Process-Oriented Language Meets Demands of Distributed Processing". Electronics, Mac Graw Hill (November 1982).

[Win84] Winskel, G.: "Categories of Models for Concurrency" LNCS.197 Seminar on Concurrency. Carnegie-Mellon University, Pittsburgh July 9-11, 1984. Springer-Verlag (1985), 246-267.

The Category of Milner Processes is Exact*

David B. Benson
Computer Science Department
Washington State University
Pullman, WA 99164–1210
USA

csnet: benson@cs1.wsu.edu

0. Introduction

Processes may be implemented by automata which interact by an exchange or an agreement upon a communication. These implementing automata are just transition systems. Following Park and Milner we agree to consider two such equivalent when there is a bisimulation between them. A central lesson of category theory is to focus attention on the morphisms preserving the structures of interest. The result in this case is what we shall call the category of Milner processes. This category is exact and exactness enables us to recreate an algebra of processes. The attempt fails to completely capture the intuitions of CCS. At the last, we add the missing ingredient of *rooted* bisimulation to claim that the intuitions of CCS are correctly embodied

A more complete version of this conference paper is planned. There we expect to complete the program begun in [5] of semantically well-grounding process algebras, [10], in automata theory by demonstrating the passage from categories of rooted Milner processes to some variant of process algebra. Exactly which variant is unclear as of this writing.

Speaking slightly imprecisely, a Milner process is a bisimulation equivalence class of automata. From [5] we have that such a class is pure epi-connected for strong bisimulation and from [2, 6] we know that these results hold for (weak) bisimulation as well. For either strong or weak bisimulation there is a unique (up to isomorphism) representing automaton for each class, so we may say that a Milner process is such a representing automaton. A morphism of Milner processes is a function from pure states to either pure states or zero, compatibly with the action in each Milner process.

In considering the forgetful functor from the category of Milner processes to the category of pointed sets and point-preserving functions between pointed sets, $\mathbf{Set}_0$, one notices that all the abnormal epimorphisms* have been swept up in forming the pure epi-connected classes. Thus the interesting forgetful factor is to $\mathbf{Neset}_0$, the largest subcategory of $\mathbf{Set}_0$ in which every epimorphism is a normal epimorphism.

---

*Research supported in part by NSF grant MCS–8402305.

Now $\mathbf{Set}_0$ is normal and it comes as no surprise therefore that $\mathbf{Neset}_0$ is both normal and conormal, which following [12] we take as the definition of exact.** This property is easily seen to lift to categories of Milner processes. Therefore categories of Milner processes has "especially nice factorization properties," to quote [12].

Another central lesson of category theory is to study morphisms with *appropriate* compositional properties. In this study we find that the pure epimorphisms are too large a class, corresponding to the well known fact that bisimulation equivalence is not a congruence. So we turn to the retractions to find that this class is too small. An intermediate class, provisionally called *good* epimorphisms, appear to be the class of morphisms whose weakly connected components are the CCS congruence classes.

As the body of the paper demonstrates, I have been influenced by the work of E. G. Manes on control categories, [19, 20, 21], as well as all the work on algebraic and categorical automata theory, typlified by [1, 17, 18]. This paper is a direct descendent of [22]. Other, less direct contributions are [15, 23].

While the results are both pleasant and surprising, I shall be even more pleased to understand how this result should be generalized to an arbitrary control category as presented in [19, 20, 21]. Nonetheless, we provide an exposition of certain process forming operators, standard from CCS, in categories of Milner processes. These are derived from ideas in [2], and we see that the structure of Milner processes developed here is just enough to define the alternative choice operator, denoted "+" in CCS, and the sequential composition operator as well as a host of related operators.

1. Milner Process Automata Framework

For a category theorist I am now going to be overly concrete. All this could be stated rather more abstractly, but I want to make the fact that I am actually doing automata theory and describing CCS processes readily apparent. What follows in this section is abstracted from [2, 5, 6].

1.1 Semirings

A set S equipped with a binary addition and a binary multiplication is a *semiring* when we have 0 a distinguished element of S and

---

* A morphism is an abnormal epimorphism if it is an epimorphism which is not a normal epimorphism.

** Unfortunately, the term *exact* has various different meanings in category theory. Our use is a concept which might also be called *binormal*. Here *exact* is used as in [11, 12].

$$(r+s)+t = r+(s+t),$$
$$r+s = s+r,$$
$$r+0 = r,$$
$$(r.s).t = r.(s.t),$$
$$r.0 = 0.r = 0,$$
$$r.(s+t) = r.s+r.t,$$
$$(r+s).t = r.t+s.t$$

for all $r, s, t, \in S$. A semiring S is *unital* if there is a distinguished element $1 \in S$ such that

$$1.s = s = s.1$$

for all $s \in S$. A semiring S is *commutative* if

$$r.s = s.r$$

for all $r, s \in S$. A semiring S is *zerosumfree*, [13], if

$$r+s=0 \qquad \text{implies} \qquad r=s=0$$

for all $r, s \in S$. A *scalar* semiring is a zerosumfree commutative unital semiring, which we will denote by $\flat$. The typical scalar semiring is the Boolean $\flat = \{0, 1\}$ with "and" as multiplication and "inclusive or" as addition. Possibilities for scalar semirings include: the natural numbers, [3], any distributive lattice as conditions or reasons, the minimization semiring.

However, the main reason for stating this theory at this level of abstraction is that these axioms are sufficient to define bisimulation, and all but the zerosumfree property are necessary simply to state the framework of transition systems.

1.2 Right S-modules

An abelian monoid $(M, +, 0)$ is a right S-module over semiring S when equipped with a *right action*

$$\mu\colon\ M \times S \longrightarrow M$$

satisfying the usual so-called scalar multiplication laws to follow. We write $m.s$ for $(m, s)\mu$ with $m \in M$ and $s \in S$. The right action laws are

$$(m+n).s = m.s+n.s,$$
$$m.(r+s) = m.r+m.s,$$
$$m.0 = 0 = 0.s,$$
$$(m.r).s = m.(r.s)$$

for all $m, n \in M$ and $r, s \in S$.

A right S-module is *unital* if the semiring S is unital and $m.1 = m$ for all $m \in M$. A *free* unital right S-module with *pure states* X is the free unital right S-module freely generated by X by the usual insertion of the generators. This object is denoted $X \cdot S$ and may be identified with a subset of the collection of all functions from X to S, the particular subset depending upon the enclosing category. If only finite sums are considered, the subset is the collection of all functions from X to S with finite support. If enough infinite sums exist this restriction of finite support may be removed; see [19, 20, 21] for suggestive ideas. To be definite, however, we shall stay with the usual finitary algebraic ideas throughout this paper. Such a function from X to S is denoted

$$\sum_{x \in X} x.s_x$$

with $s_x \in S$ being zero for all but a finite number of the $x \in X$.

When $S = \flat$ is a scalar semiring, we shall abbreviate from "free unital right $\flat$-module" to just "$\flat$-module." The $\flat$–module $X \cdot \flat$ is to be considered as the set of all $\flat$-nondeterministic distributions over the pure states X. If $\flat$ is the two element Boolean semiring, $X \cdot \flat$ is the unital semilattice of finite subsets of X with set union as the abelian addition and $\emptyset$ as the zero of the $\flat$-module.

1.3 Nondeterministic Communication Symbols

Let E be the alphabet of communication symbols, which for the purposes of this paper is fixed throughout. As usual, E^* is the set of all finite strings over E, isomorphic to the free monoid freely generated by E. As usual, string concatentation is written multiplicatively. However, the unit or null string is denoted by $1 \in E^*$. The unit has been called variously the unobservable event type or the silent communication or other, similar terms.

Let $A = E^* \cdot \flat$ be the $\flat$-module freely generated by E^*. String concatentation extends to all of A by setting

$$(\sum_{\alpha \in E^*} \alpha.r_\alpha).(\sum_{\beta \in E^*} \beta.s_\beta) = \sum_{\alpha,\beta \in E^*} (\alpha\beta).r_\alpha s_\beta$$

for all $\sum_{\alpha \in E^*} \alpha.r_\alpha$, $\sum_{\beta \in E^*} \beta.s_\beta$ in A. The $\flat$-module A is called the monoid $\flat$-algebra generated by the monoid E^*. Since the above multiplication is bilinear, we may choose to view this multiplication as a $\flat$-linear map on the tensor product $A \otimes A$, the tensor taken with respect to $\flat$. Universal considerations as in [16] guarantee this tensor exists. So we may write the $\flat$-nondeterministic concatenation as

$$\kappa \colon A \otimes A \longrightarrow A$$

with $(\alpha \otimes \beta)\kappa = \alpha\beta$ on pure states $\alpha, \beta \in E^*$. Clearly A is then a unital semiring which is zerosumfree since the underlying scalars $\flat$ are zerosumfree by fiat.

Note that under these circumstances if $X \cdot \flat$ and $Y \cdot \flat$ are $\flat$-modules over pure states X, Y respectively, then $X \cdot \flat \otimes Y \cdot \flat \approx (X \times Y) \cdot \flat$, [16].

When $\flat$ is the two element Boolean semiring the monoid $\flat$-algebra is the collection of all finite subsets of strings with set union as addition and the usual pointwise concatenation of sets as multiplication.

1.4 A-Automata

An incompletely-specified $\flat$-nondeterministic input-only automaton is a pair $(X, \zeta\colon X \cdot \flat \otimes A \to X \cdot \flat)$ where ζ is a $\flat$-linear map called the *run map* or *reachability map* of the automaton, this data subject to the requirement that the run map ζ is the right action making $X \cdot \flat$ an A-module. Specifically,

$$(u \otimes (a.b))\zeta = ((u \otimes a)\zeta \otimes b)\zeta$$

for all $u \in X \cdot \flat$ and all $a, b \in A$. Writing the right action ζ in scalar multiplication form, from the above law and from the fact that the domain of ζ is a tensor product, we have

$$\begin{aligned} u.(a.b) &= (u.a).b, \\ u.(a+b) &= u.a + u.b, \\ (u+v).a &= u.a + v.a, \\ u.0 &= 0.a = 0 \end{aligned}$$

for all $u, v \in X \cdot \flat$ and all $a, b \in A$.

The right A-module run maps are in general neither free with respect to the semiring A nor unital. If a run map is a unital right A-module, $u.1 = u$ for all $u \in X \cdot \flat$ and $1 \in A$ being the $1 \in E^*$, then all of the automaton's actions are observable. Further considerations of unital automata may be found in [1, 3, 4, 7, 8, 13, 17, 18] and in many other papers and monographs. Nonunital run maps have interesting properties, some of which are explored in [4].

Note that we use the symbol "1" with $1 \in A$, rather than the symbol "τ" found in Milner's writings. Later we shall introduce τ as a process.

1.5 Morphisms

Let $(X,\ \zeta\colon X \cdot \flat \otimes A \to X \cdot \flat)$ and $(Y,\ \xi\colon Y \cdot \flat \otimes A \to Y \cdot \flat)$ be automata as above which we notationally abbreviate to (X, ζ) and (Y, ξ), respectively, and without loss of information as $\flat$ and A are fixed.

A morphism h from (X, ζ) to (Y, ξ) in the category of automata, **Impure**(A), is an A-linear and $\flat$-linear map

$$h\colon X \cdot \flat \to Y \cdot \flat$$

such that $Xh \subseteq Y \cup \{0\}$.

This needs both explanation and justification. The requirement that h be $\flat$-linear guarantees that $\flat$-nondeterminism behaves in the expected way, cf. [16] and references cited therein. The requirement that h be A-linear is the guarantee that a morphism is a simulation of one automation by another,

$$(u.a + v.b)h = uh.a + vh.b,$$

for all $u, v \in X \cdot \flat$ and all a, b in A.

The set Xh is the image of the pure states X in $Y \cdot \flat$. The requirement that $Xh \subseteq Y \cup \{0\}$ means that pure states map under h either to pure states or else are rubbed out by mapping to zero. This requirement is going to give rise to exactness. That pure states must map to pure states is technically justified by consideration of bisimulation, [2, 5, 6]. But further, one is loath in practice to let a pure state be simulated by a distribution, the various pure states thereof offering some, but not all, of the communication symbols offered by the simulated state. For example, suppose automaton (X, ζ) has $x_0.\alpha = x_1$, $x_0.\beta = x_2$ which we diagram in something resembling CCS style as

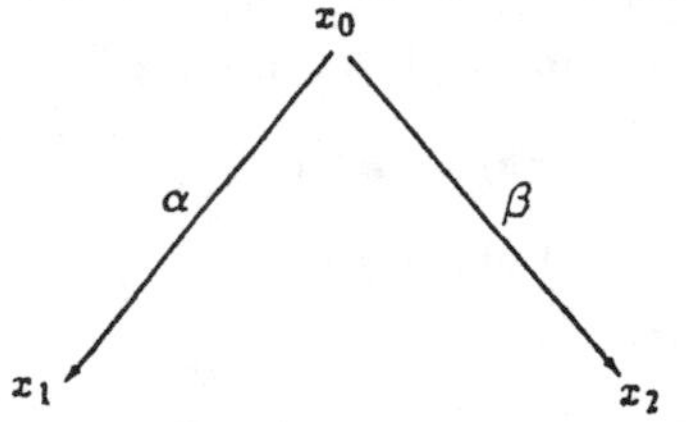

while automaton (Y, ξ) has $y_{00}.\alpha = y_1$, $y_{01}.\beta = y_2$ as its only communication offers, diagrammed as

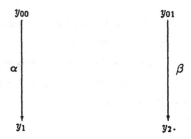

Then one might consider a map h such that $x_0h = y_{00}+y_{01}$, $x_1h = y_1$, $x_2h = y_2$ and observe that this is A-linear. The practicing software engineer objects that this not a simulation sufficiently faithful for his purposes. On these grounds, related to what are called *must* or *demonic* considerations by various authors, we have to reject maps such as this example has sketched.

However, the practicing software engineer seems—at least on occasion—to be quite satisfied with using his eraser to remove a s indicating that the result of the removal is an adequate simulation. Mathematically then, we are left with the requirement that $Xh \subseteq Y \cup \{0\}$.

The category of all such automata and all such morphisms between them is denoted **Impure**(A). We define a forgetful functor U: **Impure**(A) $\rightarrow$ **Set**$_0$ on objects (X, ζ) by

$(X,\zeta)U = X_0$ with X_0 notation for $X \sqcup \{0\}$ where 0 is the distinguished element of the objects in $\mathbf{Set}_0$, and on morphisms h: $(X,\zeta) \to (Y,\xi)$ by hU: $X_0 \to Y_0$ being the restriction of h to X_0, where we identify the zero of $Y \cdot \flat$ with the distinguished element of Y_0.

Since automata in **Impure**(A) are not, in general, *unital* right A-modules, it appears that U does not, in general, have a left adjoint. As only finite sums of pure states are in general available, smallness considerations are very small indeed. Thus it seems that free automata in **Impure**(A) are not generally available. Nonetheless, given the structure of morphisms in **Impure**(A), U is the clear choice of forgetful functor to consider.

2. Pure(A)

A morphism h: $(X,\zeta) \to (Y,\xi)$ is *pure* if $Xh \subseteq Y$. Let **Pure**(A) be the subcategory of **Impure**(A) consisting of all automata and all and only the pure morphisms between the automata. Then the forgetful functor U: **Pure**(A) $\to$ $\mathbf{Set}_0$ is defined as the composition of the inclusion functor and the forgetful functor of the previous section.

2.1 Proposition. If h is an epimorphism in **Pure**(A) which is not an isomorphism, then h is an abnormal epimorphism.

Proof: This holds for hU in $\mathbf{Set}_0$ and is clearly reflected. □

Now **Pure**(A) is an (epi, normal mono)-category and cocomplete, [2, 6]. A weakly connected component of a category consists of all objects connected in the underlying *undirected* graph of the category. An epi-connected component consists of all objects connected solely by epimorphisms. That is, objects c_0 and c_n are epi-connected when there are objects $c_1, c_2, \ldots, c_{n-1}$ and epimorphisms $c_0 \to c_1 \leftarrow c_2 \to \cdots \leftarrow c_n$. The epi–connected components of **Pure**(A) are exactly the bisimulation equivalence classes of the automata in **Pure**(A). The zerosumfree property of the scalars is necessary to our proof of this fact, and may be abandoned if one is interested in the epi-connected classes themselves rather than in bisimulation. As **Pure**(A) is cocomplete each such epi-connected component is an upward-directed class with colimiting automaton the minimal automaton in the bisimulation equivalence class, [2, 6]. Thus all the abnormal epimorphisms are swept up into the bisimulation equivalence classes.

A route to a suitable category of processes is obtained by considering the collection of colimiting, minimal automata representing the epi-connected classes. We must check that the morphisms between any pair of automata result in a corresponding morphism between the representing automata. Since **Pure**(A) is an (epi, extremal mono)-category, it suffices to consider just the extremal monomorphisms. For the following result, we denote automata by single Greek or Roman letters, possibly with subscripts and various other embellishments.

2.2 Lemma. Let $m\colon \zeta_1 \to \zeta_2$ be an extremal monomorphism. Let c_i be the colimiting object of the epi-connected component containing ζ_i, and let $e_i\colon \zeta_i \to c_i$ be the epimorphisms in the associated cocones, for $i = 1, 2$. Then there exists an extremal monomorphism $\bar{m}\colon c_1 \to c_2$ such that $e_1\bar{m} = me_2$.

Proof: All the objects and morphisms used in the proof are illustrated in the following diagram.

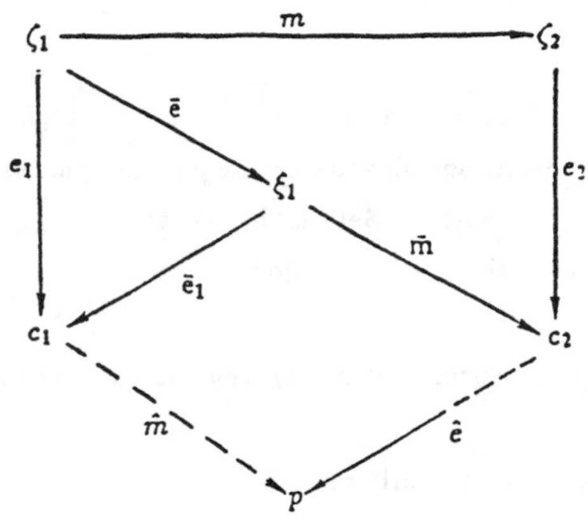

Let $(\bar{e}, \bar{m})$ be the (epi, extremal mono)-factorization of me_2, $me_2 = \bar{e}\bar{m}$. As $\bar{e}$ is an epimorphism, ξ_1 is an object of the epi-connected component containing ζ_1. Therefore, there is an epimorphism $\bar{e}_1\colon \xi_1 \to c_1$ in the cocone to c_1. Let $(\hat{m}, \hat{e}; p) \approx$ *Pushout* $(\xi_1; \bar{e}_1, \bar{m})$. Since $\bar{e}_1$ is an epimorphism, so is $\hat{e}$. As c_2 is the colimiting object of its cocone, $\hat{e}$ is an isomorphism. Therefore, $\bar{m} = \bar{e}_1\hat{m}\hat{e}^{-1}$. As $\bar{m}$ is extremal mono, $\bar{e}_1$ is an isomorphism. Therefore we may identify ξ_1 with c_1 and $\bar{m}\colon c_1 \to c_2$ is an extremal monomorphism between colimiting objects which represents $m\colon \zeta_1 \to \zeta_2$. □

Let **Bisim**(A) be the full subcategory of **Pure**(A) determined by the colimiting, representing automata as the only objects in **Bisim**(A). The above lemma establishes a functor B: **Pure**$(A) \to$ **Bisim**(A) which sends each object to its representing minimal automaton and each morphism to the associated monomorphism between representing automata. Every morphism of the category **Bisim**(A) is a normal monomorphism. Clearly **Bisim**(A) is a subcategory of **Impure**(A). The automata of **Bisim**(A) we shall call *reduced.*

Now **Bisim**(A) has a rather impovrished class of morphisms. For example, there are no morphisms whatsoever between the reduced unital A-automata diagrammed by

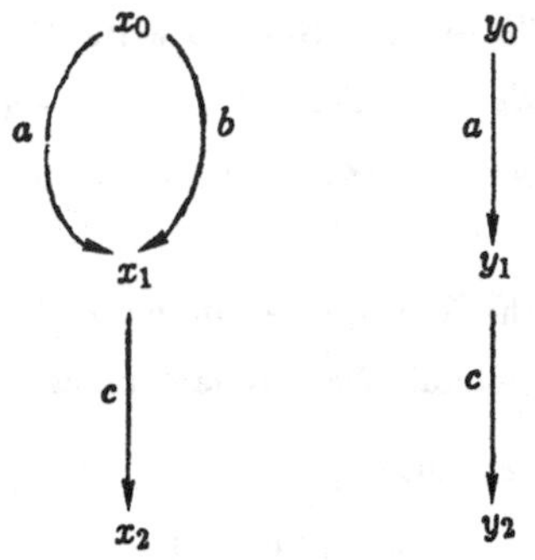

We shall go to considerable effort to find a satisfying category of processes.

3. A Category of Park-Milner Processes

We begin by noting some facts about $\mathbf{Set_0}$, a control category [19, 20, 21] isomorphic to **Pfn**.

$\mathbf{Set_0}$ is normal. That is, $\mathbf{Set_0}$ has kernels, cokernels, each of its monomorphisms is a normal monomorphism, i.e., m is a monomorphism iff $m \approx Ker(Cok(m))$, and $\mathbf{Set_0}$ is (epi, normal mono)-factorizable. To fix notation, for X_0 and Y_0 objects of $\mathbf{Set_0}$ let $X_0 \amalg Y_0 \approx (X \sqcup Y)_0$ be the coproduct in $\mathbf{Set_0}$. When f: $X_0 \to Z_0$ and g: $Y_0 \to Z_0$ are a pair of $\mathbf{Set_0}$-morphisms with common codomain, let $[f, g]$: $X_0 \amalg Y_0 \to Z_0$ be the unique mediating morphism determined by the universal property of the coproduct, $X_0 \amalg Y_0$. From now on, the bracket notation $[f, g]$ and the $\amalg$ notation always refer to the coproduct as formed in $\mathbf{Set_0}$

A $\mathbf{Set_0}$-morphism t: $X_0 \to Y_0$ is *total* if for all f: $W_0 \to X_0$, $ft = 0$ implies $f = 0$. The morphism t: $X_0 \to Y_0$ is *cototal* if for all g: $Y_0 \to Z_0$, $tg = 0$ implies $g = 0$. A morphism that is both total and cototal is called *bitotal.* An **Impure**-morphism h is a pure epimorphism iff hU is bitotal.

Every $\mathbf{Set_0}$-morphism f: $X_0 \to Y_0$ has a [kernel, total]-decomposition and a [cokernel, cototal]-decomposition. As to the former, let $K_0 \approx Ker(f)$ and $T_0 \approx (X - K)_0$ to observe that $X_0 \approx K_0 \amalg T_0$. With the coproduct injections μ_K: $K_0 \to K_0 \amalg T_0$, μ_T: $T_0 \to K_0 \amalg T_0$, the [kernel, total]-decomposition is given by $f = [0_{KY}, t_f]$ with 0_{KY}: $K_0 \to Y_0$ the zero morphism from K_0 to Y_0 and t_f: $T_0 \to Y_0$ total such that $0_{KY} = \mu_K f$ and $t_f = \mu_T f$. The [cokernel, cototal]-decomposition is obtained by considering the coinjections π_C: $C_0 \amalg R_0 \approx Y_0 \to C_0$, π_R: $C_0 \amalg R_0 \to R_0$ with $C_0 \approx Cok(f)$ and $R_0 \approx (Y - C)_0$.

3.1 Proposition. A morphism f: $X_0 \to Y_0$ is a normal epimorphism in $\mathbf{Set_0}$ if and only if there exists a set K_0 such that $X_0 \approx K_0 \amalg Y_0$ and $f \approx [0_{KY}, 1_Y]$, where 1_Y is the identity morphism on Y_0.

Proof: Consider any morphism $f: X_0 \to Y_0$ in $\mathbf{Set}_0$ to write its [kernel, total]-decomposition as $f \approx [0_{KY}, t_f]$: $K_0 \amalg T \to Y_0$ with $X_0 \approx K_0 \amalg T_0$. We have $Ker(f) \approx \mu_K$: $K_0 \to K_0 \amalg T_0$ the insertion of K_0 into X_0. Then $Cok(\mu_K) = Coeq(\mu_K, 0_{KX}) \approx [0_{KT}, 1_T]$. Thus $f \approx Cok(Ker(f))$ iff $T_0 \approx Y_0$. □

Let $[0_{KY}, g]$: $K_0 \amalg T_0 \to Y_0$ be the [kernel, total]-decomposition of $\mathbf{Set}_0$-epimorphism e. As e is an epimorphism, both e and g are cototal. We may factor e as

$$\begin{aligned} e &= [0_{KY}, g] \\ &= (1_K \amalg g)[0_{KY}, 1_Y], \end{aligned}$$

to observe that $(1_K \amalg g)$: $K_0 \amalg T_0 \to K_0 \amalg Y_0$ is bitotal. Thus every $\mathbf{Set}_0$-epimorphism is (bitotal, normal epi)-factorizable and this factorization is clearly unique up to isomorphism. As $\mathbf{Set}_0$ is an (epi, normal mono)-category, we may now say that $\mathbf{Set}_0$ is a (bitotal, normal epi, normal mono)-category.

This later property clearly lifts to **Impure**(A). The bitotal components of this factorization provide the pure epimorphisms internal to the epi-connected components of **Pure**(A). The remaining, normal components of the (bitotal, normal epi, normal mono)-factorization provide the interesting structure.

Now in $\mathbf{Set}_0$ a bitotal morphism is a retraction. This property does not lift to either **Pure**(A) or **Impure**(A) and this is the reason that **Bisim**(A) is so poorly endowed with morphisms. Therefore we begin again to consider the retraction-connected components of **Pure**(A).

By reasoning similar to that sketched in the previous section one sees that the retraction-connected components of **Pure**(A) have colimits with retractions forming the cocones. Call the colimiting objects *retracts*, or *retract automata*. Now this provides a proper refinement of the bisimulation equivalence relation. In general the retract is not the reduced automaton. In the following picture the retract is on the left and the reduced automaton is on the right, with an obvious epimorphism from the retract to the reduced automaton. This epimorphism is not a retraction.

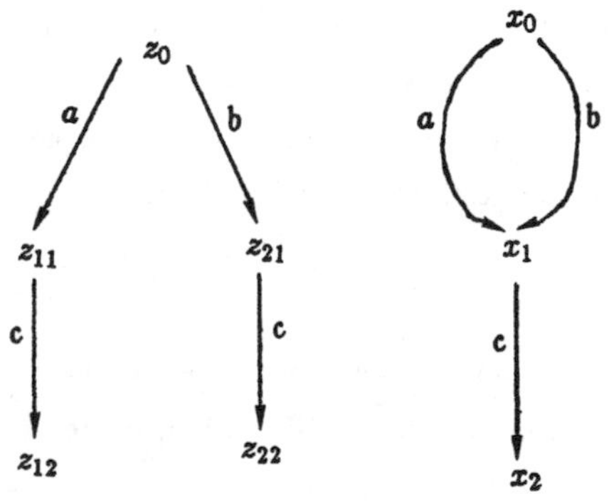

We are now ready for the definition of a category of processes based on the retract automata.

3.2 Definition. The category **Process**(A) has as its objects the retract automata representing the retraction-connected components of **Pure**(A). In this setting the retract automata may be called *processes*. The **Process**-morphisms from process p to process q are all the morphisms from p to q in **Impure**(A) which have a (normal epi, normal mono)-factorization.

In the above definition we might just as well have said that each morphism from p to q in **Impure**(A) with a (bitotal retraction, normal epi, normal mono)-factorization was a morphism from process p to process q. For as p and q are retract automata, the bitotal retraction in the factorization is always an isomorphism. What has happened is the removal of the nonretractive bitotal morphisms, exemplified by the previous picture, from further consideration. This is not a serious matter, but the issues raised by the nonretractive bitotal morphisms needs must wait for a complete version of this somewhat sketchy report. For now it perhaps suffices to note that removing the bitotal nonretractions is a means to obtain morphism composition.

Now **Process**(A) is a non-full subcategory of **Impure**(A). It remains to show that **Process**(A) is exact. We shall do so by showing that the forgetful functor U, when restricted to **Process**(A), actually carries **Proces**(A) faithfully into an exact subcategory of $\mathbf{Set}_0$.

Let $\mathbf{Neset}_0$ be the subcategory of $\mathbf{Set}_0$ with the same objects as $\mathbf{Set}_0$ and as morphisms all the $\mathbf{Set}_0$-morphisms with a (normal epi, normal mono)-factorization. Thus $\mathbf{Neset}_0$ is exact, the morphisms being functions of the form $[0,m]$ where 0 denotes a zero map and m a monomorphism, here an injective function.

Now the forgetting of **Process**(A) into $\mathbf{Set}_0$ via the functor

$$(X,\zeta)U = X_0$$

is clearly faithful. If h: $(X,\zeta) \to (Y,\xi)$ is an epimorphism of **Process**(A) then hU is an epimorphism of $\mathbf{Set}_0$. We have $hU \approx (1_K \amalg g)[0_{KY}, 1_Y]$ for some total g. But then $(1_K \amalg g)$ lifts to a morphism of **Impure**(A) with domain (X,ζ). As $(1_K \amalg g)$ is total in $\mathbf{Set}_0$, it is in fact a morphism of **Pure**(A) and as (X,ζ) is an object of **Process**(A), $(1_K \amalg g)$ is an isomorphism. Therefore $hU \approx [0_{KY}, 1_Y]$.

It remains to show that U reflects normal epimorphisms. To this end suppose h: $(X,\zeta) \to (Y,\xi)$ is such that hU, h restricted to X_0, is a normal epimorphism. From the characterization of the [kernel, domain-of-definition]-decomposition in the previous proposition, we have $X_0 \approx K_0 + Y_0$ and thus $X = K \sqcup Y$. Then

$$X\cdot\flat \cong (K \sqcup Y)\cdot\flat \cong K\cdot\flat \oplus Y\cdot\flat$$

where "$\oplus$" denotes the biproduct of unital right $\flat$-modules. Then h: $X\cdot\flat \to Y\cdot\flat$ is the direct sum of $\flat$-linear maps 0_{KY}: $K\cdot\flat \to Y\cdot\flat$ and 1_Y: $Y\cdot\flat \to Y\cdot\flat$, $h = 0_{KY} \oplus 1_Y$. A proof similar to that in the proposition demonstrates that h is a normal epimorphism of $\flat$–modules. As h is an A-linear map, it is a normal epimorphism of A-modules. This finishes the proof of the following result.

3.3 Theorem. **Process(A)** is exact and is equipped with an exact functor U to $\mathbf{Neset_0}$.

Again, we may view **Process(A)** as a non-full subcategory of $\mathbf{Neset_0}$.

4. Structure of Process(A) and Neset$_0$

In $\mathbf{Set_0}$ the projections $X_0 \times Y_0 \xrightarrow{\pi} X_0$ from the $\mathbf{Set_0}$-product are abnormal epimorphisms since $Ker(\pi) \approx \langle 0, 1 \rangle \colon Y_0 \to X_0 \times Y_0$ and π is not isomorphic to $Cok(Ker(\pi))$. Thus categorical products are conspicuously absent from $\mathbf{Neset_0}$ and hence from **Process(A)**. As $\mathbf{Neset_0}$ is a self-dual category, categorical coproducts are in general missing as well. This deficiency lifts to **Process(A)**. The category **Process(A)** does not have pullbacks as this would imply products and abelianness. Neither category has a semiadditive structure on hom-sets for similar reasons. However, both categories have equalizers formed in the usual pointwise fashion, and dually both categories have coequalizers.

The lack of coproducts is a particular hinderance to the development of a satisfying category of processes, this lack being the outcome of considering either the retraction or the bisimulation equivalence. The disappearance of coproducts in passing from $\mathbf{Set_0}$ to $\mathbf{Neset_0}$ is exemplified by copowers. While $X_0 \amalg X_0 \approx (X \sqcup X)_0$ is a coproduct object of $\mathbf{Set_0}$, the morphism $[1_X, 1_X] \colon X_0 \amalg X_0 \to X_0$ is bitotal and not an isomorphism so $X_0 \amalg X_0$ is not an object of $\mathbf{Neset_0}$. One may easily check that $X_0 \amalg Y_0$ survives in $\mathbf{Neset_0}$ just in case at least one of X, Y is the empty set. Despite these difficulties, exactness enables us to define the composition of certain processes, this being the subject of the next section.

5. Process Composition Via Fill-in Diagrams

Consider the diagram

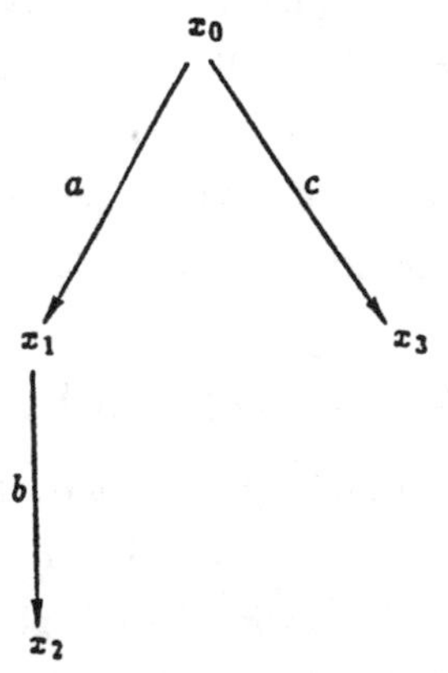

which in CCS terminology consists of first forming the process $x_0 \xrightarrow{a} x_1 \xrightarrow{b} x_2$ by sequentially composing two processes and then forming the alternative choice at the top.

Let $X = \{x_0, x_1, x_2\}$, $Y = \{x_0, x_1\}$, $Z = \{x_1, x_2\}$. The obvious normal epimorphism from X to Y and the obvious monomorphism from Z to X provide the identification of $x_1 \in X$ with x_1 in each of Y and Z. These morphisms lift to **Process**(A) providing the sequential composition of the two processes in the example.

With $X = \{x_0, x_1, x_2\}$ and $Y = \{x_0, x_3\}$, a pair of normal epimorphisms from $\{x_0, x_1, x_2, x_3\}$ to each of X and Y constructs the alternative choice.

Here is the construction the alternative choice composition of two processes: Let (X, ζ), (Y, ξ), and (Z, ψ) be objects of **Process**(A). If f: $(Z, \psi) \to (X, \zeta)$ and g: $(Z, \psi) \to (Y, \xi)$ are normal epimorphisms such that $Im(fU) = X_0$ and $Im(gU) = Y_0$, we say that the process (Z, ψ) is an (f, g)-composition of the processes (X, ζ) and (Y, ξ).

Such pairs of normal epimorphisms form a mono-source which we further remark is a natural source. The limit would be the product of (X, ζ) and (Y, ξ) which does not in general exist. Thus one is forced to find other organizing principles for the collections of (f, g)-compositions. A similar difficulty arises in the constructions generalizing the sequential composition of processes. The organizing principle in both cases is the same: diagrams of short exact sequences, familiar from homological algebra, provide the general setting for the notion of *fill-in*. Such diagrams only make sense in an exact category, and the notion of fill-in diagram is what is required to define the general composition of processes.

A ***short exact sequence*** is a sequence of five objects, the first and last of which must be the zero object, and connecting morphisms such that the image of each morphism is isomorphic to the kernel of the next morphism in sequence. Here is a diagram:

$$0 \longrightarrow \zeta_1 \xrightarrow{f} \zeta_2 \xrightarrow{g} \zeta_3 \longrightarrow 0.$$

The morphisms $0 \to \zeta_1$ and $\zeta_3 \to 0$ are zero morphisms, so f is a normal monomorphism and g is a normal epimorphism, with f characterizing the kernel of g.

View finite, acyclic processes as trees, exemplified by the tree beginning this section. Short exact sequences with a given tree as the central factor faithfully describe decompositions of the given tree into top and bottom parts of the tree. Using notation from the above diagram of a short exact sequence, let ζ_2 be the given tree. Then ζ_1 is the bottom of the tree and ζ_3 is the top of the tree. As the bottom part of a tree is a forest of trees, we should and will view each process as a forest of trees. Here is an abstracted picture of a process ζ_2 decomposed into ζ_1 and ζ_3:

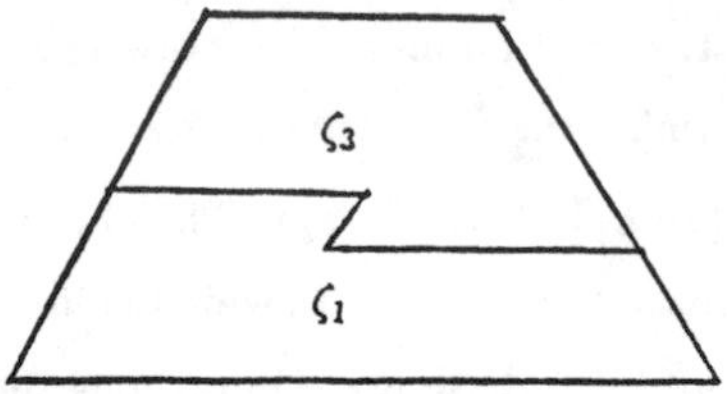

The decompositions need not preserve level and need not be proper.

We translate a standard result about short exact sequences into the view as forests of trees. The standard result is: If the diagram

$$\begin{array}{ccccccccc} 0 & \longrightarrow & \zeta_1 & \longrightarrow & \zeta_2 & \longrightarrow & \zeta_3 & \longrightarrow & 0 \\ & & & & \downarrow & & & & \\ 0 & \longrightarrow & \xi_1 & \longrightarrow & \xi_2 & \longrightarrow & \xi_3 & \longrightarrow & 0 \end{array}$$

has short exact rows while $\zeta_2 \to \xi_2$ is any morphism, then there exists a morphism $\zeta_1 \to \xi_1$ if and only if there exists a morphism $\zeta_2 \to \xi_3$.

Thinking of ζ_2 and ξ_3 as forests of trees, this result says that the choice of decompositions into bottoms ζ_1, ξ_1 which are compatible—in the sense that there is a morphism $\zeta_1 \to \xi_1$ —occurs when and only when the corresponding tops are compatible.

All of the standard results about fill-in diagrams of short exact sequences have similar purely structural interpretations. Indeed, they must, since these results hold in all exact categories and depend solely upon the exactness structure.

A standard form of process composition is the sequential composition of two processes. Thinking of forests, the bottom of the first forest in the sequential composition must be equal to the top of the second. This common portion we shall call the *join*, and denote by the letter j. Here is a picture:

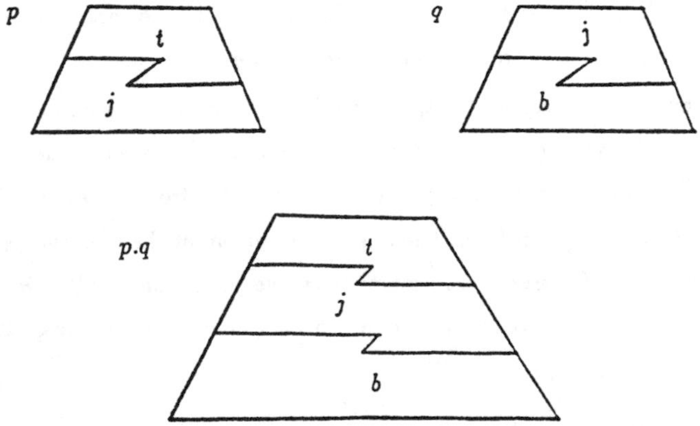

Formally, let p, q, j, t, b be processes and let

$$0 \longrightarrow j \longrightarrow p \longrightarrow t \longrightarrow 0$$

$$0 \longrightarrow b \longrightarrow q \longrightarrow j \longrightarrow 0$$

be short exact sequences. The process $p.q$ is that process, if such exists, which makes all rows and columns into short exact sequences and all squares commutative in the following diagram.

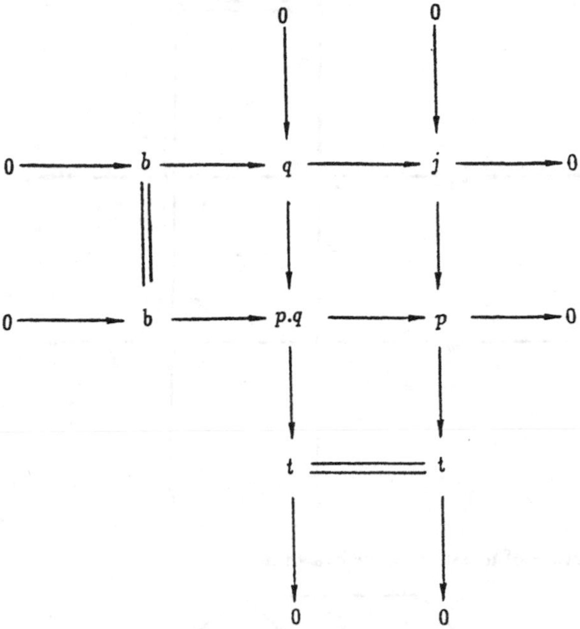

(SQ)

This is an example of a *fill-in* diagram from the given data. Of course, the notation appears to be deficient in that the particular sequential composition of processes p and q depends upon the choice of the common join, j, if indeed such a join exists at all. But, for example, with $p = t = b = q$ and $j = 0$ there is no way to complete (SQ) without coproducts. The problem is the choice of the join j and we shall return to the consideration of (SQ) a bit later on.

The so-called alternative choice of two processes p and q is denoted $p + q$. This combinator is analogous to the sequential composition operator, requiring a common join subprocess, j. Formally let

$$0 \longrightarrow b_p \longrightarrow p \longrightarrow j \longrightarrow 0$$

$$0 \longrightarrow b_q \longrightarrow q \longrightarrow j \longrightarrow 0$$

be short exact sequences. The process $p \cup q$, joined at j, is the process such that the following diagram commutes with short exact rows and columns.

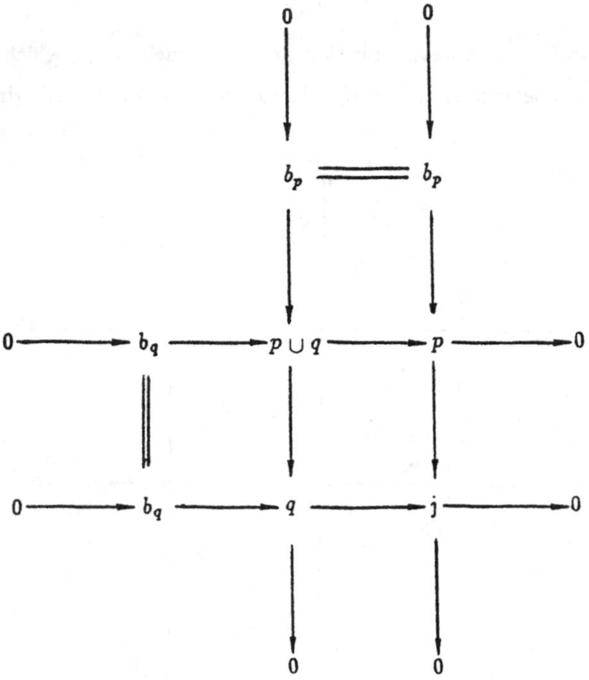

(U)

The corresponding picture of forests may be drawn as

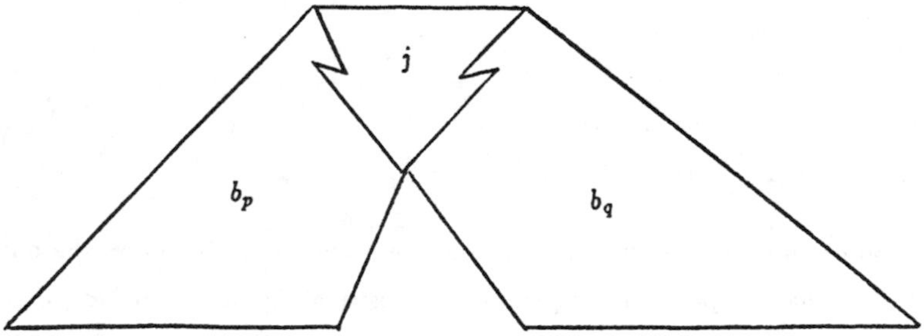

Again there is no guarantee of a (U) fill-in given the two short exact sequences of data. For example, set $b_p = p = q = b_q$ and $j = 0$ to see that the fill-in $p \cup p$ would have to be the copower of p, hence $p \cup p$ is defined by (U) iff $p = 0$. The problem here is the wrong choice of the common join j. Bisimulation clearly requires the greatest commonality and exactness comes to the rescue via [12, ¶39.10].

Given processes p and q, consider the lattices (possibly on classes) of the quotient objects of p

and q. Since 0 is a quotient object of every process, these two lattices have a non-empty intersection. Let j be the least element of the intersection of the two lattices of quotient objects. This fixes normal epimorphisms e_p: $p \to j$ and e_q: $q \to j$. Now let $b_p \to p = Ker(e_p)$ and $b_q \to q = Ker(e_q)$ to observe that

$$0 \longrightarrow b_p \longrightarrow p \longrightarrow j \longrightarrow 0$$

$$0 \longrightarrow b_q \longrightarrow q \longrightarrow j \longrightarrow 0$$

are short exact sequences.

For our next consideration we work within **Neset$_0$**, recalling that we may view each process as possessing a distinct set of pure states. Let X be the pure states of p, Y be the pure states of q, and Z be the pure states of j. From the normal epimorphisms $p \to j$ and $q \to j$ we have [kernel, total]-factorizations of x, Y as

$$X \approx K_X \amalg Z,$$

$$Y \approx K_Y \amalg Z.$$

Let $W \approx K_X \amalg Z \amalg K_y$ be equipped with the normal epimorphisms

$$[1_X, 0_{K_y}]\colon\ W \longrightarrow X,$$

$$[0_{K_X}, 1_Y]\colon\ W \longrightarrow Y$$

to observe that these latter two morphisms lift to **Process**-morphisms. Write the lift of W as $p+q$ to note that

$$\begin{array}{ccc} p+q & \longrightarrow & p \\ \downarrow & & \downarrow \\ q & \longrightarrow & j \end{array}$$

is a commuting square of normal epimorphisms, where $p+q \to p$ is the lift of $[1_X, 0_{K_Y}]$ and $p+q \to q$ is the lift of $[0_{K_X}, 1_Y]$ Then $Ker(p+q \to p) \approx Ker(e_p)$ and $Ker(p+q \to q) \approx Ker(e_q)$.

Other than the slight set-theoretical sloppiness in the foregoing, which could be made clean, this provides the definition of the alternative choice $p+q$ for general processes. We have the following facts as immediate consequences of this definition for alternative choice:

$$p+p=p,$$

$$p+q=q+p,$$

$$p+0=p,$$

$$p+(q+r)=(p+q)+r$$

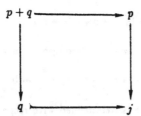

for all processes p, q, r objects of **Process**(A). If p has short exact sequence

$$0 \longrightarrow b_p \longrightarrow p \longrightarrow 1_0 \longrightarrow 0$$

then

$$p + 1_0 = p.$$

An appropriate choice of the common join for the sequential composition in (SQ) seems rather more delicate. The easiest approach is to attach the two processes as lightly as possible in something resembling the CCS sequential combinator.

Let $1 = \{x\}$ be any one-point set so that 1_0 is a two-point objec t of **Neset**$_0$. As an automaton, the action of 1_0 is given by $x.1 = x$ and $x.a = 0$ for all communication symbols a. If processes p and q have decompositions

$$0 \longrightarrow 1_0 \longrightarrow p \longrightarrow t \longrightarrow 0$$

$$0 \longrightarrow b \longrightarrow q \longrightarrow 1_0 \longrightarrow 0$$

as short exact sequences, then (SQ) always has a fill-in. The proof of this fact is tedious and is omitted.

Now if processes p and q have decompositions

$$0 \longrightarrow 1_0 \longrightarrow p \longrightarrow t_p \longrightarrow 0$$

$$0 \longrightarrow 1_0 \longrightarrow q \longrightarrow t_q \longrightarrow 0$$

as short exact sequences, then $p + q$ has a decomposition

$$0 \longrightarrow 1_0 \longrightarrow p + q \longrightarrow t_{p+q} \longrightarrow 0$$

as a short exact sequence. So if process r has a decomposition

$$0 \longrightarrow b \longrightarrow r \longrightarrow 1_0 \longrightarrow 0$$

as short exact sequence, then $(p + q).r$ is well-defined with respect to the particular point of composition determined by $1_0 \rightarrow p + q$ given above. This is a distinct process from $p.r + q.r$, unless $p = q$, being one of $p.r + q$ or $p + q.r$, depending upon the particular point of composition.

At the level of **Impure**(A) the copower automaton $r \amalg r$ exists and is equipped with the normal epimorphism $r \amalg r \rightarrow 1_0 \amalg 1_0$. So at this level we have $(p + q).(r \amalg r)$ well defined and isomorphic to $p.r + q.r$.

Here are a few other easy consequences of (SQ).

$$0.p = p.0 = p, \text{ for all process } p.$$

If p has short exact sequence

$$0 \longrightarrow 1_0 \longrightarrow p \longrightarrow t \longrightarrow 0$$

then

$$p.1_0 = p.$$

If q has short exact sequence

$$0 \longrightarrow b \longrightarrow q \longrightarrow 1_0 \longrightarrow 0$$

then

$$1_0.q = q.$$

If p has short exact sequence

$$0 \longrightarrow 1_0 \longrightarrow p \longrightarrow t \longrightarrow 0,$$

q has short exact sequences

$$\begin{array}{ccccccccc} 0 & \longrightarrow & 1_0 & \longrightarrow & q & \longrightarrow & t_q & \longrightarrow & 0 \\ & & \downarrow & & \| & & \downarrow & & \\ 0 & \longrightarrow & b_q & \longrightarrow & q & \longrightarrow & 1_0 & \longrightarrow & 0 \end{array}$$

and r has short exact sequence

$$0 \longrightarrow b \longrightarrow r \longrightarrow 1_0 \longrightarrow 0$$

then

$$(p.q).r = p.(q.r).$$

As is well-known, it is necessary to refine the bisimulation equivalence to obtain a congruence with respect to some combinators, alternative choice in particular. From our current, entirely semantical orientation, certain processes are not the fill-in diagram (U) combination of component processes. One example was given above in the discussion of $(p+q).r$. For another, assume the scalars $\mathfrak{b}$ are the Boolean semiring. Then the automaton diagram

$$x_1 \xrightarrow{1} x_2 \xrightarrow{a} x_3 \qquad (\tau a)$$

denotes no object in **Process**(A) since there is an abnormal epimorphism, which is a retraction, from (τa) to the automaton diagram

$$y_2 \xrightarrow{a} y_3 \qquad (a)$$

corresponding to the bisimulation $(\tau a) \approx (a)$. Thus the automaton diagramed as

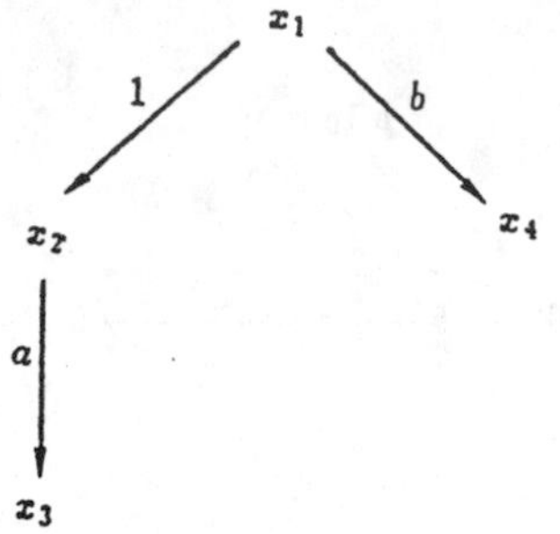

$(\tau a + b)$

is not the (U) fill-in of (τa) and (b). But this defect arises solely from the special nature of the start state, or root, in CCS related studies. We therefore consider the rooted correspondents of these same concepts.

6. Rooted Bisimulation

An *automata with start state* is an automation (X, ζ) together with a distinguished element $x_0 \in X$. The morphisms are required to preserve *and reflect* start states. The corresponding equivalence relation is called *rooted bisimulation* in [10, 9, 6].

Here we take a slightly different approach to the self-same concepts. Every automaton with start state p possesses a normal epimorphism of **Impure**(A) of the form $p \to 1_0$ where 1_0 is the one-state automaton introduced earlier. This normal epimorphism identifies the start state of p. Therefore it suffices to work in categories normal-epi-over 1_0, the various categories being denoted **R**—(A).

Recall that in any exact category, $p \xrightarrow{e} g \longrightarrow 0$ is exact iff e is a normal epimorphism. We shall use this notation to mean that e is a normal epimorphism even if the enclosing category under consideration is not exact. With this notation, the objects of **Rpure**(A) are those automata (X, ζ) of **Pure**(A) with "exact sequence"

$$(X, \zeta) \longrightarrow 1_0 \longrightarrow 0.$$

The morphisms of **Rpure**(A) are those homomorphisms $f\colon (X, \zeta) \to (Y, \xi)$ of **Pure**(A) such that

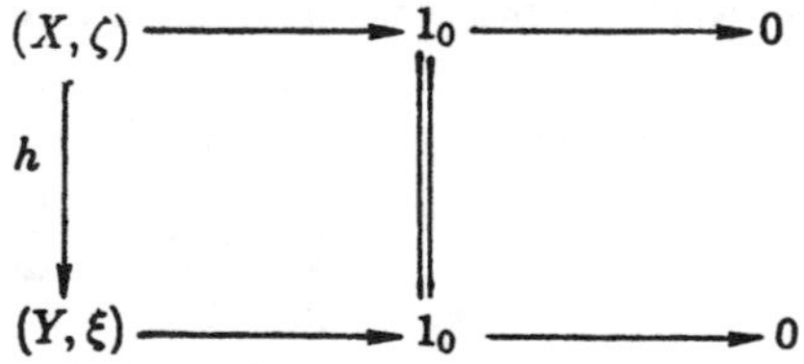

commutes. Such morphisms not only preserve but also reflect the start state.

The epi-connected classes have similar properties as in **Pure**(A), forming the rooted bisimulation equivalence classes. In a manner entirely analogous to that of prior sections of this report, **Rbisim**(A) is then defined. This category is essentially discrete as all morphisms are monomorphisms in **Bisim**(A) and the only rooted monomorphisms are isomorphisms.

Following the previous reasoning, we define the *rooted retracts* in **Rpure**(A) as the colimiting objects of the retract-connected classes of **Rpure**(A). Then **Rprocess**(A) is the non-full subcategory of **Rimpure**(A) with objects the rooted retracts and as morphisms, the (normal epi, normal mono)-factorizable morphisms of **Rimpure**(A) between the rooted retracts.

6.1 Proposition. Every morphism of **Rprocess**(A) is a rooted epimorphism of the $\mathbf{Set}_0$ form $[0, i]$ where 0 is a zero morphism in $\mathbf{Set}_0$ and i is an isomorphism in $\mathbf{Set}_0$. Further, **Rprocess**(A) is not pointed, but 1_0 is the terminal object of **Rprocess**(A).

Therefore the exact sequence ideas of previous sections must be abandoned as **Rprocess**(A) lacks monomorphisms other than isomorphisms. However, essential features of the fill-in diagrams can be maintained in this setting. To define the alternative choice of processes p and q, use the same construction: Let j be the most common quotient object of p and q as before, use the $\mathbf{Neset}_0$ level construction of object $p+q$ and normal epimorphisms so that in **Rprocess**(A), the diagram

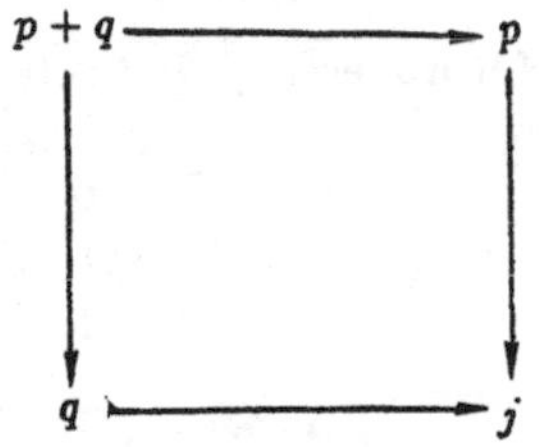

commutes. We note that $p+q \to p$ and $p+q \to q$ are rooted normal epimorphisms, so indeed morphisms of **Rprocess**(A). All of the pleasant results regarding the alternative choice $p+q$ carry into this setting, only noting that 0 is not an object of **Rprocess**(A).

Now sequential composition is not strictly definable with only rooted processes, since the root of q is neither preserved nor reflected in forming $p.q$. Inspection of (SQ) shows that one is only concerned with preservation and reflection of the root within the square

$$\begin{array}{ccc} p.q & \longrightarrow & p \\ \downarrow & & \downarrow \\ t & = & t. \end{array}$$

We then define sequential composition *at the level of* **Process**(A) and then note the rather obvious fact that if $p \to t$ is normal-epi-over 1_0, so are the remaining morphisms in the above square, these morphisms as determined via (SQ).

If we restrict attention to *terminating* processes p in the sense that p possesses short exact sequences

$$\begin{array}{ccccccccc} 0 & \longrightarrow & 1_0 & \longrightarrow & p & \longrightarrow & t & \longrightarrow & 0 \\ & & \downarrow & & \| & & \downarrow & & \\ 0 & \longrightarrow & b & \longrightarrow & p & \longrightarrow & 1_0 & \longrightarrow & 0 \end{array}$$

then without other restriction the identities

$$p + q = q + p,$$

$$p + (q + r) = (p + q) + r,$$

$$(p.q).r = p.(q.r),$$

hold for terminating processes p, q, r of **Rprocess**(A). But note that 1_0 is not a terminating process according to this definition.

7. The Milner Tau-Laws

In **Rprocess**(A), the process

$$x_0 \xrightarrow{1} x_1 \qquad (\tau)$$

is not only a retract automata but is reduced and in particular (τ) is inequivalent to 1_0. But if p is a terminating process,

$$p.\tau = p \qquad (T1)$$

since sequential composition is formed in **Process**(A). Note that while in general p and $\tau.p$ are distinct processes in **Rprocess**(A), $\tau.\tau = \tau$ since τ is a terminating process.

The remaining tau-laws are

$$p + \tau.p = \tau.p \qquad (T2)$$

$$a.(\tau p + q) = a.(\tau.p + q) + a.p \qquad (T3)$$

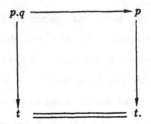

for all processes p and q and atomic action processes (a). Note that the equation

$$\tau.(\tau.p+q)=\tau.(\tau.p+q)+\tau.p$$

is a consequence of $(T1)$, $(T2)$ and previously given equations. Similarly, if $p=\tau.q$ for some process q, then

$$\begin{aligned} p+\tau.p &= \tau.q+\tau.\tau.q \\ &= \tau.q+\tau.q \\ &= \tau.q \end{aligned}$$

as a consequence of $(T1)$ and previously given equations. Therefore $(T2)$ requires further attention only in the case that p is not rooted in τ. Similarly, we may as well consider $(T3)$ only in the case of p not rooted in τ.

Up to this point in the development, any scalar semiring suffices for this theory. The equalities $(T2)$ and $(T3)$ will require the idempotency of the sums of pure states X within $X\cdot\flat$. Now if $x+x=x$ for $x\in X$, then as

$$\begin{aligned} x+x &= x.1+x.1 \\ &= x.(1+1) \end{aligned}$$

for $1\in\flat$, we have $1+1=1$ in the scalar semiring $\flat$. Therefore the addition in $\flat$ is idempotent. This idempotency suffices to obtain idempotent addition in $X\cdot\flat$. So the remainder of the development in this section is conducted with respect to any additively idempotent scalar semiring $\flat$.

Here is an example of our further considerations. Let p be the automaton

$$x_0 \xrightarrow{a} x_1 \tag{a}$$

to consider $p\cup\tau p$ diagramed as

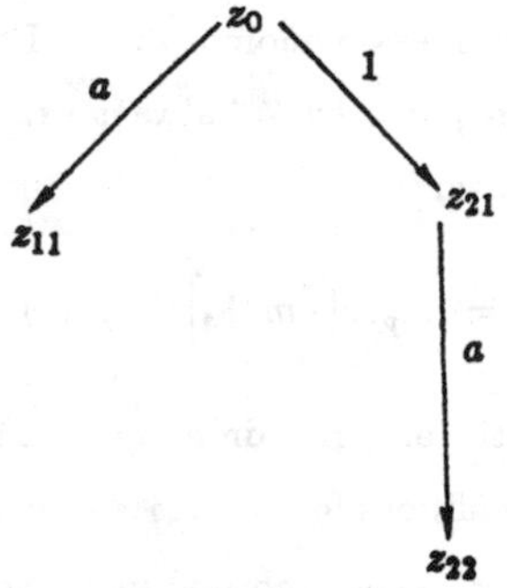

and τp with diagram

$$y_0 \xrightarrow{1} y_1 \xrightarrow{a} y_2.$$

The rooted normal epimorphisms $[1_p,0]$: $p\cup\tau p\rightarrow p$ and $[0,1_{\tau p}]$: $p\cup\tau p\rightarrow\tau p$ are clear. But in this case there is a normal monomorphism of **Process**(A), m: $p\rightarrow\tau p$, obtained by sending x_i

to y_{i+1}, $i = 0, 1$. Further, there is a rooted pure epimorphism, which is not a retraction, from $p \cup \tau p$ to τp when the scalars are idempotent. Naming this pure epimorphism b: $p \cup \tau p \to \tau p$, we have the following table of assignments for b,

b	
z_0	y_0
z_{11}	y_2
z_{21}	y_1
z_{22}	y_2

from which the reader may wish to check that b is actually a morphism of A-modules.

Therefore while $p \cup \tau p$ and τp are inequivalent as retract automata, there is a bisimulation between them. For the general case which the above exemplifies, we require a comparison between the pure epimorphism and the normal epimorphism between two retract automata, when two such exist. The following theorem is stated irrespective of the idempotence of the scalars, but appears to be vacuously true when $1 + 1 \neq 1$ in the scalars $\mathfrak{b}$.

7.1 Theorem. Let p and q be processes in **Process**(A) with join j used to form $p + q$. If there exists a pure epimorphism b: $p + q \to q$ such that

$$Cok(Ker([0, \mathbf{1}_q]) \cdot b) \approx Coeq(b, [0, \mathbf{1}_q])$$

and $Ker[0, \mathbf{1}_q] \cdot b$ is a monomorphism then with c the codomain object of $Coeq(b, [0, \mathbf{1}_q])$, there is a rooted normal epimorphism $c \to j$ and

$$q \approx c.p.$$

The proof known to me, being much too long, is omitted.

Now suppose m: $p \to q$ is a **Process**-monomorphism. Dropping to the level of $\mathbf{Set}_0$, a bitotal epimorphism with the properties required by *7.1* always exists. Let $p + q$ be given over join j as earlier described. The $\mathbf{Set}_0$-morphism

$$b = \Big[[\mathbf{1}_p, 0] \cdot m, \mathbf{1}_q\Big]: \quad p + q \to q$$

has the properties given in *7.1* so the only remaining question is whether b is actually a morphism of **Pure**(A). Here are sufficient conditions for processes over idempotent scalars.

We begin by noting that if i: $\tau \to c$ is a normal monomorphism from the process τ into some process c, then for any process r there is a commuting diagram

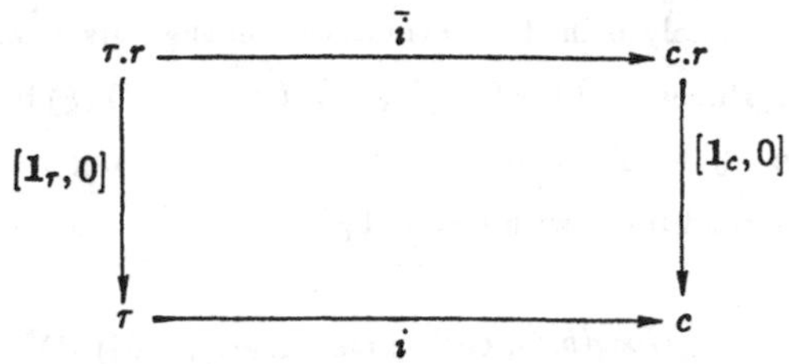

with $\bar{i}$ a normal monomorphism. Now let $p = j.r$ and $q = j.(\tau.r + s)$ for processes j, r, s to note that the sum of p and q is formed with respect to join j provided r is not rooted in τ. If s is absorbed so that $j.(\tau.r + s) = j.\tau.r = j.r$, we have $p = q$ and this instance of $(T3)$ has been previously considered. So assume s is not absorbed and r is not rooted in τ. Let $c = j.(\tau + s)$ to note that $q = c.r$ by choosing the sequential join at the end of the process τ as displayed in c.

7.2 Theorem. With the above definitions of processes, there is a normal monomorphism $m\colon p \to q$ of **Process**(A). Further, $b = \left[[1_p, 0]\cdot m, 1_q\right]\colon p+q \to q$ satisfies the hypotheses of *7.1* as a morphism of **Pure**(A).

7.3 Corollary. Both $(T2)$ and $(T3)$ are instances of this situation.

Proof: $(T3)$. For the automaton

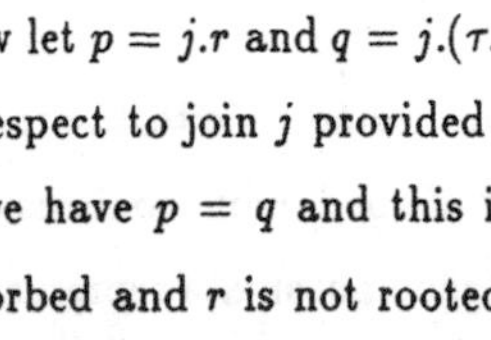

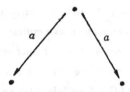

there is a pure epimorphism of the form described to the process (a). Let $c = a.(\tau + q)$ to see that the necessities for the above comparison hold for $a.p$, provided p is not rooted in τ. Thus there is a comparison $b\colon a.p + a.(\tau.p + q) \to a.(\tau.p + q)$ if p is not rooted in τ. If p is rooted in τ, since p only appears in a "context," use $(T1)$ to remove the contextual τ. Therefore $(T3)$ is a rooted bisimulation with a comparison. □

8. Yet Another Category of Milner Processes

Neither the epi-connected classes nor the retraction-connected classes in **Pure**(A) provides a satisfactory category of processes with regard to bisimulation. Thus a notion intermediate between these two is required and the analysis in the previous section suggests a candidate.

Consider morphisms in **Pure**(A) of the form $[g, 1_Y]$: $(X, \zeta) \to (Y, \xi)$ between automata. These are epimorphisms when g: $T_0 \to Y_0$ is total, where we are viewing X_0 as the **Set**$_0$-coproduct $X_0 \approx T_0 \amalg Y_0$. Under these conditions, with $b = [g, 1_Y]$,

$$Coeq(b, [0, 1_Y]) \approx Cok(Ker([0, 1_Y]) \cdot b),$$

as $Ker([0, 1_Y]) \cdot b = g$. Every retraction in **Pure**(A) is of this form, for let s: $Y_0 \to X_0$ be the dual of $[0, 1_Y]$.

We provisionally call epimorphisms isomorphic to the form $[g, 1_Y]$ *good* epimorphisms. If the direct limit of upward-directed classes of good epimorphisms exists, the colimiting objects of such are the appropriate candidates for the good automata representing processes. Lack of time prevents me from checking this condition, but I have little doubt about the matter. In such a category all the usual CCS laws can be established via the interplay between exactness and rooted morphisms established thus far.

9. Conclusions

This analysis has shown that there are several levels of ideas used in categories of Park-Milner processes. First and foremost, the theory of exact categories provides the fundamental structures. Second, the idea of rooted processes means one is attempting to work in a bipointed category. As this brief analysis shows, bipointed categories have a rather weak collection of nice properties—at least known to me. Third, additive idempotence introduces considerable additional structure, and it is here that the non-unital aspects of the A-modules play an important rôle.

References

[1] M. A. Arbib and E. G. Manes, Fuzzy Machines in a Category, Bull. Austral. Math. Soc. 13, 1975, 169–210.

[2] O. Ben–Shachar, Bisimulation of State Automata, MS thesis, Washington State University, 1986.

[3] D. B. Benson, Counting Paths: Nondeterminism as Linear Algebra, IEEE Trans. Softw. Eng. SE–10, 1984, 785–794.

[4] D. B. Benson, String Algebra and Coalgebra, Automata and Coautomata, ms.

[5] D. B. Benson and O. Ben–Shachar, Bisimulation of State Automata, IEEE Symp. Logic in Computer Science, Cambridge, MA, 1986.

[6] D. B. Benson and O. Ben–Shachar, Bisimulation of Automata, WSU Comput. Sci. Tech. Rpt. CS–87–162.

[7] D. B. Benson and I. Guessarian, Iterative and Recursive Matrix Theories, J. Algebra 86, 1984, 302–314.

[8] D. B. Benson and I. Guessarian, Algebraic Solutions to Recursion Equations, JCSS, to appear.

[9] D. B. Benson and J. Tiuryn, Fixed Points in Free Process Algebras with Silent Events, Part I., WSU Comput. Sci. Tech. Rpt. CS-86-152.

[10] J. A. Bergstra and J. W. Klop, Algebra of Communicating Processes with Abstraction, Theoret. Comput. Sci. 37, 1985, 77–121.

[11] H.–B. Brinkmann and D. Puppe, Abelsche und Exakte Kategorien; Korrespondenzen, Springer–Verlag Lecture Notes in Mathematics 96, 1969.

[12] H. Herrlich and G. E. Strecker, *Category Theory,* second edition, Heldermann-Verlag, Berlin, 1979.

[13] W. Kuich and A. Salomaa, *Semirings, Automata, Languages,* Springer–Verlag, Berlin, 1986.

[14] S. MacLane, *Categories for the Working Mathematician,* Springer–Verlag, New York, 1971.

[15] M. G. Main, Demons, Catastropies and Communicating Processes, Univ. Colorado Tech. Rpt. CU–CS–343–86.

[16] M. G. Main and D. B. Benson, Functional behavior of nondeterministic and concurrent programs, Inform. and Control 62, 1984, 144–189.

[17] E. G. Manes, ed., *Category Theory Applied to Computation and Control,* Springer–Verlag LNCS 25, 1975.

[18] E. G. Manes, A Class of Fuzzy Theories, J. Math. Analysis and Applications 85, 1982, 409–451.

[19] E. G. Manes, Additive Domains, Springer–Verlag LNCS 239, 1986, 184–195.

[20] E. G. Manes, Weakest preconditions: Categorical insights, Springer–Verlag LNCS 240, 1986, 182–197.

[21] E. G. Manes, Assertional Categories, Proc. Third Workshop on Math. Found. Program. Semantics, Tulane, April 1987, to appear.

[22] R. Milner, Calculi for Synchrony and Asynchrony, Theoret. Comput. Sci. 25, 1983, 267–310.

[23] S. B. Niefield, Adjoints to Tensor for Graded Algebras and Coalgebras, J. Pure Appl. Alg. 41, 1986, 255–261.

Relating two models of hardware

by

Glynn Winskel

University of Cambridge,
Computer Laboratory,
Corn Exchange Street,
Cambridge CB2 3QG.

The idea of this note is to show how Winskel's static-configuration model of circuits in [W] is related formally to Gordon's relational model in [G, G1]. Once so related the simpler proofs in the model in [G] can, for instance, be used to justify results in terms of the model in [W]. More importantly, we can exhibit general conditions on circuits which ensure that assertions which hold of a circuit according to the simpler model are correct with respect to the more accurate model. The formal translation makes use of a simple adjunction between (partial order) categories associated with the two models in [W] and [G], in a way reminiscent of abstract interpretation [CC]. Preliminary results suggest similar lines of approach may work for other kinds of abstraction such as temporal abstraction in reasoning about hardware (see [M]), and, more generally, make possible a formal algebraic treatment of the relationship between different models of hardware.

1. Formalising abstraction.

The models of hardware we shall relate fit into a general scheme. In many models a circuit is represented by its set of possible behaviours. So assume a circuit c denotes a subset $[\![c]\!] \subseteq B$ of *behaviours* according to a model. It will be the case that each behaviour $b \in B$ will possess structure which can be described by a *behaviour assertion*. Such an assertion A denotes a subset of behaviours $[\![A]\!] \subseteq B$ consisting of those behaviours which satisfy it. A *circuit specification* $Spec$ should pick-out those circuits which satisfy it and so we expect it to denote a subset $[\![Spec]\!] \subseteq P(B)$. There are two obvious ways a behaviour assertion A can be made into into a basic circuit specification. Firstly, we say a circuit c satisfies a circuit specification $\Diamond A$ when it has some behaviour which satisfies A, or, more formally, we can write

$$c \models \Diamond A \text{ iff } [\![c]\!] \cap [\![A]\!] \neq \emptyset.$$

Secondly, we say c satisfies $\Box A$ when all its behaviours satisfy A, *i.e.*

$$c \models \Box A \text{ iff } [\![c]\!] \subseteq [\![A]\!].$$

Of course more complicated circuit specifications can be built up by taking conjunctions, for instance, of such basic ones. In the models we shall consider behaviour assertions will have negations, with the negation $\neg A$ of an assertion A being denoted by the complement of $[\![A]\!]$. Hence we shall have

$$c \models \Diamond A \text{ iff } [\![c]\!] \not\subseteq [\![\neg A]\!].$$

Thus determining if a circuit satisfies either kind of basic circuit specification reduces to considering whether or not an inclusion

$$[\![c]\!] \subseteq [\![A]\!]$$

holds for the circuit term c and a behaviour assertion A.

The key to relating two models is to find conditions under which such an inclusion in one model implies such an inclusion in another. To this end, imagine two models, model 1 and model 2, for hardware behaviour, and that model 1 is more detailed and accurate than model 2. Assume that both models are based on their respective notions of behaviour which form sets B_1 and B_2. With luck, the fact that model 1 is more detailed than 2 will be expressed through there being an "abstraction function" from the behaviours of 1 to the behaviours of 2; it is intended that such a function shows how a more detailed behaviour can be viewed as a less detailed behaviour. Sometimes a more detailed behaviour may be outside the scope of the less detailed model so we cannot expect the function to be always defined. To reflect this, the abstraction function will be a partial function

$$abs : B_1 \rightharpoonup B_2.$$

How are the two inclusions of models 1 and 2 related by the abstraction function which exists between their associated representations of behaviour? Certainly the function *abs* extends to sets; define

$$abs_* : P(B_1) \to P(B_2) \text{ by taking}$$
$$abs_*(C) = \{abs(b_1) \mid b_1 \in C \;\&\; abs(b_1)\downarrow\}$$

for $C \in P(B_1)$. (We use $abs(b_1)\downarrow$ to mean $abs(b_1)$ is defined and $abs(b_1)\uparrow$ for $abs(b_1)$ is undefined.) Given a subset C of B_1 the function abs_* yields its image under *abs*. Accompanying the function abs_* is another function

$$abs^* : P(B_2) \to P(B_1) \text{ by taking}$$
$$abs^*(A) = \{b_1 \mid abs(b_1)\downarrow \Rightarrow abs(b_1) \in A\}$$

for $A \in P(B_2)$. Given a subset A of B_2 the function abs^* yields the largest subset of B_1 whose image under *abs* lies in A. It is easy to see that the pair of functions form an adjunction in the following sense.

1.1 Proposition. *For any $C \in P(B_1)$ and $A \in P(B_2)$,*

$$C \subseteq abs^*(A) \Leftrightarrow abs_*(C) \subseteq A,$$

a property which says the pair abs_*, abs^* forms an adjunction from $(P(B_1), \subseteq)$ to $(P(B_2), \subseteq)$ with left adjoint abs_* and right adjoint abs^*.

Further, if abs is onto (i.e. $abs_*(B_1) = B_2$) then

$$C \subseteq abs^* \circ abs_*(C) \text{ and } A = abs_* \circ abs^*(A),$$

for any $C \in P(B_1)$ and $A \in P(B_2)$.

For adjunctions like this some further facts hold which we record for later use.

1.2 Proposition. *Let $abs : B_1 \rightharpoonup B_2$ be a partial function, with abs^* defined as above. Then*

$$abs^*(B_2) = B_1, \qquad abs^*(\emptyset) = \{x \in B_1 \mid abs(x)\uparrow\},$$
$$abs^*(X \cap Y) = abs^*(X) \cap abs^*(Y),$$
$$abs^*(X \cup Y) = abs^*(X) \cup abs^*(Y),$$
$$abs^*(B_2 \setminus A) = \{x \in B_1 \mid abs(x)\uparrow\} \cup (B_1 \setminus abs^*(A)),$$

The facts expressed in 1.1 and 1.2 are fairly obvious. Still, they are significant because such an adjunction expresses how inclusion according to a more detailed model is related to inclusion in the less detailed one. Of course this is for inclusion and is just between sets, and does not involve the syntax used in the models to build circuit terms and assertions. However proposition 1.2 expresses how boolean operations of conjunction, disjunction and negation on assertions in the less detailed model translate to the more detailed one, and so will enable a smooth translation from assertions built using these connectives in model 2 to "equivalent" assertions of model 1.

In the two models we shall consider, circuit terms of the more detailed model will include those of the less detailed model, while the syntax of assertions will differ. Letting c be a circuit term and A an assertion of model 2 (the less detailed model), and using $[\![\ ]\!]_1$ and $[\![\ ]\!]_2$ for the semantic functions associating subsets of behaviours with terms and assertion in the two models, the proposition above gives

$$[\![c]\!]_1 \subseteq abs^*([\![A]\!]_2) \Leftrightarrow abs_*([\![c]\!]_1) \subseteq [\![A]\!]_2. \qquad (*)$$

As it stands this is not wholly satisfactory. Central to the models 1 and 2 are the two inclusion relations involving *terms* and *assertions*

$$[\![c]\!]_1 \subseteq [\![A_1]\!]_1 \quad \text{and} \quad [\![c]\!]_2 \subseteq [\![A_2]\!]_2,$$

where c is a circuit term and A_1 is an assertion in model 1, and A_2 is an assertion in model 2. The equivalence $(*)$ does not yet relate these inclusions directly because, obviously, $abs^*([\![A]\!]_2)$ is not an assertion and $abs_*([\![c]\!]_1)$ is not a term. However, for the two models we shall consider, for an assertion A of model 2, it is quite easy (using proposition 1.2) to construct uniformly an assertion of model 1, call it ${}^*A$, which denotes the same behaviours as $abs^*([\![A]\!]_2)$ *i.e.* so

$$[\![{}^*A]\!]_1 = abs^*([\![A]\!]_2).$$

Then asking for the two models to agree on the assertions a circuit term c satisfies amounts to requiring a condition on circuit terms c so that

$$[\![c]\!]_2 = abs_*([\![c]\!]_1),$$

because then and only then do we have

$$[\![c]\!]_1 \subseteq [\![{}^*A]\!]_1 \text{ iff } [\![c]\!]_2 \subseteq [\![A]\!]_2,$$

for any assertion A of model 2. Then, by definition, we obtain

$$c \models_1 \Box\, {}^*A \text{ iff } c \models_2 \Box\, A$$

directly and, as we shall see, a similar result holds for circuit specifications $\Diamond A$. Thus a condition on circuits ensures the two models agree. One can, in addition, ask for weaker conditions on circuits which ensure that assertions established for a circuit in one model guarantees that the corresponding assertion holds in the other, but not necessarily the converse. For the two models we shall consider the conditions will amount to simple and intuitively reasonable restrictions on the way circuit terms are built up.

We make some remarks about the present state of hardware verification. Up till now a great deal of hardware verification has focussed on establishing that circuits c meet specifications of the form $\Box\, A$—as we have seen this amounts to showing $[\![c]\!] \subseteq [\![A]\!]$—and ignored the question of whether or not a circuit satisfies specifications $\Diamond A$. However it appears both kinds of modal formulae should be considered as circuit specifications, and the effect of insisting a circuit satisfies a specification $\Diamond A$ is sometimes achieved by imposing some requirement expressed in higher-order logic (see *e.g.* section 10 of [G1]). After all a short circuit obtained by connecting power to ground denotes the emptyset in the model of [G, G1] and so satisfies any specifications of the form $\Box\, A$. This indicates the lack of expressiveness of specifications purely of the form $\Box\, A$. This problem was discussed by Mike Fourman in [F] though his proposal on how to extend specifications was less specific. The proposal here to introduce two kinds of modal assertions as basic specifications is mathematically obvious and, while much simpler, follows the same lines as used in other areas of semantics; the powerdomains of denotational semantics can be seen as spaces whose basic open sets are described by such modal formulae (see [W, R]).

2. Circuit terms.

We simplify the language in [W], ignoring the components responsible for charge storage and resistance because these are not addressed in the model in [G, G1]. Terms for circuits have the following form.

$$c ::= Pow\ (\alpha) \mid Gnd\ (\alpha) \mid wre(\alpha,\beta) \mid ntran(\alpha,\beta,\gamma) \mid ptran(\alpha,\beta,\gamma) \mid c \bullet c \mid c \setminus \alpha.$$

This assumes a set of points (point names) $\alpha, \beta, \gamma, \cdots \in \Pi$, and we assume α, β, γ are distinct point names in $ntran(\alpha,\beta,\gamma)$, $ptran(\alpha,\beta,\gamma)$, $wre(\alpha,\beta)$. The term $wre(\alpha,\beta)$ will denote a wire connecting α and β, the term $ntran(\alpha,\beta,\gamma)$ an n-type transistor with gate γ, the term $ptran(\alpha,\beta,\gamma)$ a p-type transistor with gate γ, while the composition $c_0 \bullet c_1$ joins two circuits c_0 and c_1 at their common points and the hiding operation $c \setminus \alpha$ insulates from the environment the point α.

We define the sort of term by structural induction.

$$\begin{aligned}
&\mathrm{sort}(Pow\ (\alpha)) = \mathrm{sort}(Gnd\ (\alpha)) = \{\alpha\}\\
&\mathrm{sort}(wre(\alpha,\beta)) = \{\alpha,\beta\}\\
&\mathrm{sort}(ntran(\alpha,\beta,\gamma)) = \mathrm{sort}(ptran(\alpha,\beta,\gamma)) = \{\alpha,\beta,\gamma\}\\
&\mathrm{sort}(c \bullet d) = \mathrm{sort}(c) \cup \mathrm{sort}(d)\\
&\mathrm{sort}(c \setminus \alpha) = \mathrm{sort}(c) \setminus \{\alpha\}.
\end{aligned}$$

3. A relational model.

We explain the model in [G, G1] used by Mike Gordon and others. The presentation is a little different from usual, because we concentrate more on the model and do not present circuits as just special kinds of assertions. However the equivalence with the model in [G, G1] is clear.

We assume the set of points Π, and distinct values H and L, standing for high and low, and define, for $\Lambda \subseteq \Pi$,

$$F[\Lambda] = \{V \mid V : \Lambda \to \{H, L\}\}.$$

Write $F = \bigcup_{\Lambda \subseteq \Pi} F[\Lambda]$, and say $V \in F$ has *sort* Λ if $V \in F[\Lambda]$.

A circuit term of sort Λ will denote a subset of $F[\Lambda]$, following the idea that a circuit imposes a relation between values at points.

3.1 Notation. Let $k \in \{H, L\}$. Let $\alpha \in \Pi$. For $V \in F[\Lambda]$ define $V[k/\alpha] \in F[\Lambda \cup \{\alpha\}]$ by taking

$$V[k/\alpha](\beta) = \begin{cases} V(\beta) & \text{if } \beta \neq \alpha,\\ k & \text{if } \beta = \alpha,\end{cases}$$

for $\beta \in \Lambda$.

In this model assertions for expressing the properties of circuits have the following syntax:

Variables: We assume a set of variables

$$\{v_\alpha \mid \alpha \in \Pi\}.$$

Value terms: Terms, denoting values in $\{H, L\}$ have the form

$$t ::= v_\alpha \mid H \mid L.$$

Behaviour assertions: The set of G assertions is generated by

$$A ::= t_0 = t_1 \mid A_0 \wedge A_1 \mid A_0 \vee A_1 \mid \neg A \mid \mathbf{tt} \mid \mathbf{ff} \mid \exists v_\alpha . A \mid \forall v_\alpha . A.$$

We shall use assertions such as $A \rightarrow B$ (A implies B) with the understanding that this abbreviates $\neg A \vee B$, and $A \leftrightarrow B$ for $A \rightarrow B \wedge B \rightarrow A$.

Semantically, a value term denotes a partial function $[\![t]\!] : F[\Lambda] \rightharpoonup \{H, L\}$; we take

$$[\![v_\alpha]\!]V = \begin{cases} V(\alpha) & \text{if } \alpha \in \mathrm{sort}(V), \\ \text{undefined} & \text{otherwise,} \end{cases}$$

$$[\![H]\!]V = H, \quad \text{and} \quad [\![L]\!]V = L$$

for any $V \in F$.

For an assertion A we define $\mathrm{G}[\![A]\!] \in P(F)$ by induction on the structure of A:

$$\begin{aligned}
&\mathrm{G}[\![t_0 = t_1]\!] = \{V \in F \mid [\![t_0]\!]V \downarrow \;\&\; [\![t_1]\!] \downarrow \;\&\; [\![t_0]\!]V = [\![t_1]\!]V\} \\
&\mathrm{G}[\![A_0 \wedge A_1]\!] = \mathrm{G}[\![A_0]\!] \cap \mathrm{G}[\![A_1]\!], \\
&\mathrm{G}[\![A_0 \vee A_1]\!] = \mathrm{G}[\![A_0]\!] \cup \mathrm{G}[\![A_1]\!], \\
&\mathrm{G}[\![\mathrm{tt}]\!] = F, \qquad \mathrm{G}[\![\mathrm{ff}]\!] = \emptyset, \\
&\mathrm{G}[\![\neg A]\!] = F \setminus \mathrm{G}[\![A]\!], \\
&\mathrm{G}[\![\exists v_\alpha . A]\!] = \{V \in F \mid \exists k \in \{H, L\}.\ V[k/\alpha] \in \mathrm{G}[\![A]\!]\}, \\
&\mathrm{G}[\![\forall v_\alpha . A]\!] = \{V \in F \mid \forall k \in \{H, L\}.\ V[k/\alpha] \in \mathrm{G}[\![A]\!]\}.
\end{aligned}$$

Of course we have

$$\begin{aligned}
&\mathrm{G}[\![\exists v_\alpha . A]\!] = \{V \in F \mid V[H/\alpha] \in \mathrm{G}[\![A]\!] \text{ or } V[L/\alpha] \in \mathrm{G}[\![A]\!]\}, \text{ and} \\
&\mathrm{G}[\![\forall v_\alpha . A]\!] = \{V \in F \mid V[H/\alpha] \in \mathrm{G}[\![A]\!] \;\&\; V[L/\alpha] \in \mathrm{G}[\![A]\!]\}.
\end{aligned}$$

We write $\mathrm{G}[\![A]\!]_\Lambda$ for $\mathrm{G}[\![A]\!] \cap F[\Lambda]$, where $\Lambda \subseteq \Pi$.

Semantics of circuit terms: In line with the model in [G, G1], we denote a circuit term c of sort Λ by a subset $\mathrm{G}[\![c]\!]$ of $F[\Lambda]$. The semantics as we describe it follows that of [C] closely. We first define operations $\bullet$ and $\backslash\alpha$ on elements of F.

Let $V_0 \in F[\Lambda_0]$, $V_1 \in F[\Lambda_1]$. Define

$$V_0 \bullet V_1 = \begin{cases} V_0 \cup V_1 & \text{if } V_0 \lceil \Lambda_1 = V_1 \lceil \Lambda_0, \\ \text{undefined} & \text{otherwise.} \end{cases}$$

Let $V \in F[\Lambda]$. Let $\alpha \in \Pi$. Write

$$V \setminus \alpha = V \lceil (\Lambda \setminus \{\alpha\}).$$

We extend the operations to subsets of F. For $R, R_0, R_1 \in P(F)$ define

$$\begin{aligned}
&R_0 \bullet R_1 = \{V_0 \bullet V_1 \mid V_0 \in R_0 \;\&\; V_1 \in R_1 \;\&\; V_0 \bullet V_1 \downarrow\} \\
&R \setminus \alpha = \{V \setminus \alpha \mid V \in R\}.
\end{aligned}$$

Now we define the semantics of circuit terms:

$$\begin{aligned}
&\mathrm{G}[\![Pow\,(\alpha)]\!] = \{V \in F[\{\alpha\}] \mid V(\alpha) = H\} \\
&\mathrm{G}[\![Gnd\,(\alpha)]\!] = \{V \in F[\{\alpha\}] \mid V(\alpha) = L\} \\
&\mathrm{G}[\![wre(\alpha, \beta)]\!] = \{V \in F[\{\alpha, \beta\}] \mid V(\alpha) = V(\beta)\} \\
&\mathrm{G}[\![ntran(\alpha, \beta, \gamma)]\!] = \{V \in F[\{\alpha, \beta, \gamma\}] \mid V(\gamma) = H \Rightarrow V(\alpha) = V(\beta)\} \\
&\mathrm{G}[\![ptran(\alpha, \beta, \gamma)]\!] = \{V \in F[\{\alpha, \beta, \gamma\}] \mid V(\gamma) = L \Rightarrow V(\alpha) = V(\beta)\} \\
&\mathrm{G}[\![c_0 \bullet c_1]\!] = \mathrm{G}[\![c_0]\!] \bullet \mathrm{G}[\![c_1]\!] \\
&\mathrm{G}[\![c \setminus \alpha]\!] = \mathrm{G}[\![c]\!] \setminus \alpha.
\end{aligned}$$

It follows that

$$\mathrm{G}[\![c_0 \bullet c_1]\!] = \{V \in F[\Lambda_0 \cup \Lambda_1]) \mid V\lceil\Lambda_0 \in \mathrm{G}[\![c_0]\!] \;\;\&\;\; V\lceil\Lambda_1 \in \mathrm{G}[\![c_1]\!]\},$$

where $\Lambda_0 = \mathrm{sort}(c_0)$ and $\Lambda_1 = \mathrm{sort}(c_1)$. In other words, the behaviours of the composition of two circuits are precisely those which restrict to behaviours of the component circuits.

3.2 Proposition.

(i) Let c be a circuit term of sort Λ. Let A be an assertion such that

$$G[\![c]\!] = G[\![A]\!]_\Lambda.$$

Then

$$G[\![c \setminus \alpha]\!] = G[\![\exists v_\alpha . A]\!]_{\Lambda\setminus\{\alpha\}}.$$

(ii) Let c_0 be a circuit term of sort Λ_0, and c_1 be a circuit term of sort Λ_1. Let A_0 and A_1 be assertions such that

$$G[\![c_0]\!] = G[\![A_0]\!]_{\Lambda_0} \quad \text{and} \quad G[\![c_1]\!] = G[\![A_1]\!]_{\Lambda_1}.$$

Then

$$G[\![c_0 \bullet c_1]\!] = G[\![A_0 \wedge A_1]\!]_{\Lambda_0 \cup \Lambda_1}.$$

The denotations of basic components are readily expressed as assertions, *e.g.*

$$\mathrm{G}[\![ntran(\alpha,\beta,\gamma)]\!] = \mathrm{G}[\![v_\gamma = H \rightarrow v_\alpha = v_\beta]\!]_{\{\alpha,\beta,\gamma\}}.$$

Consequently, in this simple model, the proposition implies we can replace circuit operations by logical ones, the course followed in [G, G1].

Inclusion on $P(F)$ induces a semantic entailment:

$$A_0 \models^G A_1 \text{ iff } \mathrm{G}[\![A_0]\!] \subseteq \mathrm{G}[\![A_1]\!],$$

where A_0, A_1 are assertions. As in the introduction, basic circuit specifications in this model have the form $\Diamond\, A$ and $\Box\, A$ where A is an assertion. At present the work using this model has concentrated on specifications of the form $\Box\, A$ with

$$c \models^G \Box\, A \text{ iff } \mathrm{G}[\![c]\!] \subseteq \mathrm{G}[\![A]\!],$$

where c is a circuit term and A is an assertion. In the approach of [G, G1], no separate syntax is given for circuit terms. Instead circuits are translated directly into assertions in the manner of proposition 3.2 so that showing a circuit meets a specification amounts to showing an entailment, and so an implication, holds between two assertions, and this is purely a matter of logic. Restricting attention to specifications of the form $\Box\, A$ has led to the paradox that a short-circuit *Pow* $(\alpha) \bullet$ *Gnd* (α) satisfies any specification because $\mathrm{G}[\![Pow\,(\alpha) \bullet Gnd\,(\alpha)]\!] = \emptyset$, but this is no longer the case, of course, when specifications of the form $\Diamond\, A$ are permitted too.

4. The static configurations model.

For some purposes the model in [G, G1] is inadequate, for example, it fails to deal with some resistance and capacitance effects used in hardware design, and with the fact that sometimes the value at a point is not purely high or low. The work in [W] attempts to find a model and logic for circuits without these inadequacies. Essentially it takes ideas of Bryant (eg. [B]), on which several hardware simulators are based, and uses them to provide a semantics and proof system for a language for circuits which includes resistances and capacitances.

In [W] the value at a point is assumed to lie in the set

$$\mathbf{V} = \{H, L, X, Z\}.$$

A point assumes the value Z if it is not connected to any source, value H if it is connected to power but not to ground, the value L if it is connected to ground but not to power, and the value X if it is connected to both the power and ground. It is useful to order $\mathbf{V}$:

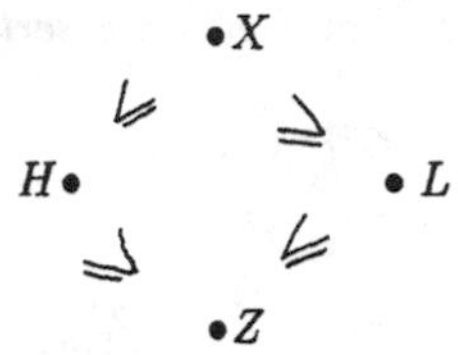

Thus X is the least upper bound of H and L with respect to the order $\leq$, and, in general, we shall use ΣW for the least upper bound of a set W of values and write the least upper bound of a pair as $w_0 + w_1$.

The static configuration model is described and motivated in detail in [W] and [W1]. As was pointed out there, if resistance and capacitance effects are ignored the definition of state (called static configuration) is simplified. Certainly one component of the state of a circuit should be a value function V which assigns a value in $\mathbf{V}$ to every point in the sort of the circuit because it is with these values that circuits do calculations. Because we want to account for the behaviour of circuits in environments where, for example, a high voltage is placed on the gate of a transistor in order to ensure that the values on its other two points are the same, we take the value function to give the value at points in an environment in which other components including sources may be present. (We did this earlier for Gordon's relational model.) We want our model to be compositional in the sense that the value functions associated with a circuit term are determined by those of its proper subterms. This cannot be achieved with value functions alone. We need to keep track of that contribution to the value function which comes from sources within the circuit and how points are connected by wires and transistors to give a satisfactory treatment of hiding. Then we can obtain a compositional model (see [W1] for a detailed argument for the necessity of this extra complexity).

4.1 Definition.

Let Λ be a subset of the points Π. A *static configuration* of sort Λ is a structure

$$\langle V, I, \sim \rangle,$$

where $V : \Lambda \to \mathbf{V}$, the *value* function, $I : \Lambda \to \mathbf{V}$, the *internal-value* function and $\sim$ is an equivalence relation on Λ, the *connectivity* relation, which satisfy

$$I(\alpha) \leq V(\alpha),$$
$$\alpha \sim \beta \Rightarrow V(\alpha) = V(\beta) \;\; \& \;\; I(\alpha) = I(\beta),$$

for all points α, $\beta \in \Lambda$.

Write Sta[Λ] for the static configurations of sort Λ, and Sta for static configurations of any sort $\Lambda \subseteq \Pi$.

Assertions in [W] for expressing the properties of circuits have the following syntax:

Variables: We assume a set of variables Var which range over points Π, and have typical members $x, y, z, \cdots$. (Note in this model variables range over points *not* values.)

Value terms: Terms, denoting values in $\{H, L, Z, X\}$ have the form

$$t ::= V(\pi) \mid I(\pi) \mid H \mid L \mid Z \mid X,$$

where π is a point term, *i.e.* an element of Π or a variable in Var.

Behaviour assertions: The set of W-assertions is generated by

$$\phi ::= t_0 = t_1 \mid t_0 \leq t_1 \mid \pi_0 = \pi_1 \mid \pi_0 \sim \pi_1 \mid \phi_0 \wedge \phi_1 \mid \phi_0 \vee \phi_1 \mid \neg\phi \mid \mathrm{tt} \mid \mathrm{ff} \mid \exists x.\phi \mid \forall x.\phi$$

where t_0, t_1 are value terms and π_0, π_1 are point terms. As with G-assertions, we shall regard $\phi \to \theta$ as abbreviating $\neg\phi \vee \theta$ and $\phi \leftrightarrow \theta$ as abbreviating $\phi \to \theta \wedge \theta \to \phi$.

The semantics of assertions: We give the semantics of closed assertions. But first we have to treat value terms which are denoted by partial functions Sta $\rightharpoonup$ **V**, as follows:

$$\begin{aligned}
&W[\![H]\!]\sigma = H \text{ for all } \sigma \in \mathrm{Sta}\\
&W[\![L]\!]\sigma = L \text{ for all } \sigma \in \mathrm{Sta}\\
&W[\![V(\alpha)]\!]\sigma = \begin{cases} V(\alpha) & \text{if } \sigma \in \mathrm{Sta} \ \& \ \alpha \in \mathrm{sort}(\sigma),\\ \text{undefined} & \text{otherwise.}\end{cases}\\
&W[\![I(\alpha)]\!]\sigma = \begin{cases} I(\alpha) & \text{if } \sigma \in \mathrm{Sta} \ \& \ \alpha \in \mathrm{sort}(\sigma),\\ \text{undefined} & \text{otherwise.}\end{cases}
\end{aligned}$$

Each W-assertion is denoted by the subset of static configurations at which it is true, defined by the following induction on length:

$$\begin{aligned}
&W[\![t_0 = t_1]\!] = \{\sigma \in \mathrm{Sta} \mid W[\![t_0]\!]\sigma\downarrow \ \& \ W[\![t_1]\!]\sigma\downarrow \ \& \ W[\![t_0]\!]\sigma = W[\![t_1]\!]\sigma\}\\
&W[\![t_0 \leq t_1]\!] = \{\sigma \in \mathrm{Sta} \mid W[\![t_0]\!]\sigma\downarrow \ \& \ W[\![t_1]\!]\sigma\downarrow \ \& \ W[\![t_0]\!]\sigma \leq W[\![t_1]\!]\sigma\}\\
&W[\![\alpha_0 = \alpha_1]\!] = \{\sigma \in \mathrm{Sta} \mid \alpha_0 \in \mathrm{sort}(\sigma) \ \& \ \alpha_1 \in \mathrm{sort}(\sigma) \ \& \ \alpha_0 = \alpha_1\}\\
&W[\![\alpha_0 \sim \alpha_1]\!] = \{\sigma \in \mathrm{Sta} \mid \alpha_0 \in \mathrm{sort}(\sigma) \ \& \ \alpha_1 \in \mathrm{sort}(\sigma) \ \& \ \alpha_0 \sim \alpha_1\}\\
&W[\![\phi_0 \wedge \phi_1]\!] = W[\![\phi_0]\!] \cap W[\![\phi_1]\!]\\
&W[\![\phi_0 \vee \phi_1]\!] = W[\![\phi_0]\!] \cup W[\![\phi_1]\!]\\
&W[\![\neg\phi]\!] = (\mathrm{Sta} \setminus W[\![\phi]\!])\\
&W[\![\mathrm{tt}]\!] = \mathrm{Sta}\\
&W[\![\mathrm{ff}]\!] = \emptyset\\
&W[\![\exists x.\phi]\!] = \{\sigma \mid \exists \alpha \in \mathrm{sort}(\sigma).\ \sigma \in W[\![\phi[\alpha/x]]\!]\}\\
&W[\![\forall x.\phi]\!] = \{\sigma \mid \forall \alpha \in \mathrm{sort}(\sigma).\ \sigma \in W[\![\phi[\alpha/x]]\!]\}
\end{aligned}$$

Write $\sigma \models \phi$ if $\sigma \in W[\![\phi]\!]$.

In order to define the semantics of circuits we introduce composition and hiding operations on static configurations.

4.2 Notation. We have used $V \setminus \alpha$ to stand for the restriction of a function $V : \Lambda \to \mathbf{V}$ to domain $\Lambda \setminus \{\alpha\}$. In addition we write $\Lambda \setminus \alpha$ for $\Lambda \setminus \{\alpha\}$, and in the case where $\sim \subseteq \Lambda \times \Lambda$ we write $\sim \setminus\alpha$ for its restriction to the relation $\sim \cap[(\Lambda \setminus \alpha) \times (\Lambda \setminus \alpha)]$.

4.3 Definition. Let $\sigma_0 = \langle V_0, I_0, \sim_0\rangle$ be a static configuration of sort Λ_0 and $\sigma_1 = \langle V_1, I_1, \sim_1\rangle$ be a static configuration of sort Λ_1. Define their *composition* to be

$$\sigma_0 \bullet \sigma_1 = \begin{cases} \langle V, I, \sim\rangle & \text{if } V_0\lceil\Lambda_1 = V_1\lceil\Lambda_0\\ \text{undefined} & \text{otherwise}\end{cases}$$

where

$$\begin{aligned}
V &= V_0 \cup V_1,\\
\sim &= (\sim_0 \cup \sim_1)^* \quad \text{and}\\
I(\alpha) &= \Sigma\{I_0(\beta) \mid \beta \in \Lambda_0 \ \& \ \beta \sim \alpha\} + \Sigma\{I_1(\beta) \mid \beta \in \Lambda_1 \ \& \ \beta \sim \alpha\}
\end{aligned}$$

for any $\alpha \in \Lambda_0 \cup \Lambda_1$.

Thus it is only possible to compose static configurations and get a defined result when their values agree on points that they have in common.

4.4 Definition. Let $\sigma = \langle V, I, \sim\rangle$ be a static configuration of sort Λ and α a point. Define *hiding*

$$\sigma \setminus \alpha = \begin{cases} \langle V \setminus \alpha,\ I \setminus \alpha,\ \sim \setminus \alpha\rangle & \text{if } \alpha \notin \Lambda \text{ or} \\ & \quad V(\alpha) = I(\alpha) + \Sigma\{V(\beta) \mid \beta \in \Lambda \setminus \alpha \ \&\ \beta \sim \alpha\} \\ \text{undefined} & \text{otherwise.} \end{cases}$$

Thus it is only possible to hide, or insulate, a point from the environment and get a defined result when the value at the point will not be disturbed; this is the case when the value at the point is due to the combined effect of internal sources and values due to unhidden points.

We extend the operations to sets of static configurations. For $S, S_0, S_1 \in P(\mathrm{Sta})$ define

$$S_0 \bullet S_1 = \{\sigma_0 \bullet \sigma_1 \mid \sigma_0 \in S_0 \ \&\ \sigma_1 \in S_1 \ \&\ \sigma_0 \bullet \sigma_1 \downarrow\}$$
$$S \setminus \alpha = \{\sigma \setminus \alpha \mid \sigma \in S \ \&\ \sigma \setminus \alpha \downarrow\}.$$

We define the denotation $[\![c]\!]$ of a circuit term c to be a subset of $P(\mathrm{Sta})$ by structural induction:

$$\mathrm{W}[\![Pow\ (\alpha)]\!] = \{\sigma \in \mathrm{Sta}[\alpha] \mid I(\alpha) = H\}$$

$$\mathrm{W}[\![Gnd\ (\alpha)]\!] = \{\sigma \in \mathrm{Sta}[\alpha] \mid I(\alpha) = L\}$$

$$\mathrm{W}[\![wre(\alpha, \beta)]\!] = \{\sigma \in \mathrm{Sta}[\alpha, \beta] \mid I(\alpha) = Z \wedge I(\beta) = Z \wedge \alpha \sim \beta\}$$

$$\begin{aligned}\mathrm{W}[\![ntran(\alpha, \beta, \gamma)]\!] = \{\sigma \in \mathrm{Sta}[\alpha, \beta, \gamma] \mid\ & I(\alpha) = Z \wedge I(\beta) = Z \wedge I(\gamma) = Z \wedge \\ & \neg(\gamma \sim \alpha) \wedge \neg(\gamma \sim \beta) \wedge \\ & (V(\gamma) = H \rightarrow \alpha \sim \beta) \wedge (V(\gamma) = L \rightarrow \neg(\alpha \sim \beta))\}\end{aligned}$$

$$\begin{aligned}\mathrm{W}[\![ptran(\alpha, \beta, \gamma)]\!] = \{\sigma \in \mathrm{Sta}[\alpha, \beta, \gamma] \mid\ & I(\alpha) = Z \wedge I(\beta) = Z \wedge I(\gamma) = Z \wedge \\ & \neg(\gamma \sim \alpha) \wedge \neg(\gamma \sim \beta) \wedge \\ & (V(\gamma) = L \rightarrow \alpha \sim \beta) \wedge (V(\gamma) = H \rightarrow \neg(\alpha \sim \beta))\}\end{aligned}$$

$$\mathrm{W}[\![c \bullet d]\!] = \mathrm{W}[\![c]\!] \bullet \mathrm{W}[\![d]\!]$$

$$\mathrm{W}[\![c \setminus \alpha]\!] = \mathrm{W}[\![c]\!] \setminus \alpha.$$

In the definition above we have used a set-expression $\{\sigma \in \mathrm{Sta}[\Lambda] \mid \phi\}$, where ϕ is W-assertion, to mean the subset of static configurations of sort Λ which satisfy the assertion ϕ.

Inclusion in the model induces notions of entailment, writing

$$\phi \models^{W} \theta \text{ iff } \mathrm{W}[\![\phi]\!] \subseteq \mathrm{W}[\![\theta]\!],$$

where θ, ϕ are W-assertions, and as we have seen

$$c \models^{W} \Diamond\, \theta \text{ iff } \mathrm{W}[\![c]\!] \cap \mathrm{W}[\![\theta]\!] \neq \emptyset,$$
$$c \models^{W} \Box\, \theta \text{ iff } \mathrm{W}[\![c]\!] \subseteq \mathrm{W}[\![\theta]\!],$$

where c is a circuit term and θ is a W-assertion. In [W], an extension of assertions above is used to provide a sound and complete proof system to verify when a circuit term satisfies any specification of the form $\Box\, \theta$.

5. The adjunction between the models in [G, G1] and [W].

Now we have two models, those in [G, G1] and [W], it is important to understand how they are related. This is not simply a matter forgetting about resistance and capacitance terms in the language for circuits because the notions of state used in two models differ; states, as formalised in [W] do include a value function but it is not assumed that the value at a point is either high (H) or low (L), and, in addition, states in [W] have other components expressing, for instance, the connectivity between points and the effects of internal sources. Still, there is an abstraction function from the states/behaviours of [W] to the states/behaviours of [G, G1] and via this we obtain a relation between the models and proof systems of [G, G1] and [W].

As in the introduction, the two notions of inclusion are related by an adjunction between the partial orders $(P(\mathrm{Sta}), \subseteq)$ and $(P(F), \subseteq)$. The adjunction is determined by an abstraction function between the states of the two models. There is clearly a partial function

$$abs : \mathrm{Sta} \rightharpoonup F$$

which acts so

$$abs(\langle V, I, \sim \rangle) = \begin{cases} V & \text{if } V \in F, \\ \text{undefined} & \text{otherwise,} \end{cases}$$

for a static configuration σ; thus $abs(\sigma)$ is undefined when the value function V_σ of the static configuration σ attributes value Z or X to some point. As explained in the introduction, this induces a left adjoint

$$abs_* : P(\mathrm{Sta}) \to P(F),$$

given by

$$abs_*(S) = \{V_\sigma \in F \mid \sigma \in S\}$$

for $S \in P(\mathrm{Sta})$, which has a right adjoint

$$abs^* : P(F) \to P(\mathrm{Sta})$$

given by

$$abs^*(R) = \{\sigma \in \mathrm{Sta} \mid V_\sigma \in F \Rightarrow V_\sigma \in R\},$$

for $R \in P(F)$. The adjunction is expressed by the property

$$S \subseteq abs^*(R) \Leftrightarrow abs_*(S) \subseteq R,$$

for $S \in P(\mathrm{Sta})$, $R \in P(F)$.

As in the introduction, we would like

$$\mathrm{W}[\![c]\!] \subseteq \mathrm{W}[\![{}^*A]\!] \text{ iff } \mathrm{G}[\![c]\!] \subseteq \mathrm{G}[\![A]\!], \text{ or equivalently, } c \models^W \Box\, {}^*A \text{ iff } c \models^G \Box\, A,$$

where A is a G-assertion and ${}^*A$ is some translation of it into a W-assertion. Comparing this with the property expressing the adjunction, which gives

$$\mathrm{W}[\![c]\!] \subseteq abs^*(\mathrm{G}[\![A]\!]) \Leftrightarrow abs_*(\mathrm{W}[\![c]\!]) \subseteq \mathrm{G}[\![A]\!],$$

we obtain this if

$$\mathrm{W}[\![{}^*A]\!] = abs^*(\mathrm{G}[\![A]\!]) \text{ and } \mathrm{G}[\![c]\!] = abs_*(\mathrm{W}[\![c]\!]),$$

the latter being an abstract expression for the required condition on c. More generally, we can ask for conditions on c such that

$$c \models^W \Box\, {}^*A \Rightarrow c \models^G \Box\, A, \text{ or}$$
$$c \models^W \Box\, {}^*A \Leftarrow c \models^G \Box\, A.$$

The translation $A \mapsto {}^{*}A$ from G to W assertions, so $\mathrm{W}[\![{}^{*}A]\!] = abs^{*}(\mathrm{G}[\![A]\!])$, is easily defined by structural induction. First define a W-assertion

$$D \equiv \forall x.\ V(x) = H \vee V(x) = L.$$

5.1 Proposition. *Let σ be a static configuration. Then*

$$\begin{aligned} \sigma \models D &\text{ iff } V_\sigma : \mathrm{sort}(\sigma) \to \{H, L\} \\ &\text{ iff } V_\sigma \in F \\ &\text{ iff } abs(\sigma)\downarrow. \end{aligned}$$

Now, for a G-(value)term t define define its translation into a W-term ${}^{*}t$ by:

$${}^{*}v_\alpha \equiv V(\alpha), \quad {}^{*}H \equiv H, \quad {}^{*}L \equiv L.$$

For a G-assertion A define a W-assertion ${}^{*}A$ by structural induction:

$$\begin{aligned} &{}^{*}(t_0 = t_1) \equiv (D \to {}^{*}t_0 = {}^{*}t_1), \\ &{}^{*}\mathrm{tt} \equiv \mathrm{tt}, \quad {}^{*}\mathrm{ff} \equiv \neg D, \\ &{}^{*}(A_0 \wedge A_1) \equiv {}^{*}A_0 \wedge {}^{*}A_1, \\ &{}^{*}(A_0 \vee A_1) \equiv {}^{*}A_0 \vee {}^{*}A_1, \\ &{}^{*}(\neg A) \equiv (D \to \neg {}^{*}A), \\ &{}^{*}(\exists v_\alpha.\ A) \equiv {}^{*}(A[H/v_\alpha] \vee A[L/v_\alpha]), \\ &{}^{*}(\forall v_\alpha.\ A) \equiv {}^{*}(A[H/v_\alpha] \wedge A[L/v_\alpha]). \end{aligned}$$

By structural induction on A, using proposition 1.2, we can prove:

5.2 Lemma. *For A a G-assertion, $abs^{*}(\mathrm{G}[\![A]\!]) = \mathrm{W}[\![{}^{*}A]\!]$.*

Now we consider the following conditions on circuits c:

$$\begin{aligned} &(a) \quad \mathrm{G}[\![c]\!] \subseteq abs_{*}(\mathrm{W}[\![c]\!]), \\ &(b) \quad \mathrm{G}[\![c]\!] \supseteq abs_{*}(\mathrm{W}[\![c]\!]), \\ &(c) \quad \mathrm{G}[\![c]\!] = abs_{*}(\mathrm{W}[\![c]\!]). \end{aligned}$$

If (a) holds we deduce

$$\begin{aligned} c \models^{W} \Box\, {}^{*}A &\Rightarrow \mathrm{W}[\![c]\!] \subseteq \mathrm{W}[\![{}^{*}A]\!] && \text{by definition} \\ &\Rightarrow \mathrm{W}[\![c]\!] \subseteq abs^{*}(\mathrm{G}[\![A]\!]) && \text{by the above lemma} \\ &\Rightarrow abs_{*}(\mathrm{W}[\![c]\!]) \subseteq \mathrm{G}[\![A]\!] && \text{by the adjunction} \\ &\Rightarrow \mathrm{G}[\![c]\!] \subseteq \mathrm{G}[\![A]\!] && \text{by assumption (a)} \\ &\Rightarrow c \models^{G} \Box\, A && \text{by definition.} \end{aligned}$$

Thus (a) implies $c \models^{W} \Box\, {}^{*}A \Rightarrow c \models^{G} \Box\, A$. In fact, because a circuit term c has finite sort, it is not hard to show that (a) holds iff $c \models^{W} \Box\, {}^{*}A \Rightarrow c \models^{G} \Box\, A$, for all G-assertions A. Similarly, for a circuit term c, condition (b) holds iff $c \models^{G} \Box\, A$ implies $c \models^{W} \Box\, {}^{*}A$, for all assertions A, while condition (c) holds iff $c \models^{G} \Box\, A$ iff $c \models^{W} \Box\, {}^{*}A$, for all assertions A.

From these results on $\Box$ -assertions we can deduce how $\Diamond$-assertions are preserved and reflected under the assumptions (a), (b) or (c). Suppose, for instance, that (a) holds of a circuit c. Then

$$\begin{aligned} c \models^G \Diamond A &\Rightarrow c \not\models^G \Box \neg A \\ &\Rightarrow c \not\models^W \Box\, {}^*(\neg A) \\ &\Rightarrow c \not\models^W \Box\, (D \rightarrow \neg^* A) \\ &\Rightarrow c \not\models^W \Box \neg(D \wedge {}^*A) \\ &\Rightarrow c \models^W \Diamond\, (D \wedge {}^*A) \end{aligned}$$

Thus $c \models^G \Diamond\, A \Rightarrow c \models^W \Diamond\, D \wedge {}^*A$. Similarly, for a circuit term c, if condition (b) holds then $c \models^W \Diamond\, D \wedge {}^*A$ implies $c \models^G \Diamond\, A$, for all assertions A, while if condition (c) holds then $c \models^G \Diamond\, A$ iff $c \models^W \Diamond\, D \wedge {}^*A$, for all assertions A.

The strongest condition (c) would follow for any circuit term c if abs_* were a homomorphism preserving the behaviour of the basic components and the operations $\bullet$ and $\backslash\alpha$ for any $\alpha \in \Pi$. The behaviour of the basic components is preserved by abs_*:

5.3 Lemma.

(i) $abs_*(W[\![Pow\,(\alpha)]\!]) = G[\![Pow\,(\alpha)]\!]$,

(ii) $abs_*(W[\![Gnd\,(\alpha)]\!]) = G[\![Gnd\,(\alpha)]\!]$,

(iii) $abs_*(W[\![wre(\alpha,\beta)]\!]) = G[\![wre(\alpha,\beta)]\!]$,

(iv) $abs_*(W[\![ntran(\alpha,\beta,\gamma)]\!]) = G[\![ntran(\alpha,\beta,\gamma)]\!]$,

(v) $abs_*(W[\![ptran(\alpha,\beta,\gamma)]\!]) = G[\![ptran(\alpha,\beta,\gamma)]\!]$.

The function abs_* does preserve $\bullet$:

5.4 Lemma. *Let $S_0 \in P(Sta[\Lambda_0])$ and $S_1 \in P(Sta[\Lambda_1])$. Then*

$$abs_*(S_0 \bullet S_1) = abs_*(S_0) \bullet abs_*(S_1).$$

But, abs_* does not preserve $\backslash\alpha$; both inclusion (a) and (b) can fail to hold, essentially, for the reasons illustrated in the example below.

5.5 Example.

(i) To show the inclusion (a) can fail, let $S = \{\sigma\}$, a static configuration of sort Λ, where $\sigma \setminus \alpha \uparrow$ and $V_\sigma(x) \in \{H, L\}$ for all $x \in \Lambda$. Then $(abs_* S) \setminus \alpha = \{V_\sigma \setminus \alpha\}$, while $abs_*(S \setminus \alpha) = \emptyset$, so $(abs_* S) \setminus \alpha \not\subseteq abs_*(S \setminus \alpha)$.

(ii) To show the inclusion (b) can fail, let $S = \{\sigma\}$, a static configuration of sort Λ, where $V_\sigma(\alpha) = X$(or Z) with $V_\sigma(x) \in \{H, L\}$, for all $x \in \Lambda \setminus \alpha$ and $\sigma \setminus \alpha \downarrow$. Then $abs_*(S \setminus \alpha) = \{V_\sigma \setminus \alpha\}$, while $(abs_* S) \setminus \alpha = \emptyset \setminus \alpha = \emptyset$, so $abs_*(S \setminus \alpha) \not\subseteq (abs_* S) \setminus \alpha$.

By banning such examples we obtain sufficient conditions to ensure (a), (b), and their conjunction (c).

To ensure (a) we have:

5.6 Lemma. *Let $S \in P(Sta)$. If*

$$\forall \sigma \in S.\ (abs(\sigma) \downarrow \Rightarrow \sigma \setminus \alpha \downarrow) \tag{a1}$$

then $(abs_*(S)) \setminus \alpha \subseteq abs_*(S \setminus \alpha)$.

To ensure (b) we shall use:

5.7 Lemma. *Let $S \in P(Sta)$. If*

$$\forall \sigma \in S.\ (\sigma \setminus \alpha \downarrow \ \&\ \ abs(\sigma \setminus \alpha) \downarrow \Rightarrow abs(\sigma) \downarrow) \tag{b1}$$

then

$$abs_*(S \setminus \alpha) \subseteq (abs_* S) \setminus \alpha.$$

The conditions (a1) and (b1) used in lemmas 5.6 and 5.7 are expressible as W-assertions. We have already seen, writing

$$D \equiv \forall x.\ V(x) = H \vee V(x) = L,$$

that $abs(\sigma)\downarrow$ iff $\sigma \models D$, for a static configuration σ. Similarly writing

$$D_\alpha \equiv \forall x.\ \neg(x = \alpha) \rightarrow (V(x) = H \vee V(x) = L),$$

we get $abs(\sigma \setminus \alpha)\downarrow$ iff $\sigma \models D_\alpha$ and $\sigma \setminus \alpha \downarrow$, for $\sigma \in$ Sta. Expressing the definedness of hiding, as in [W], by

$$\begin{aligned} G_\alpha \equiv &[H \leq V(\alpha) \rightarrow (H \leq I(\alpha) \vee \exists x. \neg(x = \alpha) \wedge H \leq V(x)\ \ \&\ \ x \sim \alpha)] \\ &\wedge [L \leq V(\alpha) \rightarrow (L \leq I(\alpha) \vee \exists x. \neg(x = \alpha) \wedge L \leq V(x)\ \ \&\ \ x \sim \alpha)] \end{aligned}$$

we have $\sigma \setminus \alpha \downarrow$ iff $\sigma \models G_\alpha$, for any $\sigma \in$ Sta.

Suppose for each subterm of c of the form $c' \setminus \alpha$ we have that c' satisfies $\Box\, D \rightarrow G_\alpha$. Then using the facts in 5.3, 5.4, 5.6 and the monotonicity of $\bullet$ and $\setminus\alpha$, by structural induction on c, we can show that c meets the condition (a1) and so (a) above. From which we obtain the following:

5.8 Corollary. *Let c be a circuit term such that for all subterms of the form $c' \setminus \alpha$*

$$c' \models^W \Box\, D \rightarrow G_\alpha. \tag{a2}$$

Then $c \models^W \Box\ {}^{*}A$ implies $c \models^G \Box\, A$, and $c \models^G \Diamond\, A$ implies $c \models^W \Diamond\, D \wedge {}^{*}A$, for any G-assertion A.

A similar argument, this time with respect to conditions (b) and (b1), shows:

5.9 Corollary. *Let c be a circuit term such that for all subterms of the form $c' \setminus \alpha$*

$$c' \models^W \Box\, G_\alpha \wedge D_\alpha \rightarrow D. \tag{b2}$$

Then $c \models^G \Box\, A$ implies $c \models^W \Box\ {}^{*}A$, and $c \models^W \Diamond\, D \wedge {}^{*}A$ implies $c \models^G \Diamond\, A$, for any G-assertion A.

By combining the conditions in 5.8 and 5.9 we obtain:

5.10 Corollary. *Let c be a circuit term. If for all subterms $c' \setminus \alpha$*

$$c' \models^W \Box\, D_\alpha \rightarrow (G_\alpha \leftrightarrow D) \tag{c2}$$

then $c \models^G \Box\, A$ iff $c \models^W \Box\, {}^{*}A$, and $c \models^G \Diamond\, A$ iff $c \models^W \Diamond\, D \wedge {}^{*}A$, for any G-assertion A.

Remark. In the case where $\alpha \in$ sort(c') for all subterms $c' \setminus \alpha$ of c the condition (b2) in 5.9 can clearly be replaced by

$$c' \models^W \Box\, D_\alpha \rightarrow (G_\alpha \rightarrow (V(\alpha) = H \vee V(\alpha) = L)).$$

The proviso that α is in the sort of σ is necessary because otherwise $V(\alpha) = H$ and $V(\alpha) = L$ are taken to be false. Similarly, when $\alpha \in$ sort(c') for all subterms $c' \setminus \alpha$ of c the condition (c2) in 5.10 can be replaced by

$$c' \models^W \Box\, D_\alpha \rightarrow (G_\alpha \leftrightarrow (V(\alpha) = H \vee V(\alpha) = L)).$$

Corollary 5.10 provides a sufficient condition, *viz.* (c2), on terms c to ensure agreement between the models [G, G1] and [W]. If for all subterms $c' \setminus \alpha$ of a circuit c, c' meets the condition of (c2), the semantic treatment of circuits in [G, G1] will only lead to correct specifications being shown to hold of c, at least as far as the model in [W] is concerned. In this sense, if a circuit term satisfies the conditions of 5.10 the treatment of hiding given in [G, G1] is safe.

The condition of 5.10 can be expressed in a more intuitive fashion. For a static configuration σ, provided $\alpha \in \mathrm{sort}(\sigma)$, we have the equivalence

$$\sigma \models D_\alpha \to (G_\alpha \leftrightarrow D) \text{ iff } \sigma \models (\forall x.\ \neg x = \alpha \to (V(x) = H \vee V(x) = L)) \to \\ ((\exists x.\ \neg x = \alpha \wedge x \sim \alpha) \vee I(\alpha) = H \vee I(\alpha) = L \vee \\ (I(\alpha) = Z \wedge V(\alpha) = X)).$$

Hence:

5.11 Corollary. *Let c be a circuit term. If for all subterms $c' \setminus \alpha$ we have $\alpha \in \mathrm{sort}(c')$ and*

$$c' \models^W \Box\, (\forall x.\ \neg x = \alpha \to (V(x) = H \vee V(x) = L)) \to \\ ((\exists x.\ \neg(x = \alpha) \wedge x \sim \alpha) \vee I(\alpha) = H \vee I(\alpha) = L \vee \\ (I(\alpha) = Z \wedge V(\alpha) = X)) \qquad (c3)$$

then $c \models^G \Box A$ iff $c \models^W \Box\, {}^*A$, and $c \models^G \Diamond A$ iff $c \models^W \Diamond D \wedge {}^*A$, for any G-assertion A.

Strengthening (c3) we obtain:

5.12 Corollary. *Let c be a circuit term. If for all subterms $c' \setminus \alpha$ we have $\alpha \in \mathrm{sort}(c')$ and*

$$c' \models^W \Box\, (\forall x.\ \neg x = \alpha \to (V(x) = H \vee V(x) = L)) \to \\ ((\exists x.\ \neg(x = \alpha) \wedge x \sim \alpha) \vee I(\alpha) = H \vee I(\alpha) = L) \qquad (c4)$$

then $c \models^G \Box A$ iff $c \models^W \Box\, {}^*A$, and $c \models^G \Diamond A$ iff $c \models^W \Diamond D \wedge {}^*A$, for any G-assertion A.

Superficially, it would seem corollary 5.12 is less widely applicable than 5.11. In fact, on the contrary, it can be shown for circuit terms, built-up according the rules of section 2, that (c3) holds iff (c4) does and thus no generality is lost in using condition (c4) in 5.12.

As an example, consider the circuit term

$$c \equiv (ntran(\alpha, \beta, \gamma) \bullet ptran(\alpha, \beta, \gamma)) \setminus \gamma.$$

Taking $c' \equiv ntran(\alpha, \beta, \gamma) \bullet ptran(\alpha, \beta, \gamma)$ it is not the case that c' satisfies $\Box\, G_\gamma \wedge D_\alpha \to D$, the condition (b2) of 5.9, and for this reason it is no surprise that we have $c \models^G \Box\, (v_\alpha = v_\beta)$ and yet $c \not\models^W \Box\, (V(\alpha) = V(\beta))$.

The conditions on circuits we have introduced are only sufficient and not necessary to guarantee agreement between the two models. For example, for any G-assertion A,

$$c \models^W \Box\, {}^*A \text{ iff } c \models^G \Box\, A, \text{ and}$$
$$c \models^W \Diamond D \wedge^* A \text{ iff } c \models^G \Diamond A$$

when c is the circuit term $ntran(\beta, \gamma, \alpha) \setminus \alpha$ even though $ntran(\beta, \gamma, \alpha)$ fails to satisfy (c4) of 5.12.

The condition (c4) of 5.12 is intuitively appealing and seems to be met in practice when hiding points in a good many circuits. There may be some syntactic constraint on circuit terms which is not too restrictive and yet ensures this condition is met when hiding. If so, for such circuit terms, the simple relational model of [G, G1] would suffice for verification. Of course, such verification could not apply when charge sharing or ratioing of resistances were used in the design—for them more detailed models like [B] and [W] would be needed. And then there is the problem of transistor thresholds. However for CMOS it seems a more conservative model of transistors suffices to take account of thresholds—see section 6. Indeed there is the likelihood that for a wide range of CMOS designs we can *prove* that a simple relational model is adequate as the basis for verification.

6. Further problems to explore.

Both models we have presented idealise the behaviour of transistors to forms of switches. In reality transistors do not always behave as switches. Because of switching thresholds an n-type transistor may not connect when its gate is high and both its source and drain are high. One solution proposed by Mike Fourman, and followed up by Mike Gordon, is to use a more conservative model of transistors which, in terms of the model [W], would allow $\neg\alpha \sim \beta$ when all points α, β, γ of an n-type transistor $ntran(\alpha, \beta, \gamma)$ were high [F1, G2]. (For a model like that in [G] this can only make a difference if the assumption that values are either H or L is dropped.) While inadequate for NMOS this does not appear too restrictive for CMOS designs. It looks as if the results here go over with a few small changes when relating two models in this more conservative regime. This should be checked. To cope with NMOS it seems an extension of the models in [B] and [W] is needed to measure drops or gains in voltage due to threshold effects. Unfortunately this complicates the models further.

To my knowledge no model at the level of abstraction seen here copes adequately with sequential circuits with static memory. While simulators like Bryant's generate one possible sequence of behaviour it is hard to find a compositional model which predicts all possible behaviours and no more—see [W] for a little more discussion. One can postulate components which have the desired behaviour but deriving their behaviour from basic components like transistors, capacitances, wires and resistances seems difficult without encumbering the model with all sorts of details about delays across wires and gates and transient capacitance effects. Can some more abstract model be found, in the spirit of those presented here, which copes with feedback loops of the sort necessary to explain flipflops for static memory? Perhaps extra structure in the form of relations expressing causal dependency will do it—such a trick is needed in the real-time programming language Esterel [BC]. On the positive side it seems the model in [W] generalises well when memory is purely dynamic (due to charge storage).

When are directional models as sometimes used by Mike Gordon and others justified? Maybe it's useful to mix assertions in the term language here in order to impose constraints on the environment. Circuit terms would include *e.g.* resistances.

Abstraction based on I instead of V. What happens?

How generally applicable are ideas like the above *e.g.* to temporal abstraction in [M]. (Literature on abstract interpretation may be useful.)

The techniques of this note can be used to highlight the problem with the crude model relative to a more detailed one. How useful are the results above in the practice of verifying circuits? The model [G, G1] should be used wherever possible because it is so much simpler than that in [W]. But how easy is it in practice to show a circuit meets conditions sufficient to ensure the simpler model may be used? After all such conditions are generally expressed in terms of the more complicated model. Is there some reasonable syntactic constraint on circuit terms which ensures a condition like (c4) in 5.12 is met when hiding?

Work needs to be done on formalising the relationship between qualitative models like those presented here, which, for instance, presuppose a clear understanding of a voltage being high or low, and quantitative physical models. The relationship can be very subtle but must be understood if we are ever to prove a design correct assuming only the correctness of its layout.

Hiding can be understood as existential quantification in both models (in the W-model it is a left adjoint to $(-\setminus\alpha)^*$) and in the G-model composition is conjunction though not in the W-model. The indexed-category view of models has been stressed by Mike Fourman based on his experience with categorical logic. Can it lead to a more uniform presentation of the models and proof systems for circuits, and if so how is composition to be understood?

References.

[B] Bryant, R.E., A switch-level model and simulator for MOS digital systems. IEEE Transactions on Computers C-33 (2) pp. 160-177, February 1984.

[BC] Berry, G., and Cosserat, L., The Esterel synchronous programming language and its mathematical semantics. In the proceedings of the joint US-UK seminar on the semantics of concurrency, July 1984, Carnegie-Mellon University, Springer-Verlag Lecture Notes in Comp. Sc. 197, 1984.

[C] Cardelli, L., An algebraic approach to hardware description and verification. Ph.D. thesis, Comp.Sc.Dept., University of Edinburgh, 1982.

[CC] Cousot, P., and Cousot, R., Abstract interpretation: a unified lattice model for static analysis of programs by constructions or approximations of fixpoints. POPL 1977.

[F] Fourman, M.P., Verification using higher-order specifications and transformations. Department of Electrical Engineering, Brunel University, 1986.

[F1] Fourman, M.P., Verbal communication, Leeds Workshop 1986.

[G] Gordon, M.J.C., Why higher order logic is a good formalism for specifying and verifying hardware. Report no.77 of the Computer Laboratory, Cambridge University 1985.

[G1] Camilleri, A.J., Gordon, M.J.C., and Melham, T.F., Hardware verification using higher-order logic. Proceedings of IFIP International Working Conference "From H.D.L. Descriptions to Guaranteed Correct Circuit Designs", Grenoble, 1986.

[G2] Gordon, M.J.C., Switch models of CMOS. In preparation 1987.

[M] Melham, T.F., Abstraction mechanisms for hardware verification. In "VLSI Specification, Verification and Synthesis", G.M. Birtwistle and P.A. Subrahmanyam, eds, Kluwer Press 1987.

[R] Robinson, E., Powerdomains, modalities and the Vietoris monad. Report no.98 of the Computer Laboratory, University of Cambridge, 1986.

[W] Winskel, G., Lectures on models and logic of MOS circuits. Proceedings for the Marktoberdorf Summer School, July 1986, published by Springer, 1987.

[W1] Winskel, G., A compositional model of MOS circuits. To appear in the book: VLSI Specification, Verification and Synthesis, G.M. Birtwistle and P.A. Subrahmanyam(eds) Kluwer Press 1987.

[W2] Winskel, G., A note on powerdomains and modality. In the May issue of Theoretical Computer Science, 1985.

Foundations of Equational Deduction: A Categorical Treatment of Equational Proofs and Unification Algorithms

D.E. Rydeheard and J.G. Stell
University of Manchester

Abstract. *We provide a framework for equational deduction based on category theory. Firstly, drawing upon categorical logic, we show how the compositional structure of equational deduction is captured by a 2-category. Using this formulation, algorithms for solving equations are derived from general constructions in category theory. The basic unification algorithm arises from constructions of colimits. We also consider solving equations in the presence of term rewriting systems and the combination of unification algorithms.*

Introduction

Equational deduction is the process of replacing like for like using substitution and the equivalence properties of equality. It has a simple compositional structure which allows us to introduce ideas from category theory: concepts in equational deduction correspond to those in category theory, whilst algorithms for solving equations may be derived from categorical constructions.

The basic ideas of equational deduction are now fairly well understood (see, for instance, [Huet,Oppen 80]) and there is an extensive literature on general techniques of equation solving and decision procedures for equational theories. This work is presented as the processing of terms (symbolic expressions) with occasional reference to abstract properties of relations. The paper by Huet [1980] is a fine exposition of this material showing, in particular, the role of abstract properties of relations.

We describe a categorical framework for these results which allows us to move away from the details of the structure of terms to the compositional properties involved. Not only does this provide an arrow-theoretic language for describing equational systems, but also, unlike the relational treatment, constructions at the abstract level (purely categorical constructions) specialize to algorithms for solving equations.

The two ingredients of equational deduction are rewriting (subterm replacement) and substitution. Both rewrite steps and substitutions may be composed. When we take account of the 'types' of these things, we end up with a structure known as a 2-category.

We first describe this, showing how a proof construction system may be based on the 2-category structure. This is a particular example of a general categorical treatment of deductive systems.

We then select topics from the theory of equational deduction, translating them into a categorical framework. We consider

- the unification of terms—solving equations in the empty theory—and show how general constructions of colimits capture the compositional structure of unification algorithms,
- solving equations in equational theories and the results of Fay [1979] and Hullot [1980] concerning confluent theories,
- a categorical setting for combining unification algorithms for different theories.

The categorical constructions involved here can be realized as computer programs using techniques described in [Rydeheard,Burstall 88].

The background to this work is the constructivity of category theory and its role in the design of computer programs. Proofs in category theory tend to be constructive. Moreover, the constructions are often of a mechanical, symbol-processing nature and can be coded as computer programs which have a high degree of abstraction, being abstracted upon categories. They are therefore 'data independent' in the sense that, provided with suitable categories as arguments, they specialize to algorithms on various data types. This is discussed at length in the above reference and finds application here where categorical constructions, such as those of colimits, specialize to algorithms for equational deduction.

This paper is a progress report on what is intended eventually to be a comprehensive treatment. Here, we have selected only a few topics from equational deduction and, moreover, towards the end of the paper, we give only outline results. To read this paper, some familiarity with basic category theory is required, such as provided by the first few chapters of a standard text (perhaps [Goldblatt 79] is at a suitable level). However, most of the key concepts are defined.

1 Categories and Equational Deduction

We begin by linking categorical logic with the work on the behaviour of equational systems. The intention is to point out some key ideas in the categorical treatment of deductive systems; ideas which may not be familiar to those working on computational aspects of deduction. This section is therefore somewhat tutorial in nature.

An entailment $A \vdash B$ can be given a 'label'—an expression which is a proof of $A \vdash B$. Lambek and Scott [1986] say:

> "Logicians should note that a deductive system is concerned not just with unlabelled entailments or sequents $A \vdash B$ (as in Gentzen's proof theory), but with deductions or proofs of entailments. In writing $f : A \to B$ we think of f as the 'reason' why A entails B."

There may be more than one proof of an entailment so we extend the idea of entailment as a relation to that of a graph. Moreover, since we may compose proofs sequentially, we may form a category whose objects are formulae (symbolic expressions) and whose arrows $f : A \to B$ are proofs of $A \vdash B$. For this to be a category, we have to impose the natural requirements that the composition of proofs is associative and has identities.

Particularly simple are deductive systems involving only equational deduction. There is no variable binding. Such systems, based upon term-rewriting, have been extensively studied from a computational viewpoint. The basic structures are terms—symbolic expressions built from variables and operation names—and (term) substitutions which are mappings from variables to terms. A substitution (in the right variables) may be applied to a term replacing the variables in the term by their images under the substitution. From application we derive composition: $(f.g)(x) = g(f(x))$.

A standard way (e.g. [Huet 80]) of treating this compositional structure involves fixing a global set of variables and forming a category whose objects are terms (formulae) built from these variables and whose arrows are certain substitutions. This category is sometimes reduced to a pre-order ($s \lesssim t$ if t is a substitution instance of s) and, from this, a lattice (the lattice of subsumptions) may be constructed.

An alternative and, for us, more productive view is provided by Kleisli categories [Kleisli 65] and the related algebraic theories of Lawvere [1963]. We form a category whose objects are sets ('sets of variables') and whose arrows $f : X \to Y$ are substitutions, mapping variables in X to terms whose variables are drawn from Y.

In this case variables are *localized.* No longer is there a global set of variables. This is a key observation, making possible the specialization of categorical constructions to algorithms for equational deduction.

Keeping track of variables often complicates the treatment of term-rewriting. Variables are created, identified, separated and renamed as necessary. Siekmann [1984] prefaces his discussion with:

> "...we assume we have a box of symbols, GENSYM, at our disposal out of which we can take an unlimited number of "new" symbols ... We shall adopt the computational proviso that whenever GENSYM is referenced it is subsequently 'updated' ..."

When global sets of variables are used, we often need to restrict attention to subsets encompassing free variables in terms. Hullot [1980], in defining 'narrowing', says:

> "Let M be a term and V a finite set of variables containing $\mathcal{V}(M)$. ... σ is a narrowing substitution of M away from V ..."

By localizing the variables and maintaining them as sources and targets of substitutions, their explicit handling is avoided: The identification and separation of variables arise from the limit and colimit constructions being employed. Coproducts are important here as they correspond to separating variables or creating new variables distinct from those already present (like the the non-functional 'gensym' of Lisp). Pushouts, on the other hand, correspond to identifying selected variables.

The localization of variables is also important in formulating a sound deductive system for many-sorted equational logic.

Algebraic theories, as Lawvere [1963] originally proposed, provide a skeletonal treatment of (the dual of) this category of (finite) sets and term substitutions. Objects are natural numbers and arrows $n \to m$ are dual to substitutions i.e. are m-tuples of terms in n variables. Composition $n \to m \to p$ is by substitution of terms at tuple positions for numbered variables. This provides a positional notation for substitution of terms and avoids the multiplicity of names arising when variables are used. More generally, algebraic theories are categories with distinguished products and algebras are product-preserving functors from theories. We shall see later the importance of coproducts in the dual category.

1.1 Equational Deduction and 2-categories

In discussing equational deduction, we introduced two compositional structures—categories of formulae and proofs; and categories of sets of variables and substitutions. These fit together nicely in a structure known as a 2-category in which there are arrows between arrows, called 2-cells. 2-cells compose in two different ways linked by an 'interchange' law. For equational deduction, our picture is of a category whose objects (0-cells) are sets, arrows (1-cells) are term substitutions and whose 2-cells are generated by rewrite rules corresponding to oriented equations. The two compositions capture the way that term substitution interacts with the subterm replacement of rewriting.

Definition 1 *A* 2-category *is a category* **A** *together with arrows between arrows in the same homset. Objects in* **A** *are called 0-cells, arrows are called 1-cells and arrows between arrows are 2-cells. Let $f, g : a \to b$, denote a 2-cell by $\alpha : f \Rightarrow g$ or, more fully, $\alpha : f \Rightarrow g : a \to b$ or, pictorially, as:*

$$a \underset{g}{\overset{f}{\genfrac{}{}{0pt}{}{\longrightarrow}{\longrightarrow}}} b \quad (\Downarrow\alpha)$$

There are two compositions of 2-cells: If $f, g, h : a \to b$ and $\alpha : f \Rightarrow g$, $\beta : g \Rightarrow h$, there is a composition $\alpha.\beta : f \Rightarrow h$ which makes each homset into a category. We call this vertical composition. For 2-cells $\alpha : f \Rightarrow f' : a \to b$ and $\beta : g \Rightarrow g' : b \to c$ there is an associative composition $\alpha \circ \beta : fg \Rightarrow f'g'$. We call this horizontal composition and may write it as $\alpha\beta$. These two compositions satisfy the so-called interchange law:

$$(\alpha.\alpha') \circ (\beta.\beta') = (\alpha \circ \beta).(\alpha' \circ \beta')$$

Finally, identities of vertical composition are those of horizontal composition, in the sense that if $i_a : a \to a$ is the identity on a and $i_{i_a} : i_a \Rightarrow i_a$ is the identity of '.', then it is also the identity of '$\circ$'.

Let's look further at the compositional structure of 2-categories. We can define left and right compositions of 1-cells with 2-cells as $f \circ \alpha$ (or simply $f\alpha$) $= i_f \circ \alpha$ and likewise for right composition. Moreover, horizontal composition can be defined in terms of these compositions of 1-cells with 2-cells by, for $\alpha : f \Rightarrow g : a \to b$ and $\beta : f' \Rightarrow g' : b \to c$,

$$\alpha \circ \beta = (f\beta).(\alpha g') \quad (= (\alpha f').(g\beta)). \qquad (1)$$

This was pointed out to me by Malcolm Bird, who drew 'tadpoles' to illustrate these compositions.

Notice also that, for $f : X \to Y$, the map $f \circ - : \mathbf{A}(Y, Z) \to \mathbf{A}(X, Z)$ is functorial and that for $f, f' : X \to Y$ and $\alpha : f \Rightarrow f'$ the map $\alpha \circ -$ is a natural transformation, the naturality being the interchange law. It has been observed that equational deduction has this functorial nature.

For more details of 2-categories consult [Kelly 82], [Kelly, Street 74] or [Street 76].

The following example shows how the language of 2-categories (with coproducts) provides a syntax for building equational proofs.

Example: 2-Categories for Term-Rewriting

We first define a category (a 1-category) whose objects are sets and whose arrows are 'term substitutions'.

An *operator domain* is a set of operator symbols indexed by their arities (natural numbers). If Ω is an operator domain, denote by Ω_n the set of operators in Ω whose arity is the natural number n.

The *terms* in a set X over an operator domain Ω, the set of which we denote by $T_\Omega(X)$, are syntactic objects defined recursively by:

$$x \in X \Rightarrow \langle x \rangle \in T_\Omega(X),$$

$$\rho \in \Omega_n,\ t_1, t_2, \ldots, t_n \in T_\Omega(X) \Rightarrow \rho(t_1, t_2, \ldots, t_n) \in T_\Omega(X).$$

We shall drop angle brackets from variables when no confusion arises and write a constant (nullary operator) a as a term (rather than as the term $a()$).

A *(term) substitution* from set X to set Y, $f : X \to Y$ is a function, $f : X \to T_\Omega(Y)$, mapping variables to terms. Thus if $X = \{x_1, x_2, x_3\}$ and $Y = \{x, y, z\}$ and f is a unary operator and g a binary operator, then an example term substitution from X to Y is:

$$\{x_1 \mapsto g(x, y), x_2 \mapsto g(f(x), g(x, y)), x_3 \mapsto f(f(z))\}.$$

Notice that substitutions can be applied to terms: if $f : X \to Y$ is a substitution, define the application of f to a term in X by:

$$f(\langle x \rangle) = f(x) \text{ for } x \in X,$$

$$f(\rho(t_1, t_2, \ldots, t_n)) = \rho(f(t_1), f(t_2), \ldots, f(t_n)).$$

We define composition using application: $(gf)(x) = g(f(x))$. Also, for each set X the identity substitution is defined by $i_X(x) = \langle x \rangle$ (unless X is empty in which case the identity is the empty function).

Now define the category $\mathbf{T}_\Omega$ to have sets as objects and term substitutions as arrows. This is indeed a category under the composition and identities just defined. This category is an example of a Kleisli category [Kleisli 65] and some of the material that follows generalizes to arbitrary Kleisli categories. Moreover, $\mathbf{T}_\Omega$ is isomorphic to the category of free Ω-algebras and homomorphisms. The full subcategory of finite sets is denoted ${\mathbf{T}_\Omega}^{Fin}$.

Rewrite rules in Ω—oriented equations—induce a 2-category structure on $\mathbf{T}_\Omega$. 2-cells in this 2-category consist of sets of (labelled) rewrite rules.

We illustrate with an example. Let Ω be an operator domain containing a constant e, a unary operator inv and a binary operator $\times$.

An example of a rewrite rule in Ω is the following:

$$commute : x \times y \Rightarrow y \times x \quad in \ \{x, y\}$$

This corresponds to a 2-cell:

$$\{z\} \ \underset{z \mapsto y \times x}{\overset{z \mapsto x \times y}{\Downarrow commute}} \ \{x, y\}$$

The 2-cell derived from a rewrite rule is explicitly tied to the variables used to express the rule. An alternative treatment is to interpret rewrite rules as schema which, when supplied with variables, generate instances of the rules. However, here we are concerned with the symbolic manipulation involved in equational proofs. Changing variables is made explicit via a substitution and the substitution instances of rules are generated by closure under composition. Often in pen-and-paper exercises these substitutions are elided, whereas here they are explicit. This makes for somewhat unwieldy, but formally precise, notation. The z which we have introduced to form the 2-cell above can be thought of as a 'place-holder' for left composition.

Now consider right composition of the above 2-cell with a substitution,

$$h = (\{x, y\}, x \mapsto inv(u), y \mapsto v, \{u, v\}).$$

This yields a new 2-cell,

$$\{z\} \ \underset{z \mapsto v \times inv(u)}{\overset{z \mapsto inv(u) \times v}{\Downarrow commute \circ h}} \ \{u, v\}$$

corresponding to a substitution instance of commutativity. We write,

$$commute \circ h : inv(u) \times v \Rightarrow v \times inv(y)$$

and say *commute* $\circ$ h is a *proof* of the rewrite. Right composition says that validity is preserved by substitution instances.

Left composition of the *commute* 2-cell with a substitution,

$$g = (\{z'\}, z' \mapsto z \times inv(z), \{z\})$$

yields a 2-cell,

$$g \circ \mathit{commute} : (x \times y) \times \mathit{inv}(x \times y) \Rightarrow (y \times x) \times \mathit{inv}(y \times x)$$

where commutativity is applied to subterms of the expression.

The vertical composition of 2-cells is the sequential composition of the proofs. Consider the following 2-cells,

$$\{z\} \; \begin{array}{c} \xrightarrow{\quad z \mapsto x \times y \quad} \\ \Downarrow \mathit{commute} \\ \xrightarrow[\quad z \mapsto y \times x \quad]{} \end{array} \; \{x, y\}$$

$$\{z\} \; \begin{array}{c} \xrightarrow{\quad z \mapsto z \times e \quad} \\ \Downarrow \mathit{right\_ident} \\ \xrightarrow[\quad z \mapsto z \quad]{} \end{array} \; \{z\}$$

and the substitution,

$$k = (\{x, y\}, x \mapsto e, y \mapsto z, \{z\})$$

then the vertical composition $(\mathit{commute} \circ k).\mathit{right\_ident}$ is a proof of $e \times z \Rightarrow z$.

Finally, the horizontal composition of 2-cells goes as follows. Consider the *commute* rule above and the rule:

$$\{z\} \; \begin{array}{c} \xrightarrow{\quad z \mapsto z \times z \quad} \\ \Downarrow \mathit{idempotent} \\ \xrightarrow[\quad z \mapsto z \quad]{} \end{array} \; \{z\}$$

The horizontal composition of these is the proof:

$$\mathit{idempotent} \circ \mathit{commute} : (x \times y) \times (x \times y) \Rightarrow y \times x$$

The interchange law, like the associativity of the compositions, imposes an equivalence upon derivations (*cf.* [Lambek and Scott 86]). For example, as 2-cells of type $e \times (\mathit{inv}(x) \times x) \Rightarrow x$, we identify

$$(\mathit{commute} \circ k).(\mathit{right\_ident}) \circ ((\mathit{commute} \circ k').\mathit{right\_inverse})$$

with

$$((\mathit{commute} \circ k) \circ (\mathit{commute} \circ k')).(\mathit{right\_ident} \circ \mathit{right\_inverse})$$

where

$$\mathit{right\_inverse} : x \times \mathit{inv}(x) \Rightarrow x \quad \mathit{in} \;\; \{x\}$$

and the substitutions are:

$$k = (\{x, y\}, x \mapsto e, y \mapsto z, \{z\})$$

$$k' = (\{x, y\}, x \mapsto \mathit{inv}(x), y \mapsto x, \{x\}).$$

We may be tempted to identify a proof consisting of the sequential composition of two *commute* rules with an identity proof. However, we consider proofs to be strictly syntactic, apart from the laws holding in 2-categories (with coproducts), and so refrain from making these extra identifications. There is a connection here with general proof theory; this equational treatment of proofs lives in a 3-category!

The entirety of 2-cells in $\mathbf{T}_\Omega$ generated by a set of rewrite rules $\mathcal{R}$ is the closure of $\mathcal{R}$ under horizontal and vertical composition together with coproducts. We look at this in detail below, beginning with coproducts in 2-categories.

Definition 2 *A* coproduct *of objects a_1 and a_2 in a 2-category $\mathbf{A}$ is an object $a_1 + a_2$ of $\mathbf{A}$ together with a pair of arrows $j_1 : a_1 \to a_1 + a_2$, $j_2 : a_2 \to a_1 + a_2$ which induces an isomorphism of categories,*

$$\mathbf{A}(a_1 + a_2, b) \equiv \mathbf{A} \times \mathbf{A}(\langle a_1, a_2\rangle, \langle b, b\rangle).$$

In elementary terms, this means that for each pair $f_1 : a_1 \to b$, $f_2 : a_2 \to b$, there is a unique $f : a_1 + a_2 \to b$ (written $[f_1, f_2]$) such that $j_1 f = f_1$ and $j_2 f = f_2$, and also for each pair of 2-cells $\alpha_1 : f_1 \Rightarrow g_1 : a_1 \to b$, $\alpha_2 : f_2 \Rightarrow g_2 : a_2 \to b$, there is a unique 2-cell $[\alpha_1, \alpha_2] : [f_1, f_2] \Rightarrow [g_1, g_2] : a_1 + a_2 \to b$ such that $j_1[\alpha_1, \alpha_2] = \alpha_1$ and $j_2[\alpha_1, \alpha_2] = \alpha_2$.

More general limits and colimits in 2-categories may be defined, see for instance [Street 76].

Example In the category $\mathbf{T}_\Omega$ coproducts are constructed as in **Set**, that is, they are disjoint sums. This inheritance is more general, applying to arbitrary Kleisli categories.

Turning to a 2-category structure on $\mathbf{T}_\Omega$, if $\alpha_1 : f_1 \Rightarrow g_1 : a_1 \to b$ and $\alpha_2 : f_2 \Rightarrow g_2 : a_2 \to b$ are two sets of rewrite rules, $[\alpha_1, \alpha_2]$ is simply the disjoint union of the sets of rewrite rules.

Coproducts of rules allow us to apply rules to selected subterms. For instance, in the example above, we have a proof:

$$p \circ [i_q, \mathit{commute}] : x \times (x \times y) \Rightarrow x \times (y \times x)$$

where $p = (\{v\}, v \mapsto x' \times z', \{x', z'\}), \{x', z'\})$ is the coproduct (disjoint union) of $\{x\}$ and $\{z\}$ and q the inclusion $\{x\} \hookrightarrow \{x, y\}$.

1.2 On Presentations and Relations

To give an account of the 2-category structure arising from (finite) sets of rewrite rules, we introduce presentations of 2-categories. Let $\mathcal{R}$ be a set of 2-cells in a (1-)category $\mathbf{A}$ with (distinguished) coproducts. We generate a 2-category structure on $\mathbf{A}$ by closing $\mathcal{R}$ under horizontal and vertical composition and coproducts. That is 2-cells in $\mathbf{A}$ extend those in $\mathcal{R}$ by the following syntactic forms:

1. $i_f : f \Rightarrow f : a \to b$,
2. $\alpha \circ \beta : fg \Rightarrow f'g' : a \to c$ where $\alpha : f \Rightarrow f' : a \to b$ and $\beta : g \Rightarrow g' : b \to c$ are 2-cells in $\mathbf{A}$,

3. $\alpha.\beta : f \Rightarrow g$ where $f, g, h : a \to b$ and $\alpha : f \Rightarrow g, \beta : g \Rightarrow h$ are 2-cells in $\mathbf{A}$,

4. $[\alpha_1, \alpha_2] : [f_1, f_2] \Rightarrow [g_1, g_2] : a_1 + a_2 \to b$ where $\alpha_1 : f_1 \Rightarrow g_1 : a_1 \to b$ and $\alpha_2 : f_2 \Rightarrow g_2 : a_2 \to b$ are 2-cells in $\mathbf{A}$.

These are identified under the equations holding in 2-categories with coproducts: specifically, associativity of the two compositions, identity and interchange laws, and the following equations for coproducts (the second of which is the uniqueness of the universal arrow):

$$j_i[\alpha_1, \alpha_2] = \alpha_i \quad i = 1, 2$$

$$\alpha : f \Rightarrow g : a_1 + a_2 \to b = [j_1\alpha, j_2\alpha] : [j_1 f, j_2 f] \Rightarrow [j_1 g, j_2 g] : a_1 + a_2 \to b$$

where $j_1 : a_1 \to a_1 + a_2 \leftarrow a_2 : j_2$ is a coproduct.

Consequences of these are the equations linking coproducts with compositions:

$$[\gamma_1 + \gamma_2] \circ [\alpha_1, \alpha_2] \circ \gamma = [\gamma_1 \circ \alpha_1 \circ \gamma, \gamma_2 \circ \alpha_2 \circ \gamma]$$

$$[\alpha_1, \alpha_2].[\beta_1, \beta_2] = [\alpha_1.\beta_1, \alpha_2.\beta_2]$$

In any 2-category, the existence of 2-cells induces a relation on arrows $\Rightarrow$ defined by $f \Rightarrow g$ iff there is a 2-cell $\alpha : f \Rightarrow g$. Let $\equiv$ be the equivalence relation defined by: $f \equiv g$ iff $f \Rightarrow g$ and $g \Rightarrow f$. Then $\equiv$ induces a quotient functor on $\mathbf{A}$ (considered as a 1-category), $Q : \mathbf{A} \to \mathbf{C}$. We call Q, or sometimes $\mathbf{C}$, the *equational theory* of the 2-category $\mathbf{A}$.

Now let the 2-category (with distinguished coproducts) $\mathbf{A}$ be generated by the set of 2-cells $\mathcal{R}$. Define a relation $\rightsquigarrow$ on arrows in $\mathbf{A}$ as follows:

1. $f \rightsquigarrow g$ if there is an $\alpha : f \Rightarrow g$ in $\mathcal{R}$,

2. if $f \rightsquigarrow g : b \to c$ then for all $h : a \to b$ and $k : c \to d$, $hfk \rightsquigarrow hgk$,

3. if $f \rightsquigarrow g : a_1 \to b$ and $h : a_2 \to b$, then $[f, h] \rightsquigarrow [g, h] : a_1 + a_2 \to b$ (and symmetrically).

The relation $\rightsquigarrow$ corresponds to one invocation of a rule in $\mathcal{R}$. In a 2-category (with distinguished coproducts) generated by $\mathcal{R}$, the transitive reflexive closure of $\rightsquigarrow$, $\rightsquigarrow^*$, is the same relation as $\Rightarrow$. This follows from the equation (1) expressing horizontal composition in terms of vertical composition and composition with arrows, together with the observation that for $f_1 \rightsquigarrow g_1 : a_1 \to b$ and $f_2 \rightsquigarrow g_2 : a_2 \to b$,

$$[f_1, f_2] \rightsquigarrow [g_1, f_2] \rightsquigarrow [g_1, g_2].$$

We relate this to term-rewriting by taking the category $\mathbf{A}$ to be $\mathbf{T}_\Omega$ and $\mathcal{R}$ to be a set of rewrite rules. Then both relations $\Rightarrow$ and $\rightsquigarrow$ in the 2-category generated by $\mathcal{R}$ are, in the terminology of Huet [1980], stable and compatible (i.e. preserved under application of substitutions $\sigma(t)$ and subterm replacements $t[u \leftarrow s]$). The relation $\rightsquigarrow$ is the smallest compatible stable relation containing $\mathcal{R}$.

1.3 Categorical Logic

Categorical logic treats deductive systems as indexed categories. This was proposed by Lawvere [1970] in introducing 'hyperdoctrines' (see [Seely 83]). The general idea is that, for each set of variables X, we form the collection of terms in X and deductions between these terms. This forms a category for each set X. A term substitution $f : X \to Y$ induces a functor between these categories corresponding to the application of the substitution. In the case where the deductive system has quantifiers, these are interpreted as adjoint to this substitution functor.

For equational deduction there are no explicit connectives or quantifiers and the categories indexed by sets of variables are just the homsets, equipped with 2-cells as arrows, and so the indexed structure reduces to a 2-category structure. In a similar vein, Seely [1987] observes that computational systems (λ-calculi) form 2-categories.

This general algebraic framework for deduction is interesting from the point of view of automated proof-support systems. The structure of proofs is to a large extent independent of the underlying logic. Proof-support systems should be generic, based not on a particular logic but on an abstract description of logic. The Edinburgh LF is a type theory for expressing logics and building proof-support systems [Harper et al. 1987].

2 A Categorical Unification Algorithm

The unification of terms is the process of solving equations in the empty theory—that with no equations. It is decidable; the earliest algorithm for computing the most general unifier was described by Herbrand [1930]. Here we show how algorithms for unification can be derived from general constructions in category theory. This hinges upon two observations. Firstly, unification may be considered as an instance of something more abstract—as a colimit in a suitable category. Secondly, general constructions of colimits provide recursive procedures for computing the unification of terms.

The material of this section is joint work of Rod Burstall with one author (DER) of this paper. It has been published elsewhere[1].

2.1 The Unification of Terms

Unification is a basic operation on symbolic expressions. As an example, consider the following equation between two terms:

$$f(w, g(h(y)), h(z)) = f(g(x), z, h(w))$$

Here f, g, h are operator symbols with the evident arities and w, x, y, z are variables. The task of unification is to replace the variables with terms so that both sides of the equation become the same term. Such a substitution is called a *unifier*. For instance:

$$w \mapsto g(h(v)),\ x \mapsto h(v),\ y \mapsto v,\ z \mapsto g(h(v)),$$

[1] The material of this section is published in LNCS 240. It is an extended abstract. A full account of the material of this chapter may be found in internal reports of the Dept. of Computer Science in the universities of Manchester and Edinburgh

where v is a variable makes both sides of the equation equal to

$$f(g(h(v)), g(h(v)), h(g(h(v)))).$$

Not only is this a unifier, it is, in fact, the 'most general unifier' in that any other unifier factors through it.

Unifiers need not always exist for an equation. We can distinguish two cases when they fail to exist. As examples:

- *Clash* $g(x) = h(y)$
- *Cyclic* $x = g(x)$

In the 'clash' case no substitution can possibly make the two sides equal. However, in the 'cyclic' case unifiers do exist if we allow infinite terms.

In programming, unification occurs, for instance, in computational logic [Robinson 65], in polymorphic type-checking [Milner 78] and in implementing programming languages which are based upon pattern-matching such as Prolog [Colmerauer et al. 73] and Planner [Hewitt 71]. A good general survey of term rewriting and unification is [Huet,Oppen 80]. Efficient unification algorithms have been proposed, for instance those of [Paterson,Wegman 78] and of [Martelli,Montanari 82]. Unification admits several generalizations including higher-order unification [Huet 75] and unification in equational theories [Huet,Oppen 80], [Siekmann 84].

Manna and Waldinger [1980], Paulson [1985] and Eriksson [1984] have considered derivations and proofs of unification algorithms. The reader is invited to compare this categorical version with theirs. The algorithm we are to derive is a general recursive counterpart of the non-deterministic algorithm in [Martelli,Montanari 82] which there serves as a starting point for the development of an efficient algorithm. It is an open question as to whether such efficient evaluation strategies can be understood in this categorical framework. The emphasis of this work is on the compositional structure of unification algorithms rather than on deriving efficient algorithms.

2.2 Unification as a Coequalizer

In this section we show that unification can be interpreted as a coequalizer in the category $\mathbf{T}_\Omega$ (defined earlier, section 1.1). This was pointed out to us by Joseph Goguen. Coequalizers are examples of colimits and may be defined in terms of a universal property (the definition is given below).

An *equation* in set X over Ω is simply a pair of terms (s,t), which we write as $s = t$. A substitution $f : X \to Y$ is said to *unify* a set of Ω-equations in X, $\{s_i = t_i : i \in Z\}$, if $\forall i \in Z.\ f(s_i) = f(t_i)$. Such unifiers do not always exist. However, when they do exist so does a *most general unifier* defined to be a unifier $f : X \to Y$ such that for any unifier $g : X \to Y'$ there is a unique[2] substitution $h : Y \to Y'$ satisfying $f.h = g$.

[2]Uniqueness is often not demanded here but, since epis are involved, uniqueness is assured.

A set of equations in X over Ω, $\{s_i = t_i : i \in Z\}$, is equivalent to a parallel pair of arrows in $\mathbf{T}_\Omega$,

$$Z \overset{f}{\underset{g}{\rightrightarrows}} X,$$

defined by $f(i) = s_i$ and $g(i) = t_i$. Unifiers of this set of equations are arrows h such that $f.h = g.h$. Moreover the above definition of a most general unifier is exactly that of a coequalizer of f and g in $\mathbf{T}_\Omega$:

Definition 3 *A* coequalizer *of a parallel pair of arrows $f, g : a \to b$ in a category* $\mathbf{C}$ *is a arrow $q : b \to c$ with $f.q = g.q$ and such that if $q' : b \to c'$ is any arrow such that $f.q' = g.q'$ then there is a unique $u : c \to c'$ such that $q.u = q'$.*

Notice how the universal form of the definition of the most general unifier (as, in some sense, the 'best' unifier) translates directly into the universal definition of the coequalizer.

2.3 On Constructing Coequalizers

Category theory is particularly rich in ways of constructing colimits from other colimits (and similarly, of course, for limits). From a mathematical point of view, the interest in these constructions is their easy pictorial (arrow-theoretic) nature and the automatic inheritance of the universality. They are not in any sense deep but their variety is remarkable. From a programming point of view, these constructions may be implemented as programs—very general, high-level programs. The compositional nature of these constructions captures that of known algorithms as we now illustrate with the unification of terms.

Our interest here is in the construction of coequalizers. The following two theorems provide constructions, in an arbitrary category, of coequalizers in terms of other coequalizers. Amongst the many possible such theorems, these are particularly chosen so as to lead to a recursive, 'divide-and-conquer' algorithm which terminates in the category ${\mathbf{T}_\Omega}^{Fin}$ (and hence to provide unification algorithms). There are other constructions of colimits which seem appropriate; for instance, there is a calculus of pushout squares. However, those that we have considered fail to give terminating algorithms in the category ${\mathbf{T}_\Omega}^{Fin}$.

What we are doing here is separating the correctness of the operations from the termination of the reduction and from the strategy determining the order in which the operations are applied. The strategy influences the efficiency and, in general, the termination of unification algorithms. A very similar approach to unification algorithms has been considered by Kirchner [1987] and others, using operations on sets of equations.

Each of the theorems below can readily be verified by simple arrow-chasing. Alternatively, the first theorem is a special case of a general construction of colimits from other colimits (see for example [Mac Lane 71 §V.2]). The second theorem follows directly from the definition of an epi.

The first theorem considers parallel pairs of arrows whose source can be expressed as a coproduct, whilst the second theorem deals with the case when a parallel pair of arrows can be factored through a common arrow. In the case of unification (i.e. in the category

T_Ω) the first theorem corresponds to the division of a set of equations into two parts, whilst the second theorem corresponds to the division of terms into subterms.

Theorem 1 *If an arrow* $q : b \to c$ *is the coequalizer of* $f, g : a \to b$ *and* $r : c \to d$ *is the coequalizer of*

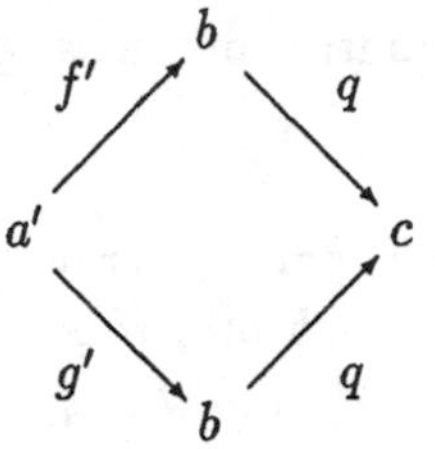

then $q.r : b \to d$ *is the coequalizer of*

$$a + a' \underset{[g,g']}{\overset{[f,f']}{\rightrightarrows}} b.$$

Here $[f, f']$ *is the unique arrow determined by the coproduct of* a *and* a' *such that* $j_a.[f, f'] = f$ *and* $j_{a'}.[f, f'] = f'$.

Theorem 2 *For all epis*[3] $h : a' \to a$, *the arrow* $q : b \to c$ *is the coequalizer of the parallel pair of arrows* $f, g : a \to b$ *iff it is the coequalizer of the parallel pair:*

$$a' \underset{h.g}{\overset{h.f}{\rightrightarrows}} b.$$

It is to be stressed that these theorems are valid for any category. However, we illustrate them in the category T_Ω. Consider the following two equations (with f, g, h, a operators and w, x, y, z variables).

$$f(w, g(h(y)), h(z)) = f(g(x), z, h(w))$$

$$h(a) = x$$

The most general unifier q of the first equation is given previously as:

$$w \mapsto g(h(v)),\ x \mapsto h(v),\ y \mapsto v,\ z \mapsto g(h(v))$$

The second equation with q applied to it is:

$$h(a) = h(v)$$

Its most general unifier r is simply

$$v \mapsto a.$$

According to theorem 1, the most general unifier of the two equations is then $q.r$ which is the following substitution:

$$w \mapsto g(h(a)),\ x \mapsto h(a),\ y \mapsto a,\ z \mapsto g(h(a))$$

[3] An arrow $e : a \to b$ is an epi iff for all pairs of arrows $f, g : b \to c$, $e.f = e.g \Rightarrow f = g$.

Theorem 2 says that, for instance, the most general unifier of the two equations above is the same as the most general unifier of the following set of equations obtained by matching subterms.

$$w = g(x)$$

$$g(h(y)) = z$$

$$h(z) = h(w)$$

$$h(a) = x$$

In choosing the above two constructions of coequalizers we have in mind a recursive algorithm based upon the expression of parallel pairs of arrows as either (letting + be a distinguished coproduct) a *coalesced sum*—

$$(a \underset{g}{\overset{f}{\rightrightarrows}} b) \oplus (a' \underset{g'}{\overset{f'}{\rightrightarrows}} b) \;=\; a + a' \underset{[g,g']}{\overset{[f,f']}{\rightrightarrows}} b$$

or as a *left composition* (with $h : a' \to a$ an epi)—

$$(a' \xrightarrow{h} a) \odot (a \underset{g}{\overset{f}{\rightrightarrows}} b) \;=\; a' \underset{h.g}{\overset{h.f}{\rightrightarrows}} b.$$

In terms of these operations, the coequalizer function ϕ taking parallel pairs to arrows satisfies the following equations (in the sense that if the right side is defined so is the left and they are equal)

$$\phi(P \oplus Q) = \phi(P).\phi(Q \odot \phi(P))$$

$$\phi(h \odot P) = \phi(P) \quad (h \text{ an epi})$$

where $(f,g) \odot h = (f.h, g.h)$ – the *right composition*. These equations are simply a rewriting of theorems 1 and 2 and provide a general structure for unification algorithms. Notice that they do not express ϕ as a homomorphism.

Let us say that a parallel pair is *irreducible* if it cannot be expressed non-trivially as a coalesced sum or as a left composite. An expression for a parallel pair P as $Q \oplus R$ is trivial if either Q or R is isomorphic to P. Likewise an expression for P as $h \odot Q$ is trivial if Q is isomorphic to P.

Now, we want to view these equations as a *definition* of a function—the coequalizer. That is we want to establish the following theorem.

Theorem 3 *Let C be the class of coequalizable parallel pairs in $\mathbf{T}_\Omega{}^{Fin}$. There is a unique (to within an isomorphism) function $\phi : C \to Arrow(\mathbf{T}_\Omega{}^{Fin})$ defined to be the coequalizer on irreducible parallel pairs and satisfying the following equations (in the sense that if the right side is defined so is the left and they are equal).*

$$\phi(P \oplus Q) = \phi(P).\phi(Q \odot \phi(P))$$

$$\phi(h \odot P) = \phi(P) \quad (h \text{ an epi})$$

Moreover, $\phi(P)$ is the coequalizer of $P \in C$.

Sketch of Proof

The universality of the arrow $\phi(P)$, if it exists, is a direct consequence of theorems 1 and 2 above and hence is at the level of general categorical skull-duggery. To establish the existence of $\phi(P)$ when P is coequalizable is more intricate—unduly intricate compared with these elegant theorems. Other authors (e.g. [Manna,Waldinger 80]) have noticed this disparity between derivation and proof of unification algorithms.

The existence of $\phi(P)$ is a termination proof and depends on defining a suitable well-founded pre-order. We give a rough argument (ignoring the partial nature of unification) of termination as follows:

> Define a well-founded pre-order on sets of equations as the lexical pre-order[4] of *(i)* the number of variables in the set of equations, *(ii)* the number of occurrences of operators and *(iii)* the number of equations.
>
> Now let E be a set of equations. Consider the construction of theorem 1. Divide E non-trivially into E_1 and E_2. The set E_1 is smaller than E in the pre-order since E_1 has no more variables or operator occurrences than E and E_1 has strictly fewer equations than E (by non-triviality). Let q be the most general unifying substitution of E_1 and let $E_2 \odot q$ be the set of equations resulting from applying q throughout E_2. There are two cases. If q is an isomorphism, then $E_2 \odot q \equiv E_2$ and so, as before, $E_2 \odot q$ is smaller than E. If q is not an isomorphism then it reduces the number of variables (an observation of Robinson [1965]) and so again $E_2 \odot q$ is smaller than E.
>
> Consider now the construction of theorem 2. In this case E is expressed non-trivially as $h \odot E'$ with h an epi. E' is smaller than E in the pre-order since both have the same number of variables but the number of operator occurrences in E' is strictly smaller than than in E or is the same but then E' contains fewer equations than E.

This proof can be cast into a more categorical form by axiomatizing suitable properties of the category $\mathbf{T}_\Omega{}^{Fin}$, principally the support of an appropriate well-founded pre-order. It has been pointed out to us that there is the possibility of a fully categorical version of this termination proof using a notion of ordinals in categorical logic.

All this can be implemented as programs as follows. The two constructions of co-equalizers yield programs which are parameterized on arbitrary categories. Categories can be considered to be a data type consisting of four functions—the source and target functions, and the composition and identity functions. These functions are just the ingredients required to write categorical algorithms. The above coequalizer constructions may be specialized to the category $\mathbf{T}_\Omega{}^{Fin}$ by, first, encoding this category and then passing it as a parameter. These two programs form the recursive part of a general, non-deterministic unification algorithm. We have implemented this in Standard ML. Categorical programming is discussed at length in [Rydeheard,Burstall 88].

---

[4]The lexical product $\lesssim$ of pre-orders $\lesssim_1$ and $\lesssim_2$ is defined by $x < y$ iff $x <_1 y$ or ($x \sim_1 y$ and $x <_2 y$), together with $x \sim y$ iff $x \sim_1 y \sim_2 x$. The lexical product of well-founded pre-orders is well-founded. The lexical pre-order determined by numerical functions is that corresponding to the usual numerical order.

3 Solving Equations in Equational Theories

We now turn to solving equations in equational theories. This is sometimes called equational unification, as unification is the solution of equations in the empty equational theory. There is an extensive literature on general techniques of solving equations and of decisions procedures in equational theories. Algorithms have been devised both for specific theories and, more generally, for theories satisfying certain properties such as 'confluence'. Here we consider the solution of equations in confluent theories and the combination of unification algorithms for different theories.

The key idea in solving an equation in an equational theory is to use equations from the theory to transform the given equation and then solve the transformed equation either directly or by further transformation. The solution of the transformed equation is then translated back to a solution of the original equation.

As an example, consider solving the equation,

$$(y + b) + c = (z + a) + c$$

in variables $\{y, z\}$, where a, b and c are constants and $+$ a binary operation. We are to solve it in the presence of a rule (oriented equation):

$$acommute : a + x \Rightarrow x + a$$

in variables $\{x\}$. As it stands the equation has no solutions—unifiers in the empty theory. However, a solution arises by observing that the LHS of the rule *acommute*, unifies against the subterm $(y + b)$ of the LHS of the equation. The relevant substitutions are

$$r = (\{y, z\}, y \mapsto a, z \mapsto z, \{z\})$$

applied to the equation and

$$s = (\{x\}, x \mapsto b, \{z\})$$

applied to the rule. We now rewrite the subterm using a substitution instance of *accommute* to get the equation $(b + a) + c = (z + a) + c$ which is to be solved. Setting z to b gives us a solution in the empty theory. This solution is composed with the substitution r to yield the following solution of the original equation:

$$(\{y, z\}, y \mapsto a, z \mapsto b, \{\}).$$

This process is known as 'narrowing'. It turns out that for certain equational theories all solutions may be obtained by successive narrowing followed by unification.

3.1 Solving Equations in 2-Categories

We have already seen how equational proofs have a 2-category structure. We now show that general techniques of solving equations, like that described above, can be interpreted in this categorical setting. The ingredients involved are colimits (for matching terms), quotient functors (induced by equational theories) and sets of solutions rather than coequalizers.

If $f, g : a \to b$ is a parallel pair of arrows in a category $\mathbf{A}$, any arrow q such that $fq = gq$ is called a *solution* of f, g. In quotients of the category $\mathbf{T}_\Omega$ this corresponds to our usual notion of the solution of a set of equations. Notice that if q is a solution of f, g so is qt for any arrow t. qt is said to be an *extension* of q. Unlike unification in the empty theory, an equation in an equational theory may have many solutions which are not all extensions of one solution. So we replace coequalizers by sets of solutions. Corresponding to the universality of coequalizers, we seek *complete sets of solutions* i.e. sets of solutions such that any solution is an extension of a solution in the set. Sets, rather than proper classes, are sufficient for the applications we have in mind.

We have seen how 'narrowing' may be used to solve equations. This can be interpreted in the setting of 2-categories.

Narrowing Let $f : a \to b$ be an arrow in a 2-category $\mathbf{A}$. Suppose that $\alpha : p \Rightarrow q : a \to c$ is a 2-cell and that arrows r and s exist to make the following square a pushout square.

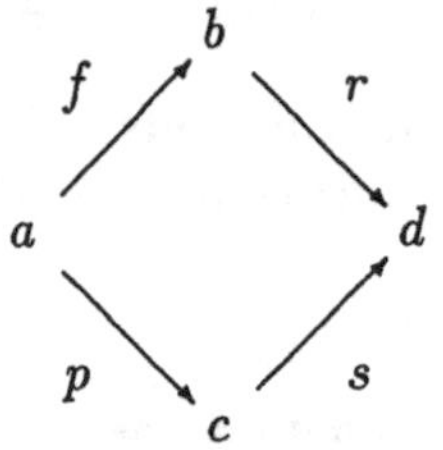

Then we say f is *narrowable* by $\alpha : p \Rightarrow q$, r is the narrowing arrow and $q.s$ the result of the narrowing.

Let $\mathbf{A}$ be a 2-category and $Q : \mathbf{A} \to \mathbf{C}$ its equational theory i.e. Q is the quotient induced by the equivalence closure $\equiv$ of the relation $\Rightarrow$. We seek solutions of parallel pairs of arrows fQ, gQ in $\mathbf{C}$, that is arrows h in $\mathbf{A}$ such that $(fh)Q = (gh)Q$ or, equivalently, $fh \equiv gh$. Since this is a common situation, let us introduce some terminology: For an equational theory $Q : \mathbf{A} \to \mathbf{C}$ of a 2-category $\mathbf{A}$, let us say that h is a *Q-solution* of a pair f, g of a $\mathbf{A}$-arrows, if $(fh)Q = (gh)Q$.

Notice that if q is a solution of f and g in $\mathbf{A}$, then q is a Q-solution of f, g. The following lemma says that narrowing in $\mathbf{A}$ transforms Q-solutions to Q-solutions.

Lemma 1 *Let $\mathbf{A}$ be a 2-category and $Q : \mathbf{A} \to \mathbf{C}$ the equational theory of $\mathbf{A}$. Let $f, g : a \to b$ in $\mathbf{A}$ with f narrowing against $\alpha : p \Rightarrow q$ with pushout square:*

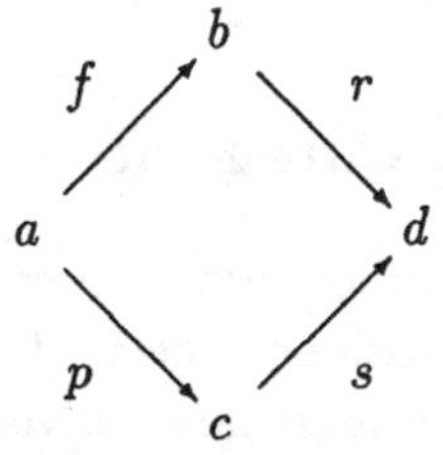

If t is a Q-solution of qs, gr, then rt is a Q-solution of f, g.

Proof

We have $fr = ps$, $p \Rightarrow q$ and $qst \equiv grt$, hence $frt = pst \equiv qst \equiv grt$ as required. □

Example Let us look again at the example an the beginning of the section.

We solve $(y+b)+c = (z+a)+c$ in $\{y,z\}$, where a, b and c are constants and $+$ a binary operation, in the presence of the rule:

$$\{v\} \underset{v \mapsto x+a}{\overset{v \mapsto a+x}{\Downarrow \; acommute}} \{x\}$$

The equation has no unifier (in the empty theory). However, a solution arises by narrowing. We use the rule *acommute*, unifying its LHS with the subterm $(y+b)$ of the LHS of the equation. That is, we narrow the substitution $(\{u\}, u \mapsto (y+b)+c, \{y,z\})$ against the rule $h \circ acommute$ where h is the substitution $(\{u\}, u \mapsto v+c, \{v\})$. The pushout gives two substitutions:

$$r = (\{y,z\}, y \mapsto a, z \mapsto z, \{z\})$$

$$s = (\{x\}, x \mapsto b, \{z\})$$

We end up with the new equation $(b+a)+c = (z+a)+c$ to be solved. Setting z to b gives us a solution in the empty theory. Thus the following composite substitution is a solution of the original equation.

$$(\{y,z\}, y \mapsto a, z \mapsto b, \{\})$$

In fact, this is the only solution.

Under certain circumstances, all solutions of parallel pairs in the equational theory of a 2-category **A** may be obtained, non-trivially, either as a solution in **A** or by transforming another solution in **A** through successive narrowing. The clue to this is 'confluence':

Definition 4 *A relation $\Rightarrow$ is said to be* confluent *if when $f \Rightarrow g$ and $f \Rightarrow h$ then there is a k such that $g \Rightarrow k$ and $h \Rightarrow k$. A 2-category is said to be confluent if the relation on arrows induced by 2-cells is confluent.*

Note that if $\equiv$ is the equivalence closure of a relation $\Rightarrow$ that is both transitive and confluent then $f \equiv g$ implies that there is an h such that $f \Rightarrow h$ and $g \Rightarrow h$.

Both solutions of parallel pairs and pushouts occur above. In a category with binary coproducts, these are linked: From a span (a pair of arrows with the same source) $f : a \to b$ and $g : a \to c$ we can construct a parallel pair $f.j_a, g.j_b$ where $j_a : a \to a+b$ and $j_b : b \to a+b$ is the coproduct. Pushouts correspond to coequalizers of the associated parallel pair. We say a span has a solution when the associated parallel pair has a solution. Notice that unification algorithms ensure that parallel pairs with solutions in the category $\mathbf{T}_\Omega$ have coequalizers.

The following proposition says that all solutions in the equational theory of a confluent 2-category arise either from those in the 2-category or from a non-trivial narrowing followed by a solution in the equational theory. Here 'non-trivial' is expressed in terms of a set of generators $\mathcal{R}$ of a 2-category and the associated relation $\leadsto$ (see section 1.2 for terminology). This observation is derived from [Fay 79] and [Hullot 80].

Proposition 1 *Let* $\mathbf{A}$ *be a confluent 2-category generated by a set of 2-cells* $\mathcal{R}$, $\leadsto$ *the induced relation on arrows and* $Q : \mathbf{A} \to \mathbf{C}$ *the equational theory of* $\mathbf{A}$. *Suppose that parallel pairs in* $\mathbf{A}$ *with solutions have coequalizers.*

Let $f, g : a \to b$ *be a parallel pair in* $\mathbf{A}$. *Every* Q*-solution of* f, g *is an arrow in one of the following two forms:*

1. *a solution of* f, g *in* $\mathbf{A}$,
2. rt *where* f *(or, symmetrically,* g*) narrows against* $p \leadsto q$ *with pushout square:*

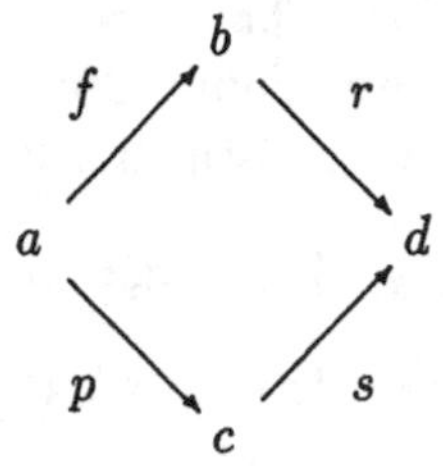

and t *is a* Q*-solution of* qs, gr.

Proof

Both forms of arrow are Q-solutions (Lemma 1 is the second case).

We show that, if $fq' \equiv gq'$, then q' is an arrow in one of the two forms above. By the confluence of $\Rightarrow$, there is an h such that $fq' \Rightarrow h$ and $gq' \Rightarrow h$. There are two cases:

1. If $fq' = h = gq'$, i.e. q' is a solution of f and g as required.
2. Suppose that $fq' \neq gq'$, then there is a k such that $fq' \leadsto k \Rightarrow h$ (or, symmetrically, for gq') because $\Rightarrow$ is the same as $\leadsto^*$. Now narrow against $fq' \leadsto k$. Since f and fq' have a solution, there is, by hypothesis, a pushout square in $\mathbf{A}$:

   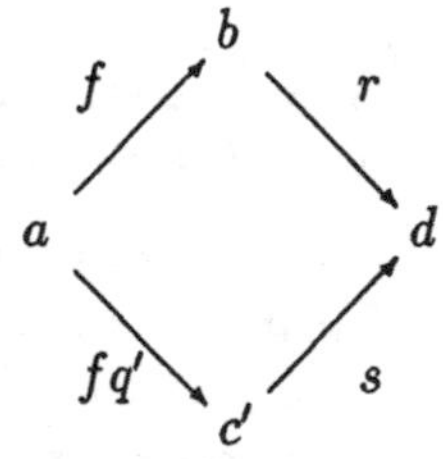

   The universal property of this pushout gives a unique arrow $t : d \to c'$ such that $q' = rt$ and $st = i_{c'}$. Since $gq' \equiv k$, we have $grt \equiv kst$. Thus q' is of the form rt where r is a narrowing arrow and t a Q-solution of gr, ks as required.

Example To show the role of confluence, consider the non-confluent rules,

$$a \Rightarrow b$$

$$a \Rightarrow c$$

where a, b and c are constants. Let us solve the equation $x = b$. Clearly, $x \mapsto b$ is a solution. Moreover, we may narrow against the first rule to obtain the solution $x \mapsto a$. However, because of the orientation of the rules, the solution $x \mapsto c$ is not obtainable by narrowing. This shows that confluence is necessary as well as sufficient.

Note that if every rule $f \Rightarrow g$ has a converse $g \Rightarrow f$ then the category is confluent and so every solution is generated by narrowing and unification. This corresponds to equational deduction in which we may use equations in either orientation.

It is possible to consider the above construction as recursively computing complete sets of solutions. However, as has been pointed out by Hullot [1980], termination is difficult to guarantee even when $\rightsquigarrow$ itself is terminating (no infinite descending chains). In this circumstance we want to consider the construction as a justification for a search through a space for solutions, providing a semi-decision procedure.

3.2 Combining Theories and Solving Equations

It seems irksome that when unification algorithms are known for, say, associative operators and for commutative operators, a new algorithm has to be devised for associative, commutative operators. Moreover, there is now a fairly well-developed theory of combining equational (and other) theories (see for instance [Burstall, Goguen 81]). What we are asking for is an analogous theory of combining unification algorithms, so that as new theories are built so are associated unification algorithms. In general, this seems to be very difficult. For the coproduct of theories (their disjoint sum) there is a partial answer in [Yelick 85], improved by [Herold 86] and [Tiden 86]. Recently, C. Kirchner [87] has considered a general framework for combining unification algorithms.

In this section we place the question within category theory, interpreting unification as a coequalizer, or more generally as a complete set of solutions, and the combination of theories as a pushout of categories. We have then resolved the problem into the construction of coequalizers in pushout categories. The task is therefore twofold: Firstly, analyzing the structure of pushout categories—a rather delicate analysis which, as far as we can see, can progress only by special cases—and, secondly, devising suitable general theorems for constructing coequalizers.

3.3 A Categorical Formulation

For the rest of this section *all functors are to be the identity on objects.* Let **Cat*** denote the category of categories with such functors (foundational questions do not concern us—we merely need some finite colimits of categories and such categories as that of finite sets

FinSet). Note that images of functors in **Cat*** are subcategories (this is not, in general, true). Consider now a pushout square in **Cat***:

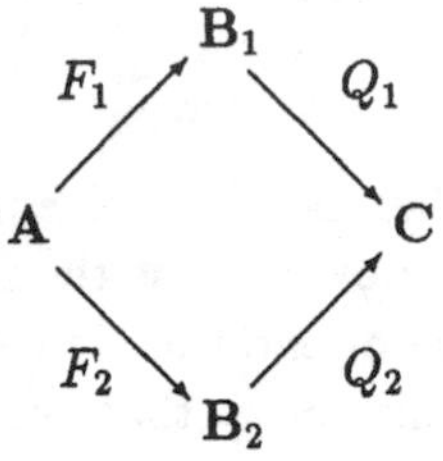

Suppose that **A**, $\mathbf{B}_1$ and $\mathbf{B}_2$ have coequalizers of certain parallel pairs of arrows and that the functors F_1 and F_2 satisfy certain conditions, then we ask: What parallel pairs of arrows in the pushout category **C** have coequalizers?

This is asking for a result like: if categories **A** and **B** have coequalizers, then so does their product $\mathbf{A} \times \mathbf{B}$. Unfortunately such a uniform result is not available in the case of the pushout of categories. The best we can do is analyze special cases.

The relevance of this to combining unification algorithms can be seen by interpreting theories as categories, namely the quotient category of $\mathbf{T}_\Omega$ induced by the equations. By suitable choices of the functors F_1 and F_2, various combinations of theories can be expressed as pushouts. For instance, if **A** is the category of finite sets **FinSet** and $\mathbf{B}_1$ and $\mathbf{B}_2$ are two equational theories, there are functors $F_1 : \mathbf{FinSet} \to \mathbf{B}_1$ and $F_2 : \mathbf{FinSet} \to \mathbf{B}_2$. The pushout of these is the coproduct of the theories—their disjoint sum. Again, if $\mathbf{A} = \mathbf{T}_\Omega$ is the empty theory (no equations) on an operator domain and $\mathbf{B}_1$ and $\mathbf{B}_2$ are two equational theories on the same operator domain with F_1 and F_2 the quotient functors then the pushout corresponds to the union of the sets of equations of the two theories. An example of this is the case of associative, commutative operators mentioned above as is also the (well studied) case where the set of equations is of the form $E \cup R$ with R a convergent (= confluent and terminating) set of rewrite rules.

3.4 The Pushout of Categories

Consider first the coproduct of categories in **Cat***. Let $\mathbf{B}_1$ and $\mathbf{B}_2$ be categories with the same objects. Their coproduct $J_1 : \mathbf{B}_1 \to \mathbf{B}_1 + \mathbf{B}_2$, $J_2 : \mathbf{B}_2 \to \mathbf{B}_1 + \mathbf{B}_2$ may be described as follows. By a path of arrows in a category we mean a non-empty string of arrows with adjacent targets and sources matching. We use "." for string concatenation and juxtaposition for composition of arrows. Objects in $\mathbf{B}_1 + \mathbf{B}_2$ are those of $\mathbf{B}_1$ (or equivalently of $\mathbf{B}_2$). Arrows in $\mathbf{B}_1 + \mathbf{B}_2$ are paths of images of the functors J_1 and J_2 (which are the disjoint sum in <u>*Set*</u>) equivalenced by the smallest equivalence relation $\sim$ such that:

1. $fJ_n.gJ_n \sim (fg)J_n \quad n = 1, 2.$
2. $s \sim s',\ t \sim t' \Rightarrow s.t \sim s'.t'.$

Pushouts may be constructed by first taking a coproduct and then a coequalizer of the resultant parallel pair as follows.

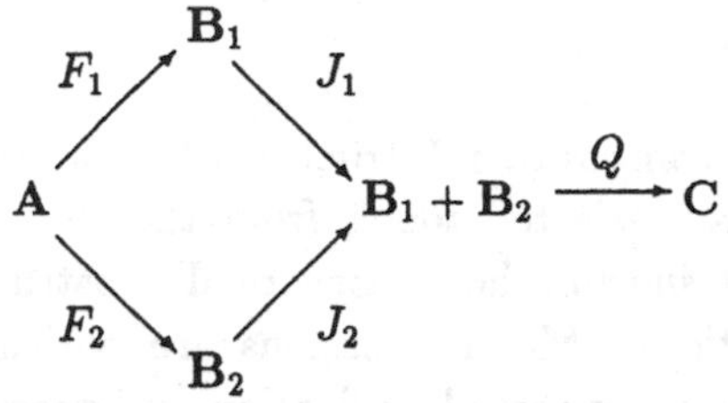

Arrows in **C** are paths of images of J_1 and J_2 equivalenced by the smallest equivalence relation $\sim$ such that:

1. $fJ_n.gJ_n \sim (fg)J_n \quad n = 1, 2.$
2. $hF_1J_1 \sim hF_2J_2.$
3. $s \sim s',\ t \sim t' \Rightarrow s.t \sim s'.t'.$

3.5 Solving Equations in Coproduct Theories

For particular cases of pushout categories, coequalizers may be constructed by devising suitable general theorems for constructing coequalizers in arbitrary categories (such theorems abound—constructing colimits in terms of colimits) and showing that they compute all coequalizers that exist. Likewise, of course, for constructing complete sets of solutions which may often be obtained from constructions of coequalizers.

As an example consider the following theorem, valid in any category, for constructing coequalizers. When extended to complete sets of solutions this theorem is a categorical counterpart of the construction of [Yelick 85].

Theorem 4 *If the following are two pushout squares:*

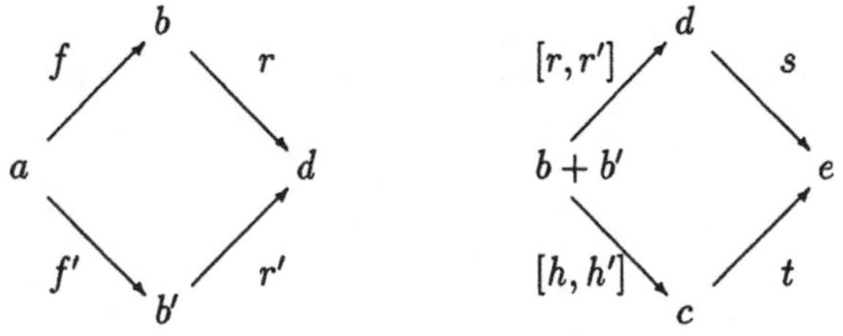

then $t : c \to e$ is a coequalizer of the parallel pair $fh, f'h' : a \to c$.

This may be proven by simple diagram chasing.

Notice that, because pushouts can be constructed in terms of coequalizers (in a category with binary coproducts), the theorem describes a construction of coequalizers in terms of coequalizers—suggesting a recursive algorithm. We have yet to establish the validity of this theorem as a recursive construction of complete sets of solutions in coproduct theories.

4 Conclusion

We have shown how the compositional structure of equational deduction may be described as a 2-category and have selected topics from the theory of equational deduction for a categorical treatment showing how categorical constructions specialize to algorithms for manipulating equations. Major omissions are critical pair completion procedures, higher-order unification, order-sorted unification and specialist unification algorithms for collections of axioms (e.g. associative, associative-commutative, Boolean rings ...).

Acknowledgements

Part of this paper (section 2 on unification algorithms) was joint work with Rod Burstall. We express our indebtedness to Rod not only for this but, more widely, for turning our attention some years ago to connections between category theory and computer programming. One author (DER) has collaborated with Rod on a project to program up basic category theory [Rydeheard,Burstall 88].

Amongst others who have helped with the material of this paper we would especially like to thank the following: Joe Goguen for bringing the categorical nature of unification to our attention and for his enthusiastic promotion of categorical aspects of programming, Horst Reichel who encouraged us to look at constructions of coequalizers for unification and Ursula Martin who helped with the proof of the unification algorithm. Others who contributed in discussion are Gordon Plotkin, David Benson, Don Sannella and Andrzej Tarlecki.

The work was carried out under grants from the Science and Engineering Research Council.

References

- LNCS = Lecture Notes in Computer Science, Springer-Verlag.
- LNM = Lecture Notes in Mathematics, Springer-Verlag.

Burstall R.M. (1980) Electronic Category Theory. Proc. Ninth Annual Symposium on the Mathematical Foundations of Computer Science. Rydzyua, Poland.

Burstall R.M. and Goguen J.A. (1981) An Informal Introduction to Specifications using Clear. In 'The correctness problem in computer science', Eds. Boyer and Moore. Academic Press, London.

Burstall R.M. and Landin P.J. (1969) Programs and Their Proofs: An Algebraic Approach. Machine Intelligence 4. Edinburgh Univ. Press. pp 17-44.

Colmerauer A. et al. (1973) Etude et realisation d'un système PROLOG. Convention de Research IRIA-Sesori No. 77030.

Eriksson L.H. (1984) Synthesis of a Unification Algorithm in a Logic Programming Calculus. Journal of Logic Programming, 1,1.

Fay M. (1979) First-Order Unification in an Equational Theory. 4th Workshop on Automated Deduction, Texas.

Goldblatt R. (1979) Topoi: The Categorial Analysis of Logic. Studies in Logic and the Foundations of Mathematics, Vol. 98, North-Holland.

Gordon M.J.C., Milner R. and Wadsworth C.P. (1979) Edinburgh LCF. LNCS 78.

Harper R., Honsell F. and Plotkin G. (1987) A Framework for Defining Logics. Proc. Symposium on Logic in Computer Science. June 22-25, 1987, Ithaca N.Y. Publ. I.E.E.E.

Herbrand J. (1930) Recherches sur la théorie de la démonstration. Thèse, U. de Paris. In: Ecrits logique de Jacques Herbrand, PUF Paris (1968).

Herold A. (1986) Combination of Unification Algorithms. In Proceedings 8th Conference on Automated Deduction, Oxford. pp 450-469. LNCS 230.

Hewitt C. (1972) Description and Theoretical Analysis (Using Schemata) of PLANNER: A Language for Proving Theorems and Manipulating Models in a Robot. Ph.D. Dept. Maths. M.I.T. Cambridge. Mass.

Huet G. (1976) Résolution d'équations dans les languages d'ordre $1,2,...,\omega$. Thèse d'etat, Specialité Maths. University of Paris VII.

Huet G. (1980) Confluent Reductions: Abstract Properties and Applications to Term Rewriting Systems. J. ACM 27,4 pp 797-821.

Huet G. and Oppen D.C (1980) Equations and Rewrite Rules: A Survey. In 'Formal Languages: Perspectives and Open Problems'. Ed. R. Book, Academic Press.

Hullot J.-M. (1980) Canonical Forms and Unification. 5th Conference on Automated Deduction, Les Arcs, France. LNCS 87, Springer-Verlag.

Jouannaud J.-P., Kirchner C. and Kirchner H. (1983) Incremental Construction of Unification Algorithms in Equational Theories. 10th ICALP, LNCS 154, Springer-Verlag.

Kelly G.M. (1982) Basic Concepts of Enriched Category Theory. Cambridge University Press.

Kelly G.M. and Street R. (1974) Review of the Elements of 2-Categories. Proc. Category Seminar, Sydney 1972/73. LNM 420. Springer-Verlag.

Kirchner C. (1987) Methods and Tools for Equational Unification. Internal report 87-R-008, Centre de Recherche en Informatique de Nancy.

Kleisli H. (1965) Every Standard Construction is Induced by a Pair of Adjoint Functors. Proc. Am. Maths. Soc. 16. pp. 544-546.

Lawvere F.W. (1963) Functorial Semantics of Algebraic Theories. Proc. Nat. Acad. of Sciences. 50. pp 869–872.

Lawvere F.W. (1970) Equality in hyperdoctrines and the comprehension schema as an adjoint functor. In: A. Heller (ed.), Proc. New York Symposium on Applications of Categorical Logic. Amer. Math. Soc. pp 1–14.

Levi G. and Sirovich F. (1975) Proving Program Properties, Symbolic Evaluation and Logical Procedural Semantics. In LNCS 32. Math. Foundations of Computer Science. Springer-Verlag.

Mac Lane S. (1971) Categories for the Working Mathematician. Springer-Verlag, New York.

MacQueen D. (1984) Modules for Standard ML. Proc. ACM Conf. on LISP and Functional Prog. Languages.

Manna Z. and Waldinger R. (1980) Deductive Synthesis of The Unification Algorithm. S.R.I. Research Report.

Martelli A. and Montanari U. (1982) An Efficient Unification Algorithm. ACM Trans. on Prog. Languages and Systems, 4, 2.

Milner R. (1978) A theory of type polymorphism in programming. J. Comp. Sys. Sci. 17, 3. pp. 348-375.

Milner R. (1984) A Proposal for Standard ML. Proc. ACM Symp. on LISP and Functional Programming.

Paterson M.S. and Wegman M.N. (1978) Linear Unification. J. Comp. Sys. Sci. 16, 2. pp. 158-167.

Paulson L.C. Verifying the Unification Algorithm in LCF. Science of Computer Programming 5.

Robinson J.A. (1965) A machine-oriented logic based on the resolution principle. J. ACM 12,1. pp. 23-41.

Robinson J.A. and Wos L.T. (1969) Paramodulation and Theorem Proving in First-Order Theories with Equality. Machine Intelligence 4. American Elsevier. pp. 135-150.

Rydeheard D.E. and Burstall R.M. (1985) The Unification of Terms: A Category-Theoretic Algorithm. Internal Report, Universities of Manchester and Edinburgh.

Rydeheard D.E. and Burstall R.M. (1986) A Categorical Unification Algorithms. Proc. Summer Conf. on Category Theory and Computer Programming 1985. LNCS 240.

Rydeheard D.E. and Burstall R.M. (1988) Computational Category Theory. To appear, Prentice-Hall.

Seely R.A.G. (1983) Hyperdoctrines, natural deduction and the Beck condition. Zeit. fur Math. Logik. 29 pp 505–542.

Seely R.A.G. (1987) Modelling Computations: A 2-Categorical Framework. Proc. Symposium on Logic in Computer Science. June 22-25, 1987, Ithaca N.Y. Publ. I.E.E.E.

Siekmann J.H. (1984) Universal Unification. In the 7th Internal. Conf. on Automated Deduction. LNCS 170.

Smyth M.B. and Plotkin G.D. (1977) The Category-Theoretic Solution of Recursive Domain Equations. Proc. Foundations of Computer Science.

Street R. (1976) Limits indexed by Category-Valued 2-Functors. J. Pure and Applied Algebra 8. pp 149–181.

Taylor P. (1987) Recursive Domains, Indexed Categories and Polymorphism. PhD. Thesis, Dept. Pure Math. and Math. Statistics, University of Cambridge.

Tiden E. (1986) Unification in combinations of collapse-free theories with disjoint sets of function symbols. In Proceedings 8th Conference on Automated Deduction, Oxford. pp 431-449. LNCS 230.

Yelick K. (1985) Combining Unification Algorithms for Confined Regular Equational Theories. In Proc. First Intern. Conf. On Rewriting Techniques and Applications. Dijon, France. LNCS 202 pp 365-380. Springer-Verlag.

A Typed Lambda Calculus with Categorical Type Constructors

Tatsuya Hagino
LFCS, Department of Computer Science, University of Edinburgh
James Clerk Maxwell Building, King's Buildings, Mayfield Road
Edinburgh EH9 3JZ, United Kingdom.

Abstract

A typed lambda calculus with categorical type constructors is introduced. It has a uniform category theoretic mechanism to declare new types. Its type structure includes categorical objects like products and coproducts as well as recursive types like natural numbers and lists. It also allows duals of recursive types, i.e. lazy types, like infinite lists. It has generalized iterators for recursive types and duals of iterators for lazy types. We will give reduction rules for this simply typed lambda calculus and show that they are strongly normalizing even though it has infinite things like infinite lists.

1 Introduction

The type structure of a simply typed lambda calculus is generally constructed from some base types using the arrow type constructor $\sigma \to \tau$. Since a pure lambda calculus is nothing but about lambda abstraction and application of lambda terms, the arrow type is more important than base types. However, if we do not have any base types, the type structure is empty and there is no point in discussing such a typed lambda calculus because there are no typed lambda terms. Therefore, we need some base types when developing a theory of a simply typed lambda calculus, but the choice of base types may vary from one calculus to another: some choose the type of natural numbers as the only base type or others, like [11], choose the type of ordinals as well. We have to be careful about choosing base types because a bad choice may ruin the whole calculus, e.g. lose the strong normalization property.

A typed lambda calculus can be regarded as a model of typed functional programming languages, so obviously the richer the type structure is, the closer to real programming languages the calculus is. Some programming languages now have quite powerful mechanisms of creating new types from existing ones, and we would naturally like to see those mechanisms in typed lambda calculi as well.

From a category theoretic point of view, the arrow type constructor is just one of the functors which can be defined by adjunctions. There is no reason why we should not have other functors in typed lambda calculi. In [4], the author introduced a category theoretic datatype declaration mechanism, by which we can define categorical objects like products and coproducts, ordinary datatypes in most of programming languages like natural numbers and lists, those datatypes which can be defined by solving recursive datatype equations, and lazy datatypes like infinite lists. All these datatypes are declared uniformly by means of generalized adjunctions. In this paper, we give a simply typed lambda calculus which incorporates this uniform categorical datatype declaration mechanism. This lambda calculus no longer needs base types. We can introduce almost all the datatypes we need in ordinary programming languages by this mechanism. Furthermore, the calculus is still strongly normalizing even though it has infinite lists and others.

In section 2, we introduce our (simply) typed lambda calculus: its type structure, its terms, its typing system and its reduction rules. In section 3, we see some types we can define in our typed lambda calculus, and, in section 4, we compare our typed lambda calculus with other typed lambda calculi. Finally, in section 5 we see the connection between our lambda calculus and the functional programming language ML.

2 A Typed Lambda Calculus with Categorical Type Constructions

2.1 Types

The type structure *Type* of an ordinary simply typed lambda calculus can be given in general by the following two rules:

$$\frac{\sigma \in \text{BaseType}}{\sigma \in \text{Type}} \qquad \frac{\sigma \in \text{Type} \qquad \tau \in \text{Type}}{\sigma \to \tau \in \text{Type}}$$

where *BaseType* is the set of base types available to this typed lambda calculus. This lambda calculus has only one way of constructing types, i.e. $\sigma \to \tau$ is the only type constructor. We can enrich the type structure by introducing new type constructors. For example, if we want the product and coproduct type constructors, we may add the following two rules:

$$\frac{\sigma \in \text{Type} \qquad \tau \in \text{Type}}{\sigma \times \tau \in \text{Type}} \qquad \frac{\sigma \in \text{Type} \qquad \tau \in \text{Type}}{\sigma + \tau \in \text{Type}}$$

Of course, we have to introduce new terms which belong to these new types and new reduction rules concerning the new terms. Although this approach is flexible to have any kind of type constructors, every time we introduce a new type constructor, we get a new typed lambda calculus and we have to prove all the properties of this calculus from the very beginning.

One way of getting out of this problem is to have a mechanism to introduce new type constructors. In domain theory, we can define domains by solving recursive domain equations and in some programming languages, e.g. ML [5], we can define datatypes (or datatype constructors as polymorphic types) in a similar manner by giving recursive datatype equations. Therefore, it is natural to introduce recursive types to a typed lambda calculus. The type structure of such a lambda calculus may be given by the following rules:

$$\frac{\rho \in \text{TVar} \qquad \rho \in \Gamma}{\Gamma \vdash \rho \in \text{Type}} \qquad \frac{\Gamma \vdash \sigma \in \text{Type} \qquad \Gamma \vdash \tau \in \text{Type}}{\Gamma \vdash \sigma \to \tau \in \text{Type}}$$

$$\frac{\rho \in \text{TVar} \qquad \Gamma \cup \{\rho\} \vdash \sigma \in \text{Type}}{\Gamma \vdash \underline{\mu}\rho.\sigma \in \text{Type}}$$

where TVar is a set of type variables and Γ is an environment under which types are checked to be well-formed. The type introduced by $\underline{\mu}\rho.\sigma$ should have terms corresponding recursively to that of $\sigma[\underline{\mu}\rho.\sigma/\rho]$, where $\sigma[\tau/\rho]$ denotes the type obtained by replacing the type variable ρ in σ by a type τ. In this way, we can get rid of some basic types. For example, the type for natural numbers can now be defined as

$$\mathrm{nat} \equiv \underline{\mu}\rho.1 + \rho$$

where 1 is the type of one element and $+$ is the coproduct type constructor. This is like in domain theory where the domain N of natural numbers is the least fixed point of $N \cong 1 + N$. Similarly we can define most of ordinary datatypes in today's programming languages in this way. However, there are still some problems about this approach. Firstly, we need to select some base types, like 1, and some type constructors, like $+$ and $\times$, to start with. Secondly, the reduction rules for this calculus may not be normalizing. Of course, it depends on the choice of terms and reduction rules in the calculus, but we would like to have a fixed point operator or something which enables us to write terms for addition, multiplication and so on. If we have a fixed point operator, reductions may not terminate. In case of an ordinary simply typed lambda calculus with the type of natural numbers, it has iterators (often denoted by J) which allows us to define these terms by primitive recursion and because we only use *primitive* recursion the reductions always terminate, but we still do not know what is the general operator for primitive recursion for $\underline{\mu}\rho.\sigma$.

From a category theoretic point of view, the types defined by $\underline{\mu}\rho.\sigma$ exactly correspond to initial T-algebras. Initial T-algebras cannot define the coproduct type constructor, but, as the author showed in [4], their extension, *initial and final* F, G*-dialgebras*, can define the coproduct type constructor as well as the product one.

Definition 2.1: Let C and D be categories and both F and G be functors from C to D. We define the category of F, G*-dialgebras* as

1. its objects are pairs $\langle A, f\rangle$ where A is a C object and f is a D morphism of $F(A) \to G(A)$, and
2. its morphisms $h: \langle A, f\rangle \to \langle B, g\rangle$ are C morphisms $h: A \to B$ such that the following diagram commutes.

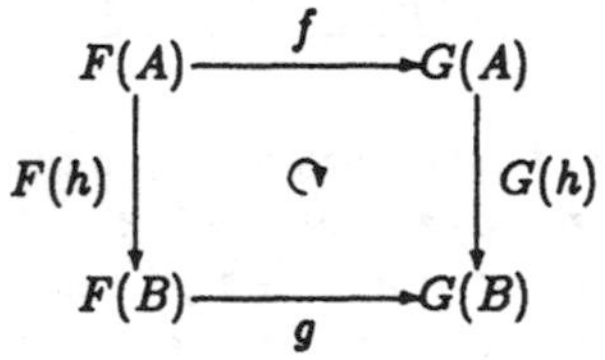

It is easy to show that it is a category; let us write $\mathbf{DAlg}(F, G)$ for it. $\square$

It is also easy to see that the definition of F, G-dialgebra is an extension of the definition of T-algebras as well as that of T-coalgebras: $\mathbf{DAlg}(T, \mathbf{I})$ is the category of T-algebras and $\mathbf{DAlg}(\mathbf{I}, T)$ is the category of T-coalgebras.

The extension is a very simple one, yet its symmetry and dividing the source category from the target one give us greater freedom. With T-algebras, we need the coproduct functor to define the domain of natural numbers, but with F, G-dialgebra we do not. Let C be any category and D be its product category $\mathrm{C} \times \mathrm{C}$. We define the functors F and G as

$$F(A) \equiv \langle 1, A\rangle \quad \text{and} \quad G(A) \equiv \langle A, A\rangle.$$

Let $\langle \text{nat}, \langle \text{zero}, \text{succ} \rangle \rangle$ be the initial F, G-dialgebra. The unique $\mathbf{DAlg}(F, G)$ morphism h from $\langle \text{nat}, \langle \text{zero}, \text{succ} \rangle \rangle$ to a $\mathbf{DAlg}(F, G)$ object $\langle A, \langle f, g \rangle \rangle$ makes the following diagram commute.

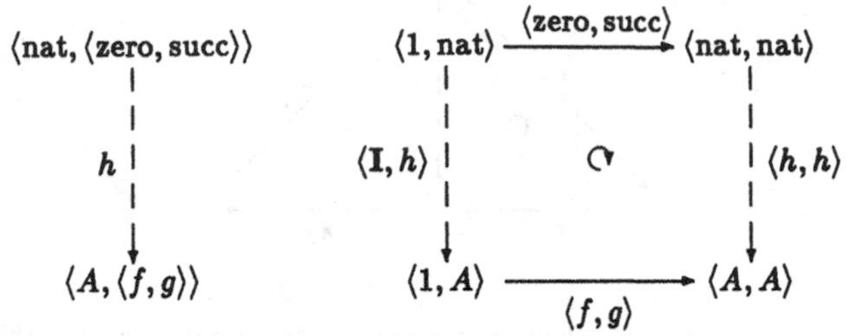

If we redraw the diagram, we get a more familiar diagram of defining 'nat' as the natural number object.

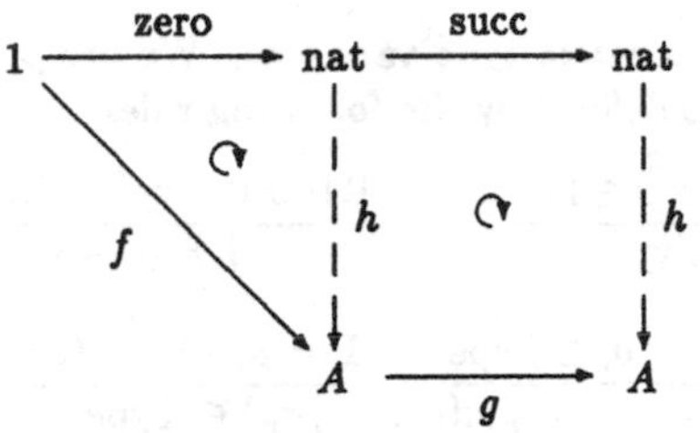

We can also demonstrate that left adjoint functors can be expressed by initial F, G-dialgebras and right adjoint functors can be expressed by final F, G-dialgebras. Let us, as an example, define the product of two C objects A and B. Remember that the product functor is the right adjoint of the diagonal functor. We set the functors F and G from C to C × C as

$$F(C) \equiv \langle C, C \rangle \quad \text{and} \quad G(C) \equiv \langle A, B \rangle.$$

Let $\langle A \times B, \langle \pi_1, \pi_2 \rangle \rangle$ be the final F, G-dialgebra. Then, from the definition, the unique $\mathbf{DAlg}(F, G)$ morphism h from a $\mathbf{DAlg}(F, G)$ object $\langle C, \langle f, g \rangle \rangle$ to $\langle A \times B, \langle \pi_1, \pi_2 \rangle \rangle$ should commute the following diagram.

$$
\begin{array}{ccc}
\langle C, \langle f, g \rangle \rangle & \quad\langle C, C \rangle \xrightarrow{\;\langle f, g \rangle\;} \langle A, B \rangle \\
\Big\downarrow h & \Big\downarrow \langle h, h \rangle \qquad\qquad \Big\downarrow \langle \mathbf{I}, \mathbf{I} \rangle \\
\langle A \times B, \langle \pi_1, \pi_2 \rangle \rangle & \langle A \times B, A \times B \rangle \xrightarrow[\langle \pi_1, \pi_2 \rangle]{} \langle A, B \rangle
\end{array}
$$

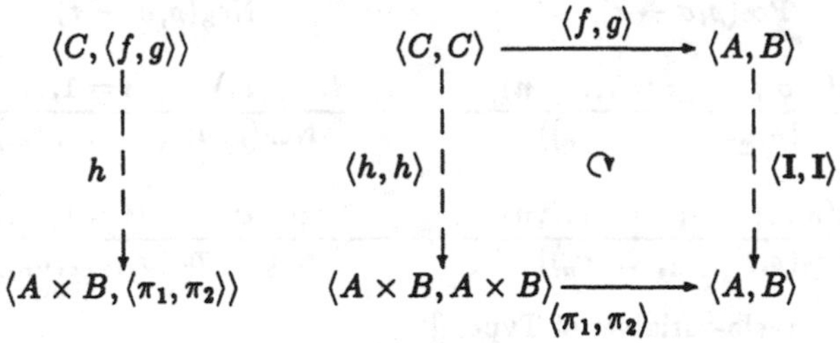

We can redraw the diagram to get an ordinary diagram for products.

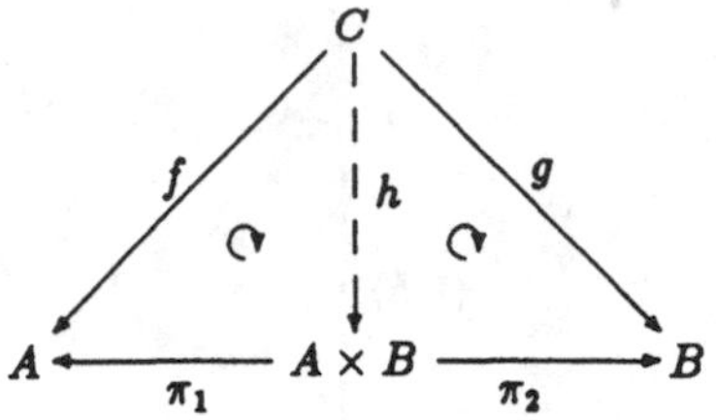

As we extended T-algebras to F, G-dialgebras, we can extend $\underline{\mu}\rho.\sigma$ into something more powerful: we use not only least fixed points but also greatest fixed points which correspond to final T-coalgebras; we allow a sequence of types instead of a single type when we take fixed points. The type structure of our lambda calculus is:

Definition 2.2: Let TVar be a set of type variables. We use $\rho, \nu, \ldots$ for the meta-variables of TVar. The set Type of types is defined by the following rules:

$$\frac{\rho \in \mathrm{TVar} \qquad \rho \in \Gamma}{\Gamma \vdash \rho \in \mathrm{Type}} \qquad \frac{\Gamma \vdash \sigma \in \mathrm{Type} \qquad \Gamma \vdash \tau \in \mathrm{Type}}{\Gamma \vdash \sigma \rightarrow \tau \in \mathrm{Type}}$$

$$\frac{\Gamma \cup \{\rho\} \vdash \sigma_i \in \mathrm{Type} \qquad \mathrm{Pos}(\rho, \sigma_i) \qquad (i = 1, \ldots, n)}{\Gamma \vdash \underline{\mu}\rho.(\sigma_1, \ldots, \sigma_n) \in \mathrm{Type}}$$

$$\frac{\Gamma \cup \{\rho\} \vdash \sigma_i \in \mathrm{Type} \qquad \mathrm{Pos}(\rho, \sigma_i) \qquad (i = 1, \ldots, n)}{\Gamma \vdash \overline{\mu}\rho.(\sigma_1, \ldots, \sigma_n) \in \mathrm{Type}}$$

where $\mathrm{Pos}(\rho, \sigma)$ is the predicate which is true when ρ occurs only positively in σ. Pos can be defined as follows together with the predicate $\mathrm{Neg}(\rho, \sigma)$ for negative occurrences of ρ in σ:

$$\mathrm{Pos}(\rho, \rho)$$

$$\mathrm{Pos}(\rho, \nu) \qquad\qquad \mathrm{Neg}(\rho, \nu)$$

$$\frac{\mathrm{Neg}(\rho, \sigma) \qquad \mathrm{Pos}(\rho, \tau)}{\mathrm{Pos}(\rho, \sigma \rightarrow \tau)} \qquad \frac{\mathrm{Pos}(\rho, \sigma) \qquad \mathrm{Neg}(\rho, \tau)}{\mathrm{Neg}(\rho, \sigma \rightarrow \tau)}$$

$$\frac{\mathrm{Pos}(\rho, \sigma_i) \qquad (i = 1, \ldots, n)}{\mathrm{Pos}(\rho, \underline{\mu}\nu.(\sigma_1, \ldots, \sigma_n))} \qquad \frac{\mathrm{Neg}(\rho, \sigma_i) \qquad (i = 1, \ldots, n)}{\mathrm{Neg}(\rho, \underline{\mu}\nu.(\sigma_1, \ldots, \sigma_n))}$$

$$\frac{\mathrm{Pos}(\rho, \sigma_i) \qquad (i = 1, \ldots, n)}{\mathrm{Pos}(\rho, \overline{\mu}\nu.(\sigma_1, \ldots, \sigma_n))} \qquad \frac{\mathrm{Neg}(\rho, \sigma_i) \qquad (i = 1, \ldots, n)}{\mathrm{Neg}(\rho, \overline{\mu}\nu.(\sigma_1, \ldots, \sigma_n))}$$

We use $\sigma, \tau, \ldots$ for the meta-variables of Type. ▯

The type $\underline{\mu}\rho.(\sigma_1, \ldots, \sigma_n)$ corresponds to the initial F, G-dialgebra where F and G are functors from C to C × ⋯ × C such that

$$F(A) \equiv \langle F_1(A), \ldots, F_n(A) \rangle \qquad \text{and} \qquad G(A) \equiv \langle A, \ldots, A \rangle$$

where $F_1, \ldots, F_n$ are functors corresponding to $\sigma_1, \ldots, \sigma_n$. On the other hand, $\overline{\mu}\rho.(\sigma_1, \ldots, \sigma_n)$ corresponds to the final G, F-dialgebra for the same F and G as above.

Note that we restricted type variables to occur only positively in $\underline{\mu}\rho.(\sigma_1, \ldots, \sigma_n)$ and $\overline{\mu}\rho.(\sigma_1, \ldots, \sigma_n)$. Therefore, we can have neither $\underline{\mu}\rho.(\rho \to \sigma)$ nor $\underline{\mu}\rho.(\rho \to \rho)$, but we can have $\underline{\mu}\rho.((\rho \to \sigma) \to \rho)$ if we want. Note also that we do not have any base types. We will show that various types we can define in this lambda calculus in section 3.

2.2 Terms and Their Types

Terms of our lambda calculus are defined as follows:

Definition 2.3: We have an enumerable set Var of variables and a set of terms is given by the following BNF expression.

$$L ::= x \mid \lambda x^{\sigma}.L \mid L_1 L_2 \mid C_{\underline{\mu}\rho.(\sigma_1,\ldots,\sigma_n),i} \mid J_{\underline{\mu}\rho.(\sigma_1,\ldots,\sigma_n),\tau} \mid$$
$$D_{\overline{\mu}\rho.(\sigma_1,\ldots,\sigma_n),i} \mid P_{\overline{\mu}\rho.(\sigma_1,\ldots,\sigma_n),\tau} \mid \sigma[L/\rho]$$

where C, J, D and P are constants and we use $x, y, z, \ldots$ for meta-variables for variables and $L, M, N, \ldots$ for meta-variables for terms. We may omit type subscripts or superscripts (e.g. $\underline{\mu}\rho.(\sigma_1, \ldots, \sigma_n)$ of $C_{\underline{\mu}\rho.(\sigma_1,\ldots,\sigma_n),i}$) if they are obvious. []

$C_{\underline{\mu}\rho.(\sigma_1,\ldots,\sigma_n),i}$ and $J_{\underline{\mu}\rho.(\sigma_1,\ldots,\sigma_n),\tau}$ are associated with $\underline{\mu}\rho.(\sigma_1, \ldots, \sigma_n)$. Remember that $\underline{\mu}\rho.(\sigma_1, \ldots, \sigma_n)$ is the initial F, G-dialgebra where F and G are

$$F(A) \equiv \langle \sigma_1[A/\rho], \ldots, \sigma_n[A/\rho] \rangle \qquad \text{and} \qquad G(A) \equiv \langle A, \ldots, A \rangle.$$

$C_{\underline{\mu}\rho.(\sigma_1,\ldots,\sigma_n),i}$ $(i = 1, \ldots, n)$ are morphisms such that

$$\langle \underline{\mu}\rho.(\sigma_1, \ldots, \sigma_n), \langle C_{\underline{\mu}\rho.(\sigma_1,\ldots,\sigma_n),1}, \ldots, C_{\underline{\mu}\rho.(\sigma_1,\ldots,\sigma_n),n} \rangle \rangle$$

is the initial F, G-dialgebra. Therefore, the type of $C_{\underline{\mu}\rho.(\sigma_1,\ldots,\sigma_n),i}$ is

$$C_{\underline{\mu}\rho.(\sigma_1,\ldots,\sigma_n),i} : \sigma_i[\underline{\mu}\rho.(\sigma_1, \ldots, \sigma_n)/\rho] \to \underline{\mu}\rho.(\sigma_1, \ldots, \sigma_n).$$

Given a term M of type $\sigma_i[\underline{\mu}\rho.(\sigma_1, \ldots, \sigma_n)/\rho]$, $C_{\underline{\mu}\rho.(\sigma_1,\ldots,\sigma_n),i} M$ constructs a term of $\underline{\mu}\rho.(\sigma_1, \ldots, \sigma_n)$. $C_{\underline{\mu}\rho.(\sigma_1,\ldots,\sigma_n),i}$ are constructors. In addition, $J_{\underline{\mu}\rho.(\sigma_1,\ldots,\sigma_n),\tau}$ is the mediating morphism such that for any morphisms $M_1, \ldots, M_n$ of type $\sigma_1[\tau/\rho] \to \tau, \ldots, \sigma_n[\tau/\rho] \to \tau$, respectively, $J_{\underline{\mu}\rho.(\sigma_1,\ldots,\sigma_n),\tau} M_1 \ldots M_n$ gives a unique morphism from $\underline{\mu}\rho.(\sigma_1, \ldots, \sigma_n)$ to τ such that the following digrams $(i = 1, \ldots, n)$ commute.

$$\begin{array}{ccc}
\sigma_i[\underline{\mu}\rho.(\sigma_1, \ldots, \sigma_n)/\rho] & \xrightarrow{C_{\underline{\mu}\rho.(\sigma_1,\ldots,\sigma_n),i}} & \underline{\mu}\rho.(\sigma_1, \ldots, \sigma_n) \\
\Big\downarrow \sigma_i[J_{\underline{\mu}\rho.(\sigma_1,\ldots,\sigma_n),\tau} M_1 \ldots M_n/\rho] & \circlearrowleft & \Big\downarrow J_{\underline{\mu}\rho.(\sigma_1,\ldots,\sigma_n),\tau} M_1 \ldots M_n \\
\sigma_i[\tau/\rho] & \xrightarrow[M_i]{} & \tau
\end{array}$$

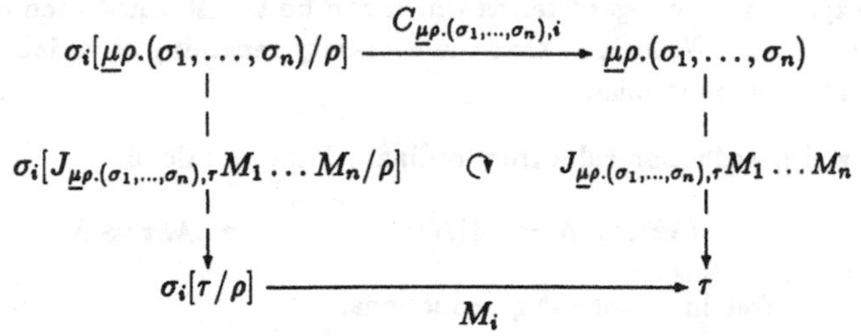

Therefore, the type of $J_{\underline{\mu}\rho.(\sigma_1,\ldots,\sigma_n),\tau}$ is

$$J_{\underline{\mu}\rho.(\sigma_1,\ldots,\sigma_n),\tau} : (\sigma_1[\tau/\rho] \to \tau) \to \cdots \to (\sigma_n[\tau/\rho] \to \tau) \to \underline{\mu}\rho.(\sigma_1, \ldots, \sigma_n) \to \tau.$$

As is well-known, $J_{\underline{\mu}\rho.(\sigma_1,\ldots,\sigma_n)}$ can be used to define primitive recursive functions.

Dually, $D_{\overline{\mu}\rho.(\sigma_1,\dots,\sigma_n),i}$ and $P_{\overline{\mu}\rho.(\sigma_1,\dots,\sigma_n),\tau}$ are associated with the type $\overline{\mu}\rho.(\sigma_1,\dots,\sigma_n)$.

$$\langle \overline{\mu}\rho.(\sigma_1,\dots,\sigma_n), \langle D_{\overline{\mu}\rho.(\sigma_1,\dots,\sigma_n),1},\dots,D_{\overline{\mu}\rho.(\sigma_1,\dots,\sigma_n),n}\rangle\rangle$$

gives the final G,F-dialgebra and $P_{\overline{\mu}\rho.(\sigma_1,\dots,\sigma_n),\tau}$ is its mediating morphism. Given a term M of type $\overline{\mu}\rho.(\sigma_1,\dots,\sigma_n)$, it can be decomposed into a term $D_{\overline{\mu}\rho.(\sigma_1,\dots,\sigma_n),i}M$ of type $\sigma_i[\overline{\mu}\rho.(\sigma_1,\dots,\sigma_n)/\rho]$ and $P_{\overline{\mu}\rho.(\sigma_1,\dots,\sigma_n),\tau}M_1\dots M_n$ can be used to construct a term of type $\overline{\mu}\rho.(\sigma_1,\dots,\sigma_n)$ from a term of type τ.

For a type σ with a free type variable ρ, a term $\sigma[M/\rho]$ denotes the result of applying the functor corresponding to σ to a term M.

Definition 2.4: Types of terms in our lambda calculus is given by the following rules:

$$\frac{x\in \mathrm{Var} \qquad x:\sigma\in\Gamma}{\Gamma\vdash x:\sigma}$$

$$\frac{\Gamma\cup\{x:\sigma\}\vdash M:\tau}{\Gamma\vdash \lambda x^{\sigma}.M:\sigma\to\tau} \qquad \frac{\Gamma\vdash M:\sigma\to\tau \qquad \Gamma\vdash N:\sigma}{\Gamma\vdash MN:\tau}$$

$$\Gamma\vdash C_{\underline{\mu}\rho.(\sigma_1,\dots,\sigma_n),i}:\sigma_i[\underline{\mu}\rho.(\sigma_1,\dots,\sigma_n)/\rho]\to\underline{\mu}\rho.(\sigma_1,\dots,\sigma_n)$$

$$\Gamma\vdash J_{\underline{\mu}\rho.(\sigma_1,\dots,\sigma_n),\tau}:(\sigma_1[\tau/\rho]\to\tau)\to\dots\to(\sigma_n[\tau/\rho]\to\tau)\to\underline{\mu}\rho.(\sigma_1,\dots,\sigma_n)\to\tau$$

$$\Gamma\vdash D_{\overline{\mu}\rho.(\sigma_1,\dots,\sigma_n),i}:\overline{\mu}\rho.(\sigma_1,\dots,\sigma_n)\to\sigma_i[\overline{\mu}\rho.(\sigma_1,\dots,\sigma_n)/\rho]$$

$$\Gamma\vdash P_{\overline{\mu}\rho.(\sigma_1,\dots,\sigma_n),\tau}:(\tau\to\sigma_1[\tau/\rho])\to\dots\to(\tau\to\sigma_n[\tau/\rho])\to\tau\to\overline{\mu}\rho.(\sigma_1,\dots,\sigma_n)$$

$$\frac{\mathrm{Pos}(\rho,\sigma) \qquad \Gamma\vdash M:\tau\to\tau'}{\Gamma\vdash\sigma[M/\rho]:\sigma[\tau/\rho]\to\sigma[\tau'/\rho]} \qquad \frac{\mathrm{Neg}(\rho,\sigma) \qquad \Gamma\vdash M:\tau\to\tau'}{\Gamma\vdash\sigma[M/\rho]:\sigma[\tau'/\rho]\to\sigma[\tau/\rho]}$$

□

2.3 Reduction rules

Let us consider the reduction rules for our typed lambda calculus. In the following we do not handle α conversions explicitly. We regard terms which can be transformed each other by α conversions are essentially the same. We assume that the necessary renaming of variables when substituting a variable by a term is done implicitly.

We have the β and η reduction rules from ordinary lambda calculi.

$$(\lambda x^{\sigma}.M)N\Rightarrow M[N/x] \qquad \lambda x^{\sigma}.Mx\Rightarrow M$$

where x needs to be free in M for the η reductions.

Since $\underline{\mu}\rho.(\sigma_1,\dots,\sigma_n)$ corresponds to the initial F,G-dialgebra for

$$F(A)\equiv\langle\sigma_1[A/\rho],\dots,\sigma_n[A/\rho]\rangle \qquad \text{and} \qquad G(A)\equiv\langle A,\dots,A\rangle,$$

for any type τ and any terms $M_i : \sigma[\tau/\rho] \to \tau$ $(i = 1, \ldots, n)$ there exists a unique morphism N which make the following diagrams $(i = 1, \ldots, n)$ commute.

$$\begin{array}{ccc}
\sigma_i[\underline{\mu}\rho.(\sigma_1, \ldots, \sigma_n)/\rho] & \xrightarrow{C_{\underline{\mu}\rho.(\sigma_1,\ldots,\sigma_n),i}} & \underline{\mu}\rho.(\sigma_1, \ldots, \sigma_n) \\
\Big\downarrow {\scriptstyle \sigma_i[N/\rho]} & \circlearrowright & \Big\downarrow {\scriptstyle N} \\
\sigma_i[\tau/\rho] & \xrightarrow[M_i]{} & \tau
\end{array}$$

N is given by the iterator as $J_{\underline{\mu}\rho.(\sigma_1,\ldots,\sigma_n),\tau}M_1 \ldots M_n$. From the commutativity, we have the following equality.

$$J_{\underline{\mu}\rho.(\sigma_1,\ldots,\sigma_n),\tau}M_1 \ldots M_n(C_{\underline{\mu}\rho.(\sigma_1,\ldots,\sigma_n),i}L) = M_i(\sigma[J_{\underline{\mu}\rho.(\sigma_1,\ldots,\sigma_n),\tau}M_1 \ldots M_n/\rho]L)$$

where L is a term of type $\sigma_i[\underline{\mu}\rho.(\sigma_1, \ldots, \sigma_n)/\rho]$. Reducing the number of constructors is a good strategy for normalizing terms, so we have a reduction rule of rewriting the left-hand side by the right-hand side.

$$J_{\underline{\mu}\rho.(\sigma_1,\ldots,\sigma_n),\tau}M_1 \ldots M_n(C_{\underline{\mu}\rho.(\sigma_1,\ldots,\sigma_n),i}L) \Rightarrow M_i(\sigma[J_{\underline{\mu}\rho.(\sigma_1,\ldots,\sigma_n),\tau}M_1 \ldots M_n/\rho]L)$$

When τ is $\underline{\mu}\rho.(\sigma_1, \ldots, \sigma_n)$ and M_i is $C_{\underline{\mu}\rho.(\sigma_1,\ldots,\sigma_n),i}$, the commutative diagram is

$$\begin{array}{ccc}
\sigma_i[\underline{\mu}\rho.(\sigma_1, \ldots, \sigma_n)/\rho] & \xrightarrow{C_{\underline{\mu}\rho.(\sigma_1,\ldots,\sigma_n),i}} & \underline{\mu}\rho.(\sigma_1, \ldots, \sigma_n) \\
\Big\downarrow {\scriptstyle \sigma_i[N/\rho]} & \circlearrowright & \Big\downarrow {\scriptstyle N} \\
\sigma_i[\underline{\mu}\rho.(\sigma_1, \ldots, \sigma_n)/\rho] & \xrightarrow[C_{\underline{\mu}\rho.(\sigma_1,\ldots,\sigma_n),i}]{} & \underline{\mu}\rho.(\sigma_1, \ldots, \sigma_n)
\end{array}$$

N should be the identity morphism, so we have the following reduction rule.

$$J_{\underline{\mu}\rho.(\sigma_1,\ldots,\sigma_n),\tau}C_{\underline{\mu}\rho.(\sigma_1,\ldots,\sigma_n),1} \cdots C_{\underline{\mu}\rho.(\sigma_1,\ldots,\sigma_n),n} \Rightarrow \lambda x^{\underline{\mu}\rho.(\sigma_1,\ldots,\sigma_n)}.x$$

The two reduction rules cannot exactly characterize $\underline{\mu}\rho.(\sigma_1, \ldots, \sigma_n)$ to be the initial F, G-dialgebra since the uniqueness condition is essentially a conditional equation, but as far as its computational aspect is concerned they seems to be enough.

Dually, for $\overline{\mu}\rho.(\sigma_1, \ldots, \sigma_n)$, we have the following two reduction rules:

$$D_{\overline{\mu}\rho.(\sigma_1,\ldots,\sigma_n),i}(P_{\overline{\mu}\rho.(\sigma_1,\ldots,\sigma_n),\tau}M_1 \ldots M_nL) \Rightarrow \sigma_i[P_{\overline{\mu}\rho.(\sigma_1,\ldots,\sigma_n),\tau}M_1 \ldots M_n/\rho](M_iL)$$

and

$$P_{\overline{\mu}\rho.(\sigma_1,\ldots,\sigma_n),\tau}D_{\overline{\mu}\rho.(\sigma_1,\ldots,\sigma_n),1} \cdots D_{\overline{\mu}\rho.(\sigma_1,\ldots,\sigma_n),n} \Rightarrow \lambda x^{\overline{\mu}\rho.(\sigma_1,\ldots,\sigma_n)}.x.$$

Finally, we have some reduction rules for functors $\sigma[M/\rho]$. Like the product functor $f \times g$ can be expressed by $\langle f \circ \pi_1, g \circ \pi_2 \rangle$, we transform $\sigma[M/\rho]$ into terms containing J, C, P and D. In the following rules, let M be a term of type $\tau \rightarrow \tau'$.

$$\rho[M/\rho] \Rightarrow M$$
$$\nu[M/\rho] \Rightarrow \lambda x^{\sigma}.x \qquad \text{(where } \rho \not\equiv \nu\text{)}$$
$$(\sigma \rightarrow \sigma')[M/\rho] \Rightarrow \lambda x^{\sigma[\tau/\rho]\rightarrow\sigma'[\tau/\rho]}.\lambda y^{\sigma[\tau'/rho]}.\sigma'[M/\rho](x\ (\sigma[M/\rho]\ y))$$
$$\underline{\mu}\nu.(\sigma_1,\ldots,\sigma_n)[M/\rho] \Rightarrow J_{\underline{\mu}\nu.(\sigma_1[\tau/\rho],\ldots),\underline{\mu}\nu.(\sigma_1[\tau'/\rho],\ldots)} M_1 \ldots M_n$$
$$\text{(where } M_i \text{ is } \lambda x^{\sigma_i[\tau/\rho][\underline{\mu}\nu.(\sigma_1[\tau'/\rho],\ldots)/\nu]}.C_{\underline{\mu}\nu.(\sigma_1[\tau/\rho],\ldots),i}(\sigma_i[\underline{\mu}\nu.(\sigma_1[\tau'/\rho],\ldots)/\nu][M/\rho]\ x))$$
$$\overline{\mu}\nu.(\sigma_1,\ldots,\sigma_n)[M/\rho] \Rightarrow P_{\overline{\mu}\nu.(\sigma_1[\tau'/\rho],\ldots),\overline{\mu}\nu.(\sigma_1[\tau/\rho],\ldots)} M_1 \ldots M_n$$
$$\text{(where } M_i \text{ is } \lambda x^{\overline{\mu}\nu.(\sigma_1[\tau/\rho],\ldots)}.\sigma_i[\overline{\mu}\nu.(\sigma_1[\tau/\rho],\ldots)/\nu][M/\rho](D_{\overline{\mu}\nu.(\sigma_1[\tau/\rho],\ldots),i}x))$$

We have some obvious propositions.

Proposition 2.5: For a term L of type σ, if there is a reduction $L \stackrel{*}{\Rightarrow} M$, M also has the type σ, where $\stackrel{*}{\Rightarrow}$ is the transitive version of $\Rightarrow$. ▯

Proposition 2.6: If ρ does not occur in σ, then $\sigma[M/\rho] \stackrel{*}{\Rightarrow} \lambda x^{\sigma}.x$ for any term M. ▯

This means that constant functors always give identities.

Proposition 2.7: For any type σ with a free variable ρ, $\sigma[\lambda x^{\tau}.x/\rho] \stackrel{*}{\Rightarrow} \lambda y^{\sigma[\tau/\rho]}.y$. ▯

This means that identities are always mapped to identities by σ, which is one of the conditions for σ being a functor.

Now, we have two important theorems about our reduction system: strong normalization and Church-Rosser. Because we only use primitive recursions, any term can be reduced to a normal term which cannot be reduced any more. In fact, any reduction leads to a normal term.

Theorem 2.8: (Strong Normalization Theorem) The reduction is strongly normalizing, that is, there is no infinite reduction sequence $L_1 \Rightarrow L_2 \Rightarrow L_3 \Rightarrow \cdots \Rightarrow L_n \Rightarrow \cdots$. ▯

Furthermore, any reduction leads to a *unique* normal term.

Theorem 2.9: (Church-Rosser Theorem) The reduction is Church-Rosser, that is, if $L \stackrel{*}{\Rightarrow} M$ and $L \stackrel{*}{\Rightarrow} N$, there exists a term K such that $M \stackrel{*}{\Rightarrow} K$ and $N \stackrel{*}{\Rightarrow} K$. ▯

Because we have the strong normalization theorem, the Church-Rosser theorem follows from the following lemma (see [6] proposition 13.1).

Lemma 2.10: If $L \Rightarrow M$ and $L \Rightarrow N$, there exists a term K such that $M \stackrel{*}{\Rightarrow} K$ and $N \stackrel{*}{\Rightarrow} K$.
Proof: All we have to do is to check any overlapping of two reduction rules. For example,

$$J_{\underline{\mu}\rho.(\sigma_1,\ldots,\sigma_n),\tau} M_1 \ldots M_n (C_{\underline{\mu}\rho.(\sigma_1,\ldots,\sigma_n),i} L) \Rightarrow M_i(\sigma[J_{\underline{\mu}\rho.(\sigma_1,\ldots,\sigma_n),\tau} M_1 \ldots M_n/\rho] L)$$

and

$$J_{\underline{\mu}\rho.(\sigma_1,\ldots,\sigma_n),\tau} C_{\underline{\mu}\rho.(\sigma_1,\ldots,\sigma_n),1} \ldots C_{\underline{\mu}\rho.(\sigma_1,\ldots,\sigma_n),n} \Rightarrow \lambda x^{\underline{\mu}\rho.(\sigma_1,\ldots,\sigma_n)}.x$$

overlap, that is, there a term to which both rules can be applied, but we can easily show that two resulting terms can be reduced to the same term.

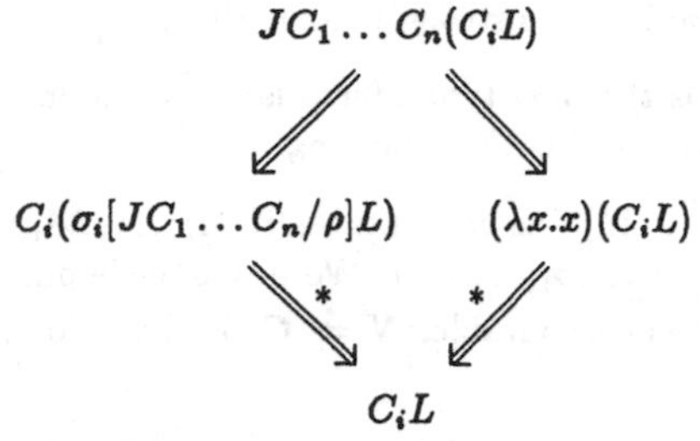

$C_i(\sigma_i[JC_1 \dots C_n/\rho]L) \stackrel{*}{\Rightarrow} C_iL$ follows from $JC_1 \dots C_n \Rightarrow \lambda x.x$ and proposition 2.7. We can check all the other overlappings similarly. []

The strong normalization theorem follows intuitively from the fact that we use only primitive recursion. In the reduction of $J_{\underline{\mu}\rho.(\sigma_1,\dots,\sigma_n),\tau}$ we reduce the number of constructors $C_{\underline{\mu}\rho.(\sigma_1,\dots,\sigma_n),i}$, whereas in the reduction of $P_{\overline{\mu}\rho.(\sigma_1,\dots,\sigma_n),\tau}$ we reduce the number of destructors $D_{\overline{\mu}\rho.(\sigma_1,\dots,\sigma_n),i}$. Therefore, there is no way we can continue reducing any terms. Formally, we prove the normalization theorem by Tait's computability method [6, 12]. First, we define the notion of *computable* terms by induction on types.

Definition 2.11: (Computability)

1. $M : \sigma \to \tau$ is computable if $MN : \tau$ is computable for any computable term $N : \sigma$
2. $M : \underline{\mu}\rho.(\sigma_1, \dots, \sigma_n)$ is computable if M is strongly normalizing and $M \stackrel{*}{\Rightarrow} C_{\underline{\mu}\rho.(\sigma_1,\dots,\sigma_n),i}L$ such that $L : \sigma_i[\underline{\mu}\rho.(\sigma_1, \dots, \sigma_n)/\rho]$ is computable.
3. $M : \overline{\mu}\rho.(\sigma_1, \dots, \sigma_n)$ is computable if M is strongly normalizing and

   $$M \stackrel{*}{\Rightarrow} P_{\overline{\mu}\rho.(\sigma_1,\dots,\sigma_n),\tau}N_1 \dots N_nL$$

   such that for any i

   $$D_{\overline{\mu}\rho.(\sigma_1,\dots,\sigma_n),i}(P_{\overline{\mu}\rho.(\sigma_1,\dots,\sigma_n),\tau}N_1 \dots N_nL) : \sigma_i[\overline{\mu}\rho.(\sigma_1, \dots, \sigma_n)/\rho]$$

   is computable. []

Note that the definition is inductive in another sense as well. The definition of computable terms of $\underline{\mu}\rho.(\sigma_1, \dots, \sigma_n)$ depends on itself. We take the least fixed point of this self recursive definition, i.e. starting with the empty set of computable terms, we increase the set by applying the definition.

$$\emptyset \subseteq S_1 \subseteq S_2 \subseteq \dots \subseteq S_n \subseteq S_{n+1} \subseteq \dots$$

where S_{n+1} is obtained by applying the definition to S_n. Since the definition is monotonic, there is a least fixed point. This process also provides the measure function over the computable terms which assigns for each computable term M an ordinal α where S_α is the first set which contains M. On the other hand, we take the greatest fixed point for the definition of computable terms of $\overline{\mu}\rho.(\sigma_1, \dots, \sigma_n)$.

It is easy to see that

Lemma 2.12: If M is computable, M is strongly normalizing. []

Therefore, all we have to prove is that any typed term is computable. This is proved by structural induction. One of the lemmas we need is the following.

Lemma 2.13: $J_{\underline{\mu}\rho.(\sigma_1,\ldots,\sigma_n),\tau}M_1 \ldots M_n N$ is computable if $M_1, \ldots, M_n$ and N are computable
Proof: N is a computable term of $\underline{\mu}\rho.(\sigma_1, \ldots, \sigma_n)$. We prove the lemma by induction on the ordinal associated with N. Because N is computable, $N \stackrel{*}{\Rightarrow} C_i L$. Therefore, any reduction sequence of $JM_1 \ldots M_n N$ should be

$$JM_1 \ldots M_n N \stackrel{*}{\Rightarrow} JM_1' \ldots M_n'(C_i L) \Rightarrow M_i'(\sigma_i[JM_1' \ldots M_n'/\rho]L).$$

Because the functor σ_i only applies $JM_1' \ldots M_n'$ to a term of $\underline{\mu}\rho.(\sigma_1, \ldots, \sigma_n)$ in L whose ordinal is smaller than that of N (formally, we have to prove this), from induction hypothesis $\sigma_i[JM_1' \ldots M_n'/\rho]L$ is computable and $M_i'(\sigma_i[JM_1' \ldots M_n'/\rho]L)$ is computable as well. Therefore, $JM_1 \ldots M_n N$ is computable. []

We have the similar lemmas for $P_{\overline{\mu}\rho.(\sigma_1,\ldots,\sigma_n),\tau}$ and $\lambda x^\sigma.M$. From these lemmas,

Lemma 2.14: Any typed term is computable. []

Hence, we have proved the strong normalization theorem.

Although we have the strong normalization theorem and the Church-Rosser theorem and, therefore, the equality of two terms is decidable, this does not mean that the equality of two computable (or more weakly, primitive recursive) functions is decidable. The equality of lambda terms are defined by "two lambda terms reduce to the same lambda term", but the reduction rules do not capture all the equality of computable (or primitive recursive) functions. Remember that we defined $\underline{\mu}\rho.(\sigma_1, \ldots, \sigma_n)$ to be the initial F, G-dialgebra and we put the commutativity of diagrams, but we did not put the uniqueness, or we could not put it. Therefore, the equality derived from the reduction is weaker. In other words, if we regard the reduction rules as the operational semantics of our lambda calculus and the initial and final F, G-dialgebras as the denotational semantics, the denotational semantics is not fully abstract.

3 Examples

In this section, we show some types which we can define in our lambda calculus.

Example 3.1: The type corresponding to the initial object is defined by $\emptyset \equiv \underline{\mu}\rho.()$. There are no constructors and $J_{\emptyset,\sigma} : \emptyset \to \sigma$ gives the unique morphism from $\emptyset$ to any type σ. Dually, the type corresponding to the terminal object is defined by $1 \equiv \overline{\mu}\rho.()$. There are no destructors and $P_{1,\sigma} : \sigma \to 1$ gives the unique morphism from σ to 1. There is an element in 1. In fact, there is only one element in 1 in some sense. Let us write $*$ for $P_{1,1\to1}\lambda x^1.x$ which is an element of 1. Actually, we can use any element of 1. The choice does not affect the computation. []

Example 3.2: The product of two types, σ and τ can be defined as $\sigma \times \tau \equiv \overline{\mu}\rho.(\sigma, \tau)$. We have two projections.

$$\pi_1 \equiv D_{\sigma\times\tau,1} \quad : \sigma \times \tau \to \sigma \qquad\qquad \pi_2 \equiv D_{\sigma\times\tau,2} \quad : \sigma \times \tau \to \tau$$

If M is a term of type σ and N is a term of type τ, we can define a term $\langle M, N\rangle$ of type $\sigma \times \tau$.

$$\langle M, N\rangle \equiv P_{\sigma\times\tau}(\lambda x^1.M)(\lambda x^1.N)* \quad : \sigma \times \tau$$

We have the following reduction.

$$\pi_1\langle M,N\rangle \equiv D_{\sigma\times\tau,1}(P_{\sigma\times\tau,1}(\lambda x.M)(\lambda x.N)*) \Rightarrow (\lambda x.x)((\lambda x.M)*) \overset{*}{\Rightarrow} M$$

Similarly, we can show that $\pi_2\langle M,N\rangle \overset{*}{\Rightarrow} N$. However, we do not have

$$\langle \pi_1 M, \pi_2 M\rangle \overset{*}{\Rightarrow} M$$

because we did not code the uniqueness condition in our reduction rules. We could have coded products specially, but then, we lose the generality of our lambda calculus. ▯

Example 3.3: Dually, the coproduct of σ and τ is defined as $\sigma+\tau \equiv \underline{\mu}\rho.(\sigma,\tau)$. Two injections are defined as follows.

$$\iota_1 \equiv C_{\sigma+\tau,1} \quad : \sigma \to \sigma+\tau \qquad\qquad \iota_2 \equiv C_{\sigma+\tau,2} \quad : \tau \to \sigma+\tau$$

$J_{\sigma+\tau,\nu}$ satisfies the following reductions.

$$J_{\sigma+\tau,\nu}MN(\iota_1 L) \equiv J_{\sigma+\tau,\nu}MN(C_{\sigma+\tau,1}L) \Rightarrow M((\lambda x.x)L) \Rightarrow ML$$

$$J_{\sigma+\tau,\nu}MN(\iota_2 L) \overset{*}{\Rightarrow} NL \qquad ▯$$

Example 3.4: Let us define the natural numbers in our lambda calculus. The type is defined by

$$\omega \equiv \underline{\mu}\rho.(1,\rho).$$

This definition is closely connected to the definition in domain theory where the domain of natural numbers is defined as the least fixed point of $N \cong 1+N$. Our $\underline{\mu}$ is the least fixed point operator. The only difference is that we use a sequence $(1,\rho)$ instead of coproduct $1+\rho$. Our approach is in this way connected to algebraic specification methods where the type of natural numbers is defined as the initial algebra of one constant and one operator. The elements are generated by zero and the successor function which are defined in our lambda calculus as follows:

$$0 \equiv C_{\omega,1}* \quad : \omega \qquad \mathrm{succ} \equiv C_{\omega,2} \quad : \omega \to \omega$$

$J_{\omega,\sigma}$ gives us almost the ordinary iterator but its type is

$$J_{\omega,\sigma} \quad : (1\to\sigma)\to(\sigma\to\sigma)\to\omega\to\sigma.$$

We can define the ordinary one by this $J_{\omega,\sigma}$ as follows.

$$\tilde{J}_\sigma \equiv \lambda x.\lambda y.\lambda n.J_{\omega,\sigma}(\lambda z.x)yn \quad : \sigma\to(\sigma\to\sigma)\to\omega\to\sigma$$

It satisfies the usual reductions:

$$\tilde{J}_\sigma MN0 \overset{*}{\Rightarrow} J_{\omega,\sigma}(\lambda z.M)N(C_{\omega,1}*) \Rightarrow (\lambda z.M)((\lambda x.x)*) \overset{*}{\Rightarrow} M$$

and

$$\tilde{J}_\sigma MN(\mathrm{succ}L) \overset{*}{\Rightarrow} J_{\omega,\sigma}(\lambda z.M)N(C_{\omega,2}L) \Rightarrow b(J_{\omega,\sigma}(\lambda z.M)NL) \approx N(\tilde{J}_\sigma MNL)$$

where $\approx$ is the equivalence relation generated by $\overset{*}{\Rightarrow}$. Using $\tilde{J}_\sigma$, we can define all the primitive recursive functions. For example, the addition function can be define as

$$\mathrm{add} \equiv \lambda n.\lambda m.\tilde{J}_\omega m\ \mathrm{succ}\ n \quad : \omega\to\omega\to\omega.$$

We can demonstrate, for example, add(succ zero)(succ zero) $\overset{*}{\Rightarrow}$ succ(succ zero). ▯

Example 3.5: As [11] and [13], we can define the type for ordinals by $\Omega \equiv \underline{\mu}\rho.(1, \omega \to \rho)$. We only check whether our definition of the iterator coincides with the ordinary one.

$$\Omega \equiv \underline{\mu}\rho.(1, \omega \to \rho)$$

$$0_\Omega \equiv C_{\Omega,1}* \quad : \Omega$$

$$\mathrm{sup} \equiv C_{\Omega,2} \quad : (\omega \to \Omega) \to \omega$$

$$J_{\Omega,\sigma} \quad : (1 \to \sigma) \to ((\omega \to \sigma) \to \sigma) \to \Omega \to \sigma$$

$$J_{\Omega,\sigma}(\lambda x.M)N0_\Omega \Rightarrow (\lambda x.a)((\lambda x.x)*) \overset{*}{\Rightarrow} M$$

$$\begin{aligned} J_{\Omega,\sigma}(\lambda x.M)N(\mathrm{sup}\, L) &\Rightarrow N((\omega \to \rho)[J_{\Omega,\sigma}(\lambda x.M)N/\rho]L) \\ &\Rightarrow N((\lambda y.\lambda z.J_{\Omega,\sigma}(\lambda x.M)N(yz))L) \Rightarrow b(\lambda z.J_{\Omega,\sigma}(\lambda x.M)N(Lz)) \end{aligned} \quad \square$$

Example 3.6: Finally, the type for finite lists can be defined by $L_\sigma \equiv \underline{\mu}\rho.(1, \sigma \times \rho)$ with

$$\mathrm{nil} \equiv C_{L_\sigma,1}* \quad : L_\sigma \qquad\qquad \mathrm{cons} \equiv C_{L_\sigma,2} \quad : \sigma \times L_\sigma \to L_\sigma$$

$$J_{L_\sigma,\tau} \quad : (1 \to \tau) \to (\sigma \times \tau \to \tau) \to L_\sigma \to \tau$$

whereas the type for infinite lists can be defined by $I_\sigma \equiv \overline{\mu}\rho.(\sigma, \rho)$ with

$$\mathrm{head} \equiv D_{I_\sigma,1} \quad : I_\sigma \to \sigma \qquad\qquad \mathrm{tail} \equiv D_{I_\sigma,2} \quad : I_\sigma \to I_\sigma$$

$$P_{I_\sigma,\tau} \quad : (\tau \to \sigma) \to (\tau \to \tau) \to \tau \to I_\sigma$$

$$\mathrm{head}(P_{I_\sigma,\tau}MNL) \overset{*}{\Rightarrow} ML \qquad\qquad \mathrm{tail}(P_{I_\sigma,\tau}MNL) \overset{*}{\Rightarrow} P_{I_\sigma,\tau}MN(NL).$$

The type for infinite lists is quite exciting to play with. The following lambda term gives us the infinite sequence of zeros

$$\mathrm{inf} \equiv P_{I_\omega,\omega}(\lambda x^\omega.x)(\lambda x^\omega.x)\mathrm{zero},$$

whereas the following gives us the infinite increasing sequence $0, 1, 2, 3, \ldots$.

$$\mathrm{inc} \equiv P_{I_\omega,\omega}(\lambda x^\omega.x)\mathrm{succ\ zero}$$

We can merge two infinite sequences by choosing elements alternatively by the following function.

$$\mathrm{comb} \equiv P_{I_\sigma,I_\sigma \times I_\sigma}(\mathrm{head} \circ \pi_1)(P_{I_\sigma \times I_\sigma, I_\sigma \times I_\sigma}\pi_2(\mathrm{tail} \circ \pi_1))$$

where $M \circ N$ is $\lambda x.(M(Nx))$. We can demonstrate, for example, that

$$\mathrm{head}(\mathrm{tail}(\mathrm{tail}(\mathrm{tail}(\mathrm{comb\ inf\ inc})))) \overset{*}{\Rightarrow} \mathrm{succ}(\mathrm{succ\ zero}) \quad \square$$

We could give many other examples: boolean, trees, automata, co-natural numbers, ...

4 Comparison with Other Lambda Calculi

While writing this paper, the author was communicated to [8, 9] where recursive types are introduced into first-order and second-order typed lambda calculi. They use least fixed points and greatest fixed points as we do, but their recursion combinator R has a different type from ours.

$$\frac{M : (\rho \to \tau) \to \sigma \to \tau}{R_{\sigma,\tau}(M[\underline{\mu}\rho.\sigma/\rho]) : \underline{\mu}\rho.\sigma \to \tau}$$

The author cannot give a clear connection between our iterator and theirs. In addition, they take fixed points over a single type expression and, therefore, they need some basic type constructors like 1 and +, whereas in our lambda calculus there are no basic type constructors.

Although we discarded the coproduct and product type constructors from basic type constructors, we still have one basic type constructor, namely the arrow type constructor $\sigma \to \tau$. Since typed lambda calculi are all about arrow types, it seems impossible to start calculi without it, but from a category theoretic point of view, the arrow type constructor can be defined as the right adjoint functor of the product type constructor so it might possible to start calculi without the arrow type constructor. In [4], the author showed that the arrow typed constructor can be defined by F, G-dialgebras, but how it can be defined in lambda calculi still has to be investigated.

The second order lambda calculus can be started without basic type constructors and it has been shown that recursive types which can be defined by least fixed points of type expressions can be defined in the calculus. The coding of recursive types is a generalization of the coding of Church numerals in untyped lambda calculus. The author does not know whether it is possible to code up greatest fixed points as well.

5 New ML?

We might say that ML is based on a simply typed lambda calculus as we might say that LISP is based on the untyped lambda calculus. The type structure of ML depends on the version of ML we are talking about. If we are talking about the original ML developed with LCF [3], it has some base types, product, disjoint sum, integer, etc. , and has ability to introduce new types via recursively defined type equations. For example, the data type for binary trees whose leaves are integers is defined as

```
absrectype btree = int + (btree # btree)
     with leaf n = absbtree(inl n)
     and node(t1,t2) = absbtree(inr(t1,t2))
     and isleaf t = isl(repbtree t)
     and leafvalue t = outl(repbtree t)
     and left t = fst(outr(repbtree t))
     and right t = snd(outr(repbtree t));;
```

Here, we need the coproduct type constructor '+' as a primitive. We cannot do without it, whereas 'int' can be defined in terms of others primitives (ML has it as a primitive type just because of efficiency).

At the next evolution of ML which yielded the current Standard ML [5], we discovered that the coproduct type constructor is no longer needed as a primitive. Standard ML has a 'datatype' declaration mechanism by which the coproduct type constructor can be defined as follows:

```
datatype 'a + 'b = inl of 'a | inr of 'b;
```

A datatype declaration lists the constructors of the defining type. An element of `''a + 'b'` can be obtained by either applying '`inl`' to an element of '`'a`' or applying '`inr`' to an element of '`'b`'. Data type declarations exactly correspond to our $\underline{\mu}\rho.(\sigma_1,\ldots,\sigma_n)$.

`datatype 'a + 'b =` `inl of 'a | inr of 'b`	$\Longleftrightarrow$	$\sigma+\tau \equiv \underline{\mu}\rho.(\sigma,\tau)$
`inl`	$\Longleftrightarrow$	$C_{\sigma+\tau,1}$
`inr`	$\Longleftrightarrow$	$C_{\sigma+\tau,2}$

We can define the data type for binary trees in Standard ML as follows.

```
datatype btree = leaf of int | node of btree * btree;
```

The symbol '`|`' is just like '+', but we shifted from the object level of the language to the syntax level. Note that we no longer need the separate definition of '`leaf`' or '`node`'. We can define the other functions using `case` statements.

```
exception btree;

fun isleaf t = case t of
                 leaf _ => true
               | node _ => false;

fun leafvalue t = case t of
                    leaf n => n
                  | node _ => raise btree;

fun left t = case t of
               leaf _ => raise btree
             | node(t1,t2) => t1;

fun right t = case t of
                leaf _ => raise btree
              | node(t1,t2) => t2;
```

We get rid of the coproduct type constructor from the primitives, but Standard ML still needs the product type constructor. From a category theoretic point of view, we can sense asymmetry in the type structure of Standard ML. Let us remember that our lambda calculus needs neither the coproduct type constructor nor the product type constructor as a primitive. We should be able to introduce the symmetry of our lambda calculus into ML. Let us imagine the next stage of the ML evolution and define Symmetric ML.

	Primitives	Declaration Mechanism
ML	`->`, `unit`, `#`, `+`	`abstype`
Standard ML	`->`, `unit`, `*`	`datatype`
Symmetric ML	`->`	`datatype`, `codatatype`
CPL in [4]		`left object`, `right object`

ML Evolution

Remember that datatype declarations correspond to $\underline{\mu}\rho.(\sigma_1,\ldots,\sigma_n)$. We list constructors for types. In order to get rid of the product type constructor from primitives, we should have a declaration mechanism which corresponds to $\overline{\mu}\rho.(\sigma_1,\ldots,\sigma_n)$. Its syntax is

```
codatatype TypeParam TypeId =
      Id is TypeExp & ... & Id is TypeExp;
```

A codatatype declaration introduces a type by listing its destructors. The product type constructor can now be defined as follows:

```
codatatype 'a * 'b = fst is 'a & snd is 'b;
```

where '`fst : 'a * 'b -> 'a`' gives the projection function to the first component and '`snd : 'a * 'b -> 'b`' gives the projection function to the second component. If the declaration is recursive, we do not take the initial fixed point of the type equation but the final fixed point. This is firstly because of symmetry and secondly because the initial fixed points are often trivial. Because of this, we can define infinite objects by codatatype declarations. For example, the following declaration gives us the data type for infinite lists.

```
codatatype 'a inflist = head is 'a & tail is 'a inflist;
```

If we took the initial fixed point, we would get the empty data type. The definition is exactly corresponds to $I_\sigma \equiv \overline{\mu}\rho.(\sigma,\rho)$.

Obviously we have destructors for co-data types because we declare them, but how can we construct data for co-data types? We had **case** statements for data types, so we have 'merge' statements as dual. Its syntax is

```
merge Destructor <= Exp & ... & Destructor <= Exp
```

For example, the function '`pair`' which makes a pair of given two elements can be defined as follows.

```
fun pair(x,y) = merge fst <= x & snd <= y;
```

As a more complicated example, we might define a function which combines two infinite lists together.

```
fun comb(l1,l2) = merge head <= head l1
                      & tail <= comb(l2,tail l1);
```

As we use pattern matching to declare functions over data types, we can also use it to declare functions over co-data types. For example, an alternative definition of 'comb' may be

```
fun head comb(l1,_) = head l1
  & tail comb(l1,l2) = comb(l2,tail l1);
```

Conclusions

We have introduced a simply typed lambda calculus with categorical type constructors and demonstrated that we can define various types which had been defined as primitives before. Therefore, our normalization theorem covers the normalization theorems for other simply typed lambda calculi: a typed lambda calculus with natural numbers, a typed lambda calculus with ordinals, and so on.

The lambda calculus we presented in this paper is a direct derivation of author's work on a Categorical Programming language [4] where a functional programming language CPL has been introduced. CPL is a categorical-combinator-style programming language which has a uniform categorical datatype declaration mechanism. CPL is more general than the lambda calculus we presented here in a sense that CPL does not need $\rightarrow$ to be a primitive type constructor. It can define it as the right adjoint of the product functor. It seems that the difference comes out from the fact that the category theory distinguish morphisms from elements of exponential types, whereas lambda calculi not. In lambda calculi (or functional programming languages based on lambda calculi), functions are always treated as closures.

An experimental version of CPL has been implemented. Whether the codatatype declaration mechanism will be adopted to ML or not remains to be seen, but the author believes that it is an elegant answer to lazy types in ML.

The connection between initial fixed points and final fixed points is quite interesting to investigate. For example, the type of natural numbers as the initial fixed point of $N \cong 1 + N$ is associated with primitive recursion, whereas the type of natural number as the final fixed point of the same equation (we call it co-natural number object) is associated with general recursion.

Neither the lambda calculus we defined here nor CPL has not yet been mixed with dependent types. This has to be investigated in the future.

Acknowledgements

The author would like to thank Furio Honsell who led me to the world of lambda calculi from the world of category theory.

References

[1] Arbib, M. A. and Manes, E. G.: The Greatest Fixed Points Approach to Data Types. In *proceedings of Third Workshop Meeting on Categorical and Algebraic Methods in Computer Science and System Theory*, Dortmund, West Germany, 1980.

[2] Curien, P-L.: *Categorical Combinators, Sequential Algorithms and Functional Programming.* Research Notes in Theoretical Computer Science, Pitman, 1986.

[3] Gordon, M. J., Milner, A. J. and Wordsworth, C. P.: *Edinburgh LCF. Lecture Notes in Computer Science*, Volume 78, 1979.

[4] Hagino, T.: *Category Theoretic Approach to Data Types.* Ph. D. thesis, University of Edinburgh, 1987.

[5] Harper, R., MacQueen, D. and Milner, R.: *Standard ML.* LFCS Report Series, ECS-LFSC-86-2. Department of Computer Science, University of Edinburgh, 1986.

[6] Lambek, J. and Scott, P. J.: *Introduction on Higher-Order Categorical Logic. Cambridge Studies in Advanced Mathematics*, Volume 7, 1986.

[7] Lehmann, D. and Smyth, M.: Algebraic Specification of Data Types – A Synthetic Approach –. *Mathematical System Theory*, Volume 14, pp. 97–139, 1981.

[8] Mendler N. P.: *First- and Second-Order Lambda Calculi with Recursive Types.* Technical Report TR 86-764, Department of Computer Science, Cornell University, 1986.

[9] Mendler N. P.: *Recursive Types and Type Constraints in Second-Order Lambda Calculus.* 1986.

[10] Smyth, M. B. and Plotkin, G. D.: The Category-Theoretic Solution of Recursive Domain Equations. *SIAM Journal of Computing*, Volume 11, 1982.

[11] Stenlund, S.: *Combinators, λ-Terms and Proof Theory*. D. Reidel, Dordrecht, 1972.

[12] Tait, W.: Intentional Interpretation of Functionals of Finite Type I. *Journal of Symbolic Logic*, 32, pp. 198–212, 1967.

[13] Troelstra, A. S.: *Mathematical Investigation of Intuitionistic Arithmetic and Analysis. Lecture Notes in Mathematics*, Volume 344, Springer-Verlag, 1973.

Final Algebras, Cosemicomputable Algebras, and Degrees of Unsolvability

Lawrence S. Moss[1,2]
Center for the Study of Language and Information

José Meseguer[1,3]
Joseph A. Goguen[1,3]
SRI International
and
Center for the Study of Language and Information

Abstract This paper studies some computability notions for abstract data types, and in particular compares cosemicomputable many-sorted algebras with a notion of finality to model minimal-state realizations of abstract (software) machines. Given a finite many-sorted signature Σ and a set V of visible sorts, for every Σ-algebra A with co-r.e. behavior and nontrivial, computable V-behavior, there is a finite signature extension Σ' of Σ (without new sorts) and a finite set E of Σ'-equations such that A is isomorphic to a reduct of the final (Σ', E)-algebra relative to V. This uses a theorem due to Bergstra and Tucker [3]. If A is computable, then A is also isomorphic to the reduct of the initial (Σ', E)-algebra. We also prove some results on congruences of finitely generated free algebras. We show that for every finite signature Σ, there are either countably many Σ–congruences on the free Σ-algebra or else there is a continuum of such congruences. There are several necessary and sufficient conditions which separate these two cases. We introduce the notion of the Turing degree of a minimal algebra. Using the results above prove that there is a fixed one-sorted signature such that for every r.e. degree $\mathbf{d}$, there is a finite set E of Σ-equations such the initial (Σ, E)-algebra has degree $\mathbf{d}$. There is a two-sorted signature Σ_0 and a single visible sort such that for every r.e. degree $\mathbf{d}$ there is a finite set E of Σ-equations such that the initial (Σ, E, V)-algebra is computable and the final (Σ, E, V)-algebra is cosemicomputable and has degree $\mathbf{d}$.

1 Introduction

Abstract data types can be specified equationally. For computer science purposes, the abstract data types in question should have appropriate computability properties. The nicest possibility is that equality between data elements be *decidable*; this leads to the notion of a **computable** abstract data type. A second possibility is that the data type be **semicomputable**; i.e., there is a semi-decision procedure for equality of data items. The initial algebra associated to an arbitrary finite set of

[1]The research reported here was supported in part by a gift from the System Development Foundation.
[2]Current address: Mathematics Department, University of Michigan, Ann Arbor MI 48109.
[3]The research reported here was supported in part by Office Of Naval Research Contracts N00414-85-C-0417 and N0014-86-C-0450.

equations is always semicomputable, and there are well known examples of such algebras that are not computable. A third possibility is that *inequality* between two data elements be semidecidable, i.e., an algorithm can be given such that if two data elements are different, we will find out after a finite number of steps, but if they are equal we may wait forever for an answer; such data types are called **cosemicomputable**.

Cosemicomputable algebras appear naturally associated with a common use of abstract data types in many programming languages. This begins of course with Simula, where the abstract data types ideas first arose, but it is present in the entire object-oriented programming tradition. According to that use, a data type is viewed as a *machine* with an internal state and a visible behavior. For instance, the elements of stack are visible, by performing stack operations, but the stack itself is a hidden state. This is of course an automaton-theoretic notion of data type. All that matters from this point of view is the *visible behavior* exhibited by the data type. Unlike initial algebra semantics where 'abstract' in 'abstract data type' means up to unique isomorphism, the relevant concept here is not that of *isomorphism* (there can be many nonisomorphic implementations of the same behavior) but rather the automaton theoretic concept of *behavioral equivalence.* Two algebras are behaviorally equivalent if their behavior relative to a designated set of visible sorts is isomorphic. Meseguer and Goguen [12] used the term **abstract machine** to designate a class of behaviorally equivalent algebras; intuitively this class represents all the implementations that are possible for a given behavior. They also showed that the initial and final realization functors of classical automaton theory (as in Goguen [6]), generalize without restriction for algebras on an arbitrary signature.

Among all the possible implementations, or realizations, of a behavior, the final or 'minimal' realization corresponds to identifying any two internal states that are behaviorally indistinguishable; abstractly this is characterized by the fact that such a realization is a final object in the class of all (reachable) algebras behaviorally equivalent to it. A set of equations can be used in this context to specify an abstract machine and one can then extract from it the final realization. Indeed, a set E of equations defines a corresponding initial algebra $T_{\Sigma,E}$ and then a set V of visible sorts defines the abstract machine of all possible implementations of the V-behavior of such an initial algebra. Thus, the pair (E, V) specifies as well the final algebra $F_{\Sigma',E,V}$ behaviorally equivalent to $T_{\Sigma,E}$. This is the *final algebra semantics* of Giarrantana, Gimona and Montanari [5] and Wand [15]; both papers build on the work of Guttag [8] (who calls the hidden sort the 'type of interest') characterizing algebras behaviorally equivalent to $T_{\Sigma,E}$ by his 'sufficient completeness' and 'consistency' conditions.

Although the final algebra is the implementation having fewest states, it need not be the most efficient implementation. In fact, if $T_{\Sigma,E}$ is computable (or even under a much weaker hypothesis), the final algebra is cosemicomputable, and we shall see in detail that it need not be computable. Perhaps the most important role of final algebras is a conceptual one, since it provides a notion of an *extensional* or 'fully abstract' model associated to a given behavior. Internal states may correspond for instance to names for functions definable in a given language, with behavior being function evaluation; then the final algebra interpretation will be extensional, i.e., two states of that final algebra will be equal iff they correspond to the same function. Since only behavior matters, one can have a computable implementation of a given behavior (e.g., for the language in question) but think of it extentionally, as if one had the (generally cosemicomputable) extensional model of that behavior given by the final algebra.

Cosemicomputable algebras are also naturally associated with a different equational specification method. The two methods specify final algebras in *different categories*; also their basic intuitions and suitability for software specification are radically different.

The other notion of finality (called BT-finality in this paper) was proposed by Bergstra and Tucker, and formalizes the intuition of *consistency.* Given a set E of equations one looks for an algebra A that satisfies not only E, but also any other equation $t = t'$ consistent with E. By "consistent with E" one means that when $t = t'$ is added to E, the corresponding initial algebra is nontrivial, i.e., does not identify all data elements. Such an A is then a final algebra in the category of all nontrivial quotients of the initial algebra defined by E. Unfortunately, the requirement that the lattice of quotients of an algebra have only one atom above the trivial quotient is very strong, and in general such an algebra A does not exist, and it is quite a delicate matter to know when it does.

For example, consider the case of a signature Σ with a single sort **Nat** together with a constant symbol 0 and a unary function s. The initial algebra is the natural numbers with a constant 0 and the usual successor. Now consider the case when E is empty. Then, e.g., $s^2(0) = 0$ is consistent with E (since it is true in the naturals mod 2) as is $s^3(0) = 0$. If both of these equations held in a single minimal algebra, then that algebra is trivial because $0 = s^3(0) = s(s^2(0)) = s(0)$. This makes BT-finality of limited use as a software specification method, since in general the BT-final algebra of a set of equations does not exist, and it is quite a delicate matter to know when it does. However, results and techniques developed by Bergstra and Tucker to study the computability of BT-final algebras are used in this paper.

When it does exist, the BT final algebra specified by a finite set of equations is cosemicomputable, and this is another link between BT finality and the finality which we study.

In this paper, we are interested in the precise relation between the algebraic and recursion theoretic concepts discussed above. For computable and semicomputable algebras the known results are as follows: Bergstra and Tucker [2] have shown that every computable data type has a finite equational specification as an initial algebra. (One is allowed to add a finite number of 'hidden functions' to the original set of function symbols. Majster [10] has shown these are necessary for some results.) In addition, they have shown in [4] that the equations can always be chosen to be confluent and terminating as rewrite rules. They studied semicomputable algebras and proved a partial converse in [2], namely that allowing an additional sort, all semicomputable algebras have an initial algebra specification by a finite set of equations. Turning to cosemicomputable algebras, Bergstra and Tucker[3] proved that any such algebra can be specified as the BT-final algebra associated to a finite set of conditional equations. They also have proved that any computable algebra has a finite equational specification that specifies it as both its initial and BT-final algebra.

In Meseguer and Goguen [12], it was conjectured that any cosemicomputable algebra having a nonunit computable V-behavior had a final algebra specification by means of a finite set of equations, provided that additional function symbols were allowed. We prove that conjecture here, making use of a theorem of Bergstra and Tucker slightly extend to suit our purposes. We also prove an associated conjecture on computable algebras, namely that for any computable algebra there is a finite set of equations specifying it under both initial and final algebra semantics, again with the help of auxiliary hidden functions.

Finally, we consider the relationship between final algebras and degrees of unsolvability. More generally, we look first at arbitrary quotients of the initial algebra and ask:

1. How many are there?

2. Do they realize all the degrees of unsolvability?

3. Do initial and final algebras realize all the r.e. degrees?

We answer (1) and (2) by giving a condition on finite signatures of operations such that: (i) when the condition is satisfied there is a continuum of quotients whose congruences realize all the degrees of unsolvability; (ii) when the condition fails there are only countably many quotients, all of them computable. Our proof does not assume the continuum hypothesis. We then answer (3) by exhibiting a simple signature Σ_0 and a fixed set V of sorts such that for every r.e. degree there is some finite set E of Σ equations such that the final Σ, E-algebra has the given degree. We also show that every r.e. degree also arises as the degree of an *initial* algebra relative to a finite set of ordinary equations.

	A_1	A_2
Carrier of sort s	$\{0,1\}$	$\{0,1\}^*$
Carrier of sort s'	$\{0,1\}$	$\{0,1\}^*$
c	0	λ
a	0	0
b	0	0
r	$x \mapsto 0$	$x \mapsto x$ concatenated with 0
l	$x \mapsto 1$	$x \mapsto x$ concatenated with 1

Figure 1: Two Σ_0-algebras.

2 Algebraic Preliminaries

This section reviews the definitions and results from many-sorted algebra that we will need; for a fuller exposition, cf., e.g., Meseguer and Goguen [12]. Readers familiar with the rudiments of many-sorted algebra will want to skip ahead to Section 3.

Let S be any set of **sorts.** A **signature** Σ over S is a set of (function) symbols, each coming with an **arity** consisting of a (possibly empty) word belonging to S^* and a **co-arity** belonging to S. (A function symbol whose arity is the empty word λ is called a **constant** symbol of the sort of its co-arity.) Usually we will ignore the dependence of Σ on S, and all the signatures in this paper will be finite or countable. A Σ**-algebra** A is a family of sets $\{A_s : s \in S\}$, known as the **carriers** of the sorts, and a family of functions $\{A_\sigma : \sigma \in \Sigma\}$ such that if σ is of arity $w = s_1 s_2 \cdots s_n$ and co-arity s, then $A_\sigma : A_{s_1} \times A_{s_2} \times \cdots \times A_{s_n}$.

For example, suppose S_0 has two sorts s and s', and Σ_0 has the following elements: a constant symbol c of sort s, two constant symbols a and b of sort s', and two unary function symbols r and l of sort s (i.e., symbols of arity s and co-arity s). In Figure 1 below, there is a tabular description of two Σ_0-algebras A_1 and A_2.

A family of maps $\phi = \{\phi_s : A_s \to B_s\}$ between (respecitve carriers of) two Σ-algebras A and B is a Σ**-homomorphism** if ϕ preserves the action of the function symbols in Σ. That is, for all function symbols of the appropriate arities and co-arities, $\phi_s(A_\sigma(a_1,\ldots,a_n)) = B_\sigma(\phi(a_1),\ldots,\phi(a_n))$. For example, a Σ_0-homomorphism ϕ from A_2 to A_1 is given by setting $\phi_{s'}$ to be the identity map on $\{0,1\}$; $\phi_s(\lambda) = 0$; and if x is not empty, $\phi_s(x) =$ the last element of x. A Σ**-isomorphism** is a bijective Σ-homomorphism; we identify Σ-isomorphic Σ-algebras.

Since the Σ-algebras and Σ-homomorphisms constitute a category, the definitions of **initial** and **final** objects apply here. That is, an initial Σ-algebra A is one such that for every Σ-algebra B, there is a unique Σ-homomorphism from A into B. Dually, A is final if for every B, there is a unique Σ-homomorphism from B into A. Initial and final objects of other categories are defined in the same way, and there is (up to unique isomorphism) at most one of each in any category.

There is a standard representation for the initial Σ-algebra consisting of the set of (ground) terms over the signature Σ with the natural interpretation of the function symbols. It is denoted T_Σ. For example, A_2 is an initial Σ_0-algebra, since $A_2 \cong T_{\Sigma_0}$. Given a Σ-algebra A, the unique Σ-homomorphism from T_Σ to A is the evaluation map ϵ_A defined in the obvious way. For example, $\epsilon_{A_2}(r(l(r(r(c))))) = 0010$. A Σ-algebra A for which the map ϵ_A is surjective is called a **minimal** Σ-algebra. These are important precisely because every element of a minimal algebra has a name; there is no 'junk'.

The canonical final Σ-algebra is the **unit** Σ-algebra $\mathbf{1}_\Sigma$. For each sort s, the carrier $(\mathbf{1}_\Sigma)_s$ is a singleton. The function symbols in Σ are interpreted as constant functions.

Given a Σ-algebra A, a **Σ-congruence** is a family $\{\equiv_s : s \in S\}$ of equivalence relations on the carriers of A such that the interpretation of every function symbol in Σ respects these equivalence relations. A Σ-congruence can be associated to every Σ-homomorphism in the natural way. Given a Σ-algebra A and a Σ-congruence $\equiv$, we can take the quotient to form a Σ-algebra $A/\equiv$. Up to isomorphism, the minimal Σ-algebras are the quotients of initial algebras, and all of the Σ-algebras in this paper will be of this form.

Let S be a set of sorts, and let Σ be a signature over S. Let $X = \{X_s : s \in S\}$ be a family of sets indexed by elements of S. We call the elements of each X_s the **variables** of sort s, and we assume that the sets of variables of distinct sorts are disjoint and that no variable is a member of Σ. We can expand Σ to a signature $\Sigma(X)$ by considering the elements of each X_s as constant symbols of sort s. Then $T_{\Sigma(X)}$ is an initial $\Sigma(X)$-algebra, and thus it also is a Σ-algebra. An **equation** is a pair of elements of $T_{\Sigma(X)}$ of the same sort together with the set of variables occurring in these terms. (It is possible to consider the case where the set of variables is a superset of the set of variables occurring in the two terms. This is essential, for example, in providing a proof theory for many-sorted equational logic (cf. Goguen and Meseguer [7]), but since we will not need a precise notion of equational deduction, we use the more restricted form of equation.) If the pair consists of elements of T_Σ (i.e., if there are no variables), then the equation is said to be a **ground** equation. A Σ-algebra A and family of maps δ_s from X_s to A_s determines a unique $\Sigma(X)$-algebra structure on A, and thus a unique $\Sigma(X)$-homomorphism $\delta : T_{\Sigma(X)} \to A$. We call δ an **assignment**. A Σ-algebra A **satisfies** the equation (t_1, t_2) if for all assignments δ defined on the variables of t_1 and t_2, $\delta(t_1) = \delta(t_2)$. Note that every equation is satisfied by the final Σ-algebra. A set E of equations gives rise to a Σ-congruence $\equiv_E$ on a Σ-algebra A by considering the least Σ-congruence extending the set

$$\{(\delta(t_1), (\delta(t_2)) : \delta \text{ is an assignment}\}$$

Then $T_\Sigma/\equiv_E$ is initial in the category of (Σ, E)-algebras — Σ-algebras in which all the equations in E hold — and Σ-homomorphisms. $T_{\Sigma,E}$ denotes this Σ-algebra. Of course, the unit Σ-algebra is also the final object of this category. Note that, up to isomorphism, every minimal Σ-algebra A is of the form $T_{\Sigma,E}$ for some, not necessarily finite, set of equations E, since one can always take E to be the congruence induced by ϵ_A .

3 Finality and Computability

The reader familiar with discussions of finality in abstract data types might know that there are several different definitions of 'final algebra semantics' in the literature. Final algebras originate with the work of Giarrantana, Gimona and Montanari [5] and Wand [15], both building upon ideas of Guttag [8]. Bergstra and Tucker [3] introduced a different notion of finality and considered computability issues for it. The alternative definitions might lead to the feeling that there is some competition between the alternatives. Actually *different intuitions* are being modelled by notions of finality in *different categories*.

Part of the intuition behind initiality is that the facts true in initial (Σ, E)-algebras are precisely those which are *deducible* from the equations in E by rules of many-sorted deduction (cf. Meseguer and Goguen [12]). These rules generalize deduction systems for one-sorted equational logic (cf. Goguen and Meseguer [7]). In fact, this *characterizes* the initial algebra among the minimal algebras. Suppose one puts a different natural condition on minimal (Σ, E)-algebras, a condition of any nature whatsoever. The problem arises as to whether this condition is characterizable by an algebraic property. We will discuss two conditions on minimal (Σ, E)-algebras which are captured by notions of *finality in appropriate categories*. Clearly, the categories must be different from the category of all (Σ, E)-algebras, since the final object in this category is trivial.

3.1 Final Algebras

Given a minimal Σ-algebra A and some fixed $V \subseteq S$, we can consider A to be a **machine** by declaring the elements of carriers whose sorts are in V to be **visible**; invisible data elements then correspond to the 'internal state' of our machine that we can imagine hidden in a 'black box.' We then define the **V–behavior** of the minimal Σ-algebra A to be the restriction to the visible sorts of the induced congruence on T_Σ. Given two minimal Σ-algebras A and B with the same V-behavior, we say that A and B are **V–behaviorally equivalent.** Intuitively this means that as machines, A and B are indistinguishable when only data in visible sorts can be observed. Now consider some set E of Σ-equations. We are interested in the 'minimal realization' of the (V-behavior of) the initial (Σ, E)-algebra, i.e., in some minimal Σ-algebra V-behaviorally equivalent to $T_{\Sigma,E}$ but using the *fewest internal states.* Such a model would be final in the category $\mathbf{Mach}(\Sigma, E, V)$ whose objects are the minimal Σ-algebras with the same V-behavior as $T_{\Sigma,E}$ and whose morphisms are the Σ-homomorphisms.

It turns out that such an algebra can be constructed explicitly. It is a quotient of the initial $\mathbf{Mach}(\Sigma, E, V)$-algebra, and since $T_{\Sigma,E}$ is not in general initial in *this* category, we give the construction of this initial algebra first. It is convenient to (temporarily) pass to a larger signature here, and to let A denote $T_{\Sigma,E}$. Let $\Sigma(A, V)$ be the signature obtained from Σ by adding new constant symbols for the elements of A of visible sort. Of course, A can be regarded as a $\Sigma(A, V)$-algebra in a natural way. Let $I_{\Sigma,E,V}$ be the S-sorted family of **V–irreducible** terms of $T_{\Sigma(A,V)}$ – i.e., those t such that for every subterm u of t, if u is of visible sort, then u is an element of A. Turn this into a Σ-algebra by interpreting the function symbols as follows: If s is not visible, let the interpretation of a function symbol f of co-arity s be the function which when applied to V–irreducible terms $t_1, \ldots, t_n$ gives the term $f(t_1, \ldots, t_n) \in I_{\Sigma,E,V}$; if s is visible, its interpretation on $t_1, \ldots, t_n$ gives $\epsilon_A(f(t_1, \ldots, t_n)) \in A$. $I_{\Sigma,E,V}$ is in this way a Σ-algebra, and in fact it is a *minimal* Σ-algebra. To see this, we describe the surjective Σ-homomorphism $\epsilon_{I_{\Sigma,E,V}}$: Given a term t of visible sort, $\epsilon_{I_{\Sigma,E,V}}(t) = \epsilon_A(t)$. If t is of invisible sort, then $\epsilon_{I_{\Sigma,E,V}}$ is the result of replacing the maximal subterms of t of visible sort by their evaluation in A. This Σ-algebra $I_{\Sigma,E,V}$ is an initial object of $\mathbf{Mach}(\Sigma, E, V)$— in fact it is initial in a larger category as well (cf. Meseguer and Goguen[12]).

The final object of $\mathbf{Mach}(\Sigma, E, V)$ is denoted $F_{\Sigma,E,V}$. It is a quotient of $I_{\Sigma,E,V}$ by the following 'Nerode' Σ-congruence $\equiv_{\mathrm{nerode}}$. Of course, $\equiv_{\mathrm{nerode}}$ must be the identity congruence on the visible carriers. To state the definition for the other carriers, we add to $\Sigma(A, V)$ a new variable $y = y_s$ for each invisible s. Let U be the set of terms of visible sort in this new signature For t_1 and t_2 of the same invisible sort s, $t_1 \equiv_{\mathrm{nerode}} t_2$ iff for all $u \in U$,

$$\epsilon_{I_{\Sigma,E,V}}(u(y \leftarrow t_1)) = \epsilon_{I_{\Sigma,E,V}}(u(y \leftarrow t_2)).$$

The notation $u(y \leftarrow t_i)$ denotes the result of substituting t_i for all occurrences of y in u. Observe the similarity of this definition with the Nerode construction in automata theory. We call $F_{\Sigma,E,V}$ the **final** (Σ, E, V)-algebra. This construction can be done more generally beginning with an arbitrary Σ-algebra A; see Meseguer and Goguen [12] for the details and a proof of finality.

3.2 BT–Final Algebras

Now we will consider the notion of finality given by Bergstra and Tucker [3] whose exposition we follow. Here one wants to model the notion of consistency –i.e., one looks for a nontrivial minimal (Σ, E)-algebra A which satisfies all ground facts satisfiable in *any* nontrivial (Σ, E)-algebra. This is the same as the set of ground equations which hold in some non-trivial minimal (Σ, E)-algebra. As noted in the introduction, such an algebra need not always exist.

Despite the fact that a nontrivial minimal algebra A satisfying exactly the equations true in all nontrivial minimal algebras need not always exist, when it does exist, the category of all non-unit

(Σ, E)-algebras has A as a final object. Conversely, when this category has a final object B, then such an algebra A exists (and $A \cong B$). In this case, we call B the **BT-final** (Σ, E)-algebra. When it exists, it can be realized as a quotient of T_Σ by the Σ-congruence $\equiv_{max}$ given by $t_1 \equiv_{max} t_2$ iff the least Σ-congruence extending $E \cup \{t_1 = t_2\}$ is not the unit congruence. It should be mentioned that the *relation* $\equiv_{max}$ can be defined for any Σ and E, but it is a nontrivial Σ-congruence iff the BT-final algebra exists.

3.3 Computability

If Σ is a finite (or countable) signature, then minimal Σ-algebras are countable sets, and we can consider them as quotients of algebras defined on sets of natural numbers. In this way recursion-theoretic questions arise concerning minimal algebras. A is a **recursive Σ-algebra** if the carriers of A are sets of natural numbers and if the interpretations of the function symbols are (restrictions of) total recursive functions. A **Gödel coding** of a Σ-algebra A is a pair $\langle B, p \rangle$ where B is a recursive Σ-algebra, and $p : B \to A$ is a Σ–isomorphism. Algebras with Gödel codings are said to be **computable**; their systematic study was initiated by Rabin [13] and also Malcev [11]. One of the elementary results there if Σ is a finite signature, then T_Σ is a computable algebra. For example, given a map Sig from the finite set $\{1, 2, \ldots, |\Sigma|\}$ onto Σ, we construct a Gödel coding of T_Σ by giving a simultaneous construction of the carriers B_s for $s \in S$: if $\sigma \in \Sigma$ is of arity $s_1, \ldots, s_n$ and co-arity s, and if $b_1, \ldots, b_n$ are elements of $B_1, \ldots, B_n$, respectively, then $< Sig^{-1}(\sigma), b_1, \ldots, b_n > \in B_s$. (Angular brackets here denote some fixed recursive tupling function.) Then each B_s is recursive as well. We will often identify initial algebras with their Gödel codings in this way. For example, we can regard terms, equations, and finite sets of equations as natural numbers.

The following result states the initiality of Gödel codings in the category of recursive Σ-algebras and recursive homomorphisms; it follows as an immediate consequence that the choice of Gödel coding is irrelevant. For a proof, see Malcev [11] or Meseguer and Goguen [12].

Theorem 1 *If $\langle B, p \rangle$ is a Gödel coding of the initial algebra T_Σ, then for any Σ-algebra A there is a unique homomorphism $\delta_A : B \to A$. If A is recursive, then so is δ_A.* ⊣

Corollary 2 *All Gödel codings of T_Σ are equivalent. That is, for each pair of Gödel codings $\langle B, p \rangle$ and $\langle C, q \rangle$ of T_Σ there is a unique family of recursive bijections $\psi (= \{\psi_s : B_s \leftrightarrow C_s\})$ such that ψ is a Σ–isomorhism and for all s and all $b \in B_s$, $p(b) = q(\psi_s(b))$.* ⊣

Given a minimal Σ-algebra A and a Gödel coding $\langle B, p \rangle$ of T_Σ, let us denote by $\equiv$ the Σ-congruence induced on B by the surjective Σ-homomorphism $\delta_A : B \to A$. We define the Σ-algebra A to be **computable** (or **semicomputable**, or **cosemicomputable**) if $\equiv$ is recursive (or r.e., or co-r.e., respectively). The equivalence property above imples that this is independent of the choice of the Gödel coding $\langle B, p \rangle$ of T_Σ.

Using the initiality of Gödel codings of T_Σ and the fact that the quotient of a recursive algebra by a recursive congruence is always isomorphic to a recursive algebra, it is not hard to check (see for instance Meseguer and Goguen [12] for details) that, a minimal algebra induces a recursive congruence on a Gödel coding of T_Σ if and only if it is isomorphic to a recursive algebra. So two notions of computable algebra given above are equivalent.

One way to determine whether a minimal algebra is computable, semicomputable, or cosemicomputable is to fix some notion of equational deduction and then to notice that the following predicate R is recursive:

$R(\Sigma, E, e, n) \Longleftrightarrow \Sigma$ is a (code of a) finite signature, E is a (code of a) finite set of equations, and e is a ground equation which follows from E in T_Σ in n steps.

Using this fact, we can see that for a finite set E of equations, the initial (Σ, E)-algebra is semicomputable. There are numerous examples from algebra and logic of cases where it is not computable (the so-called undecidable word problems). For a second example, consider the BT-final algebra for a finite set E of equations. We need to evaluate the complexity of the relation $\equiv_{\max}$. Assuming that the BT-final algebra exists, the relation $\equiv_{\max}$ is characterized as follows: $t_1 \equiv_{\max} t_2$ iff it is not the case that for all pairs (u_1, u_2) of terms of the same sort, there is some n such that $R(\Sigma, E \cup \{t_1 = t_s\}, \{u_1 = u_2\}, n)$. In the notation common when considering the arithmetic hierarchy, this looks like a Σ_2 predicate, since by driving in the negation, we obtain

$$(\exists u_1, u_2)(\forall n) \neg R(\Sigma, E \cup \{t_1 = t_2\}, \{u_1 = u_2\}, n).$$

However, we can dispense with the existential quantifier in favor a finite disjunction of universal sentences, where in place of the equations between the u's we have all of the equations between constant symbols of the various sorts. In this way, $\equiv_{\max}$ is a finite disjunction of Π_1 sentences and is thus co-r.e.

Bergstra and Tucker show that this assessment is sharp by giving a finite pair (Σ, E) such that the BT-final (Σ, E)-algebra is cosemicomputable but not computable.

Questions arise as to whether the recursion theoretic consequences imply the algebraic antecedents. That is, given a minimal semicomputable Σ-algebra A, is there a finite set E of Σ-equations such that A is the initial (Σ, E)-algebra? Given a cosemicomputable algebra, is it BT-final for some E? Given a computable algebra, is it initial and BT-final simultaneously for the same E? These questions are the so-called *adequacy problem* in abstract data type specification. It would be interesting to see a set of questions and answers relating algebraic and complexity-theoretic notions.

Bergstra and Tucker answered the second and third of these questions in [3]. In general, the answers to these are negative, but if one permits finite expansions of the signature without adding new sorts, the answers become positive. In effect, they show, e.g., that cosemicomputable algebras are *reducts* of BT-final algebras. (For the first question, a characterization of this type is obtained in Bergstra and Tucker [2] provided an expansion with new sorts is permitted.) Their paper also contains a strong theorem on cosemicomputable algebras. Since this theorem will be used in the sequel, we outline the details. Let Σ be a signature over a set S containing just one sort, and let A be a cosemicomputable Σ-algebra. Thus A is isomorphic to a quotient $R/\equiv$, where R is a recursive Σ-algebra and $\equiv$ is a co-r.e. equivalence relation on R. Suppose that A is infinite. Then

1. There is an infinite recursive subset T of R such that T is a transversal of $\equiv$, and every recursive function $f : T \to R$ extends to a total recursive function $g : R \to R$ which preserves $\equiv$;

2. There is an extension Σ' of Σ with no new sorts, an expansion R' of R, and a finite set E of conditional Σ' equations such that R' is the initial (Σ', E)-algebra. In addition, R' is recursive.

In essence, Bergstra and Tucker considered a recursive bijection between T and R. This bijection allows the actions of the function symbols in R to be projected onto T in a recursive way. Then using (1), this projection extends to all of R, and some new function symbols are interpreted by this projection. Since T is infinite, arithmetic can be simulated on it, and so the extended projection of R can be given an equational specification using Matiasevic's Theorem. Bergstra and Tucker use this result to prove their main theorem by adding one more equation to E to get a finite set E^*. Then they show that A is isomorphic to the BT-final (Σ', E^*)-algebra.

We will need to modify slightly this theorem. First we need to allow S to have more than one sort, provided that for some $s \in S$, $(\equiv)_s$ has infinitely many classes. This extension was noted by Bergstra and Tucker. The proof in the extended case is a slight generalization of the original argument; care must be taken when dealing with finite carriers. The second extension we will need concerns conclusion (1). We need to extend functions defined on cartesian products of T and the other carriers. The functions to be extended must preserve $\equiv$ of course, and this generalization of (1) is trivial (see below). Finally, it should be noted that Bergstra and Tucker worked with *conditional* equations — that is conditionals whose antecedents are conjunctions of equiations and whose consequents are single equations. However, it is possible to prove their theorem with (unconditional) equations by introducing several new conditional functions. We outline the details, assuming the reader is familiar with the original argument and has a copy of it nearby.

We first recall the basic setup of the proof of the BT Theorem. We are given a recursive Σ-algebra R together with a cosemicomputable congruence $\equiv$ on it. It is shown that there is an infinite recursive subset T of R together with a recursive map $proj : R \rightarrow$T. It should be mentioned that $proj$ preserves $\equiv$ and also that $proj(proj(x)) = proj(x)$. From this it follows that every recursive function from T to R can be extended (by composing with $proj$) to a recursive function on the whole of R which preserves $\equiv$. For example, since T is recursive, we can take an arbitrary element 0 of it and then get a map $succ_1$ on T with the properties of a successor. Then $succ_1$ extends to a recursive function $succ$ on R by $succ(x) = succ_1(proj(x))$. From this, we define on T and extend to R functions add and $mult$ which behave in the obvious way. In addition, we will need to have functions $pred$ and $monus$ for the predecessor and truncated difference functions which are defined in the usual way. For example,

$$monus(succ^m(0), succ^n(0)) = 0 \text{ if } m < n$$

$$monus(succ^m(0), succ^n(0)) = succ^{m-n}(0) \text{ if } m \geq n$$

Moreover, since T and R are both infinite and recursive, there is a recursive bijection $enum_1 : T \rightarrow R$. This extends to a recursive map $enum$ defined on all of R by $enum(x) = enum_1(proj(x))$. An important fact here is that for all $x \in R$, there is some m such that $x = enum(succ^m(0))$.

Now what Bergstra and Tucker do is to consider an expansion of Σ by new function symbols corresponding to these operations and then give a finite set E of conditional equations in the expanded signature such that the expansion of R is the initial algebra relative to the equations. In order to capture the effect of conditional equations E, we need to add new conditional operators.

On behalf of $enum$, we define two functions $cond_1$ and $cond_2$, and on behalf of each function symbol $F \in \Sigma$, we define a function $cond_F$. The definitions are by cases on the first argument:

$$cond_1(z, x, y) = \begin{cases} enum(proj(x)) & \text{if } proj(z) = 0 \\ proj(y) & \text{otherwise} \end{cases}$$

$$cond_2(z, x, y) = \begin{cases} proj(enum(proj(x))) & \text{if } proj(z) = 0 \\ enum(proj(y)) & \text{otherwise} \end{cases}$$

$$cond_F(z, x_1, \ldots, x_k, y) = \begin{cases} f(enum(proj(x_1)), \ldots, enum(proj(x_k))) & \text{if } proj(z) = 0 \\ enum(proj(y)) & \text{otherwise} \end{cases}$$

In the last of these, it is assumed that F is a k–ary function symbol interpreted in R by a function f.

Now take Σ' to be the original signature Σ augmented with new function symbols $PROJ$, $ENUM$, $SUCC$, ADD, $MULT$, $PRED$, $MONUS$, $COND_1$, $COND_2$, and $COND_F$ for each $F \in \Sigma$. Under the interpretations of these new symbols by the (lower-case) functions above, R expands to a Σ'-algebra R'. As in the original presentation, it should be noted that $\equiv$ is a Σ'-congruence on R'.

We next list the equations E. They are equations (1)–(10) of the original (the recursion equations for the arithmetic operations), together with recursion equations for the functions $PRED$ and $MONUS$:

$$PRED(0) = 0$$

$$PRED(SUCC(X)) = PROJ(X)$$

$$PRED(X) = PRED(PROJ(X))$$

$$MONUS(X, 0) = PROJ(X))$$

$$MONUS(X, SUCC(Y)) = PRED(MONUS(X, Y))$$

$$MONUS(X, Y) = MONUS(PROJ(X), PROJ(Y))$$

Note that we have also added equations which express the fact that all the functions in R' are preserved under composition with *proj*. The equations for the conditional operations are:

$$COND_1(0, X, Y) = ENUM(PROJ(X))$$

$$COND_1(SUCC(Z), X, Y) = PROJ(Y)$$

$$COND_2(0, X, Y) = PROJ(ENUM(PROJ(X)))$$

$$COND_2(SUCC(Z), X, Y) = ENUM(PROJ(Y))$$

$$COND_F(0, X_1, \ldots, X_k, Y) = F(ENUM(PROJ(X_1, \ldots, X_k)))$$

$$COND_F(SUCC(Z), X_1, \ldots, X_k, Y) = ENUM(PROJ(Y))$$

We also need equations for the function *enum*, and here we refer to the original paper, specifically to equations (11) and (12). We cannot use these equations because they are conditional, but we use the new conditional functions along with the same formal polynomials as in [3]:

$$\begin{aligned}&COND_1(ADD(MONUS(P_1(X, Y, Z_1, \ldots, Z_{k(1)}), Q_1(X, Y, Z_1, \ldots, Z_{k(1)})),\\ &\qquad MONUS(Q_1(X, Y, Z_1, \ldots, Z_{k(1)}), P_1(X, Y, Z_1, \ldots, Z_{k(1)}))),\\ &\quad X, Y)\\ &= PROJ(Y)\end{aligned}$$

$$\begin{aligned}&COND_2(ADD(MONUS(P_2(X, Y, Z_1, \ldots, Z_{k(2)}), Q_2(X, Y, Z_1, \ldots, Z_{k(2)})),\\ &\qquad MONUS(Q_2(X, Y, Z_1, \ldots, Z_{k(2)}), P_2(X, Y, Z_1, \ldots, Z_{k(2)}))),\\ &\quad X, Y)\\ &= PROJ(Y)\end{aligned}$$

$$\begin{aligned}&COND_F(ADD(MONUS(P_F(X_1, \ldots, X_k, Y, Z_1, \ldots, Z_{k(F)}), Q_F(X_1, \ldots, X_k, Y, Z_1, \ldots, Z_{k(F)})),\\ &\qquad MONUS(Q_F(X_1, \ldots, X_k, Y, Z_1, \ldots, Z_{k(F)}), P_F(X_1, \ldots, X_k, Y, Z_1, \ldots, Z_{k(F)}))),\\ &\quad X_1, \ldots, X_k, Y)\\ &= ENUM(PROJ(Y))\end{aligned}$$

For example, the first of these is the analog of (11) of the original:

$$\begin{aligned}&P_1(X, Y, Z_1, \ldots, Z_{k(1)}) = Q_1(X, Y, Z_1, \ldots, Z_{k(1)}) \rightarrow\\ &\quad ENUM(PROJ(X)) = PROJ(Y).\end{aligned}$$

We need in addition equations which express that these new conditional functions are invariant under composition with $PROJ$. It should be checked that R' satisfies all these equations. The key point here is that for all z, $succ(proj(z)) \neq 0$. The reader interested in seeing the idea should consider the long equation for $COND_1$ above. The verification that this holds in R' is uses the fact that the conditional equation immediately before this paragraph also holds in R'.

Now that we have given the equations, it remains to argue that $R' \cong T_{\Sigma',E}$. Let $\equiv'$ be the least Σ' congruence on $T_{\Sigma'}$ extending E. The key step is to verify that each $t \in T_{\Sigma'}$ is $\equiv'$-equivalent to a term of the form $ENUM(SUCC^n(0))$ for some n. The induction is *not* on the construction of terms, but on the well-founded relation $\prec$ given by $t_1 \prec t_2$ iff either t_1 is a subterm of t_2 or t_1 has fewer occurrences of conditional functions than t_2. The induction hypothesis then takes care of those terms whose leading symbol is one of the new conditional functions. For the others, the argument parallels the original closely, except that instead of using conditional equations, we use the equations involving conditional functions. The map $\phi : R' \to T_{\Sigma',E}$ given by $\phi(enum(succ^n(0))) = [ENUM(SUCC^n(0))]_{\equiv'}$ is therefore surjective. The verification that it is a Σ'-homomorphism is quite like the original argument except that we need to consider the new conditional symbols. Thus it is by induction on $\prec$.

It should be noted that these results are uniform in the sense that the extension Σ' depends only on Σ and not on A. Of course, the equations E depend on A.

Henceforth we shall refer to the modifications of (1) and (2) above as the **BT Theorem.** We will state here the precise version that will be used. Let Σ be a finite signature, and let A be an infinite cosemicomputable Σ-algebra. Thus A is isomorphic to a quotient $R/\equiv$, where R is a recursive Σ-algebra and $\equiv$ is a co-r.e. equivalence relation on R. Suppose that A is infinite, and fix a sort s such that R_s is infinite. Then for all sorts s_1 and s_2:

1. There is an infinite recursive subset T of R_s such that for all sorts s_1 and s_2, every recursive function $f : R_{s_1} \times T \to R_{s_2}$ which preserves $\equiv$ extends to a recursive function $g : R_{s_1} \times R_s \to R_{s_2}$ which also preserves $\equiv$.

2. There is an extension Σ' of Σ with no new sorts, an expansion R' of R, and a finite set E of ordinary Σ' equations such that R' is the initial (Σ', E)-algebra. In addition, R' is recursive.

Now we return to the notion of finality considered in this paper. For a minimal algebra, say $T_{\Sigma,E}$, the initial realization $I_{\Sigma,E,V}$ of its V-behavior is a minimal Σ-algebra, so we can ask whether it is computable. This depends on the original Σ and E that we started with. It is easy to see that if $T_{\Sigma,E}$ is a computable algebra, then the algebra $I_{\Sigma,E,V}$ is computable. But as we noted above, $T_{\Sigma,E}$ is not always computable. Nevertheless, $I_{\Sigma,E,V}$ is computable provided E satisfy a weaker hypothesis. Following Bergstra and Meyer [1], we need only insist that the congruence on (a Gödel coding of) T_Σ induced by the equations in E be recursive when restricted to the *visible* carriers. This immediately gives a decision procedure for telling whether two (codes of) terms of the same visible sort have the same image under the Σ-homomorphism $\epsilon_{I_{\Sigma,E,V}}$. For the invisible sorts, we describe the procedure a little more intuitively: To decide whether two (codes of) terms u and v of the same invisible sort are in the same class of the associated congruence, determine if the maximal subterms of visible sort have the same evaluations in $T_{\Sigma,E}$, and if in addition u and v are built up in the same way from these maximal subterms of visible sort.

It is clear from the definitions that if $I_{\Sigma,E,V}$ is a computable algebra, then $\equiv_{\mathrm{nerode}}$ is Π_1. By our earlier observation, this means that if $T_{\Sigma,E}$ has computable V-behavior, then the final (Σ, E)-algebra is cosemicomputable.

We can ask for this notion of finality whether this observation can be reversed. That is, for every Σ and for every cosemicomputable Σ-algebra A with computable V-behavior, can we find an expansion Σ' and equations E such that A is the final (Σ, E, V)-algebra? In general, the answer is again no. For example, V may be empty. In this degenerate case, the congruence $\equiv_{\mathrm{nerode}}$ is trivial, and the final algebra is computable. Even if $V \neq \emptyset$, if all the visible carriers of A have less than two elements, then again A cannot be the reduct of the final (Σ', E, V)-algebra for any Σ' and E. This is because the visible carriers of the final (Σ', E, V)-algebra must also be of size less than two, and the same holds for $T_{\Sigma'}$. But then once again $\equiv_{\mathbf{nerode}}$ is trivial, and the final algebra is computable, which as we show below is in general false. So we suppose that there is a visible sort v such that A_v is nonempty. with at least two elements. We say that A has **non-unit** V-behavior in this case. For algebras with non-unit computable V-behavior, Meseguer and Goguen [12] conjectured that there is an expansion of the signature as above, and we prove this in Theorem 3 below. Implicit in this conjecture is the conjecture that if A is computable, then we can find Σ' and E such that A is simultaneously a reduct of the initial (Σ, E)– and a reduct of the final (Σ', E, V)-algebra. This is shown in Corollary 4 below.

4 Every Reasonable Cosemicomputable Algebra is a Reduct of a Final Algebra

Theorem 3 *Let Σ be a finite signature over S, and let $V \subseteq S$ be a set of visible sorts. Let A be a minimal cosemicomputable Σ-algebra with non-unit computable V-behavior. Then there is a finite extension Σ' of Σ with no new sorts, and a finite set E of Σ'-equations such that A is isomorphic to the reduct of the final (Σ', E, V)-algebra.*

Proof A is minimal, so it is a quotient of T_Σ, and taking E' to be the induced congruence, A is just $T_{\Sigma,E'}$. Since A has computable V-behavior, the initial realization of its behavior $I_{\Sigma,E',V}$ is computable and is thus isomorphic to a recursive algebra R. In addition, the congruence $\equiv$ induced on R by the Σ-homomorphism from R to A is co-r.e. because it is the image under a recursive surjection of the corresponding co-r.e. congruence induced on a Gödel coding of T_Σ. Moreover, the restriction of $\equiv$ to the visible carriers is the identity because R and A have the same V-behavior. Also, since A is has non-unit V-behavior, there is some visible v^* such that A_{v^*} has size at least two. Thus the same is true of R_{v^*}, and we fix two distinct visible elements, say c_0 and c_1 of R_{v^*}.

Finally, we can assume that for some sort w^*, A_{w^*} is infinite because otherwise A is finite and the theorem has a different and easier proof. This last assumption is a hypothesis in the BT Theorem.

First we apply the BT Theorem to Σ, R, and $\equiv$, and we will obtain an expansion R^+ of R. For each invisible sort s, we introduce a new function symbol G_s of arity sw^* and co-arity v^*. We set Σ^+ to be Σ along with the new symbols, and we will fix an interpretation g_s for each new symbol in such a way that each g_s preserves the congruence $\equiv$. By the theorem, there is an infinite recursive subset T of R_{w^*}, and we need only specify the values of g_s on $R_s \times T$. In this way we get an enlarged signature Σ^+, and an expansion R^+ of R. Note that part (i) of the BT Theorem has insured that $\equiv$ is a Σ^+-congruence on R^+.

The interpretation of the new symbols G_s depends on a technical lemma concerning co-r.e. equivalence relations on recursive subsets of ω. The lemma will be discussed at the end of this Section. It states that if $\equiv$ is a co-r.e. equivalence relation on R_s (or indeed any recursive set), then there is a function $f_s : R_s \times \omega \to \{0,1\}$ such that $x \equiv y$ iff for all n, $f_s(x,n) = f_s(y,n)$. Define $g^\# : R_s \times T \to R_{v^*}$ by

$$g^\#(r,p) = c_i \text{ iff } f_s(r,n) = i \text{ and } p \text{ is the } n^{th} \text{ element of } T.$$

Now $g^\#$ is recursive, and it preserves $\equiv$ as well. Therefore by (1) of the BT Theorem. it extends to a recursive function g_s on $R_s \times R_{w^*}$ into R_{w^*}. We interpret G_s by g_s for each invisible s.

Having given an interpretation of the symbols in Σ^+, we now apply the BT Theorem again, this time to Σ^+, R^+, and $\equiv$. This time we get expansions R' and Σ' and a finite set E of Σ'-equations such that R' is the initial (Σ', E)-algebra. Again, $\equiv$ is a Σ'-congruence. We will show that $I_{\Sigma',E,V}/\equiv_{\mathbf{nerode}}$

and $R'/\equiv$ are isomorphic Σ'-algebras. Since the reduct of $R'/\equiv$ to Σ is isomorphic to the original algebra A, this will finish the proof. Since R' and $R'/\equiv$ have the same V-behavior, there is a unique surjective Σ'–homomorphsim $\phi : (R'/\equiv) \rightarrow F_{\Sigma',E,V}$. We thus need only show that ϕ is injective. By initiality, $\epsilon_{F_{\Sigma',E,V}} = \phi \circ \epsilon_{(R'/\equiv)}$. And since all these algebras are minimal, it is sufficient to show that if t and t' are V-irreducible Σ' terms of the same invisible sort s and $\epsilon_{R'}(t) \not\equiv \epsilon_{R'}(t')$, then $t \not\equiv_{\text{nerode}} t'$. So suppose the antecedent. Then for some number n,

$$f_s(\epsilon_{R'}(t), n) \neq f_s(\epsilon_{R'}(t'), n).$$

Hence if p is the nth element of the infinite set T,

$$g_s(\epsilon_{R'}(t), p) \neq g_s(\epsilon_{R'}(t'), p).$$

By initiality of R', there is some term $t_0 \in T_{\Sigma'}$ such that $\epsilon_{R'}(t_0) = p$. Let $u \in T_{\Sigma'}(\{y\})$ be the term $G_s(y, t_0)$. Note that u is of visible sort. Therefore

$$\epsilon_{I_{\Sigma',E,V}}(u(y \leftarrow t)) = \epsilon_{R'}(u(y \leftarrow t)) = g_s(\epsilon_{R'}(t), \epsilon_{R'}(t_0)) = g_s(\epsilon_{R'}(t), p).$$

The same is true with t' replacing t, and it follows that

$$\epsilon_{I_{\Sigma',E,V}}(u(y \leftarrow t)) \neq \epsilon_{I_{\Sigma',E,V}}(u(y \leftarrow t'))$$

and consequently $t \not\equiv_{\text{nerode}} t'$. ⊣

This completes the proof of the theorem modulo the lemma on co-r.e. equivalence relations. Before turning to the lemma, we note an important corollary of the theorem.

Corollary 4 *Let S be a set of sorts, V a non-empty subset of S, and Σ a signature over S. Suppose that A is a minimal computable Σ-algebra with non-unit computable V-behavior. Then there is a signature Σ' over S extending Σ and a set E of Σ'-equations such that A is isomorphic to the reduct to Σ of both the initial (Σ', E)-algebra and the final (Σ', E, V)-algebra.*

Proof A is isomorphic to a recursive number algebra R. The equality relation $=$ is a Σ-congruence on R, and R is isomorphic to $R/=$. Also, $=$ is recursive and hence co-r.e. By (the proof of) Theorem 3, we therefore get an extension Σ', an expansion R' of R to Σ', and a set E of Σ'-equations. By (ii) of the BT Theorem, R' is the initial (Σ', E)-algebra, and by the theorem, $R'/=$ is the final (Σ', E, V)-algebra. But R' and $R'/=$ are isomorphic Σ'-algebras, and A is isomorphic to the reduct R of both of these. ⊣

We now state and prove the lemma used above. It is stated in the case of co-r.e. equivalence relations on ω, but with minor changes the argument applies to such relations on any recursive subset of ω.

Lemma 5 *If $\equiv$ is a co-r.e. equivalence relation on ω, then there is a total recursive function $f : \omega \times \omega \rightarrow \{0,1\}$ such that $x \equiv y$ iff for all n, $f(x,n) = f(y,n)$.*

Proof Fix an enumeration of the complement of $\equiv$. We construct f in stages along with a finite equivalence relation which approximates $\equiv$. At stage n we will have a finite function f_n with domain $\{0, ..., n\} \times \{0, \ldots, c_{n-1}\}$ for some c_{n-1} along with an equivalence relation $\equiv_n$ on $\{0, \ldots, n\}$. The approximations will satisfy the following conditions:

1. For all $x, y \leq n$, if $x \equiv y$, then $x \equiv_n y$.

2. $x \equiv_n y$ iff for all $i < c_n$, $f_n(x,i) = f_n(y,i)$.

3. If $x \not\equiv y$, then for some n, $x \not\equiv_n y$.

4. $c_n < c_{n+1}$.

The construction will give an increasing sequence $f_0 \subseteq f_1 \subseteq \cdots \subseteq f_n \subseteq \cdots$ of functions, and f will be the union. Thus f is recursive, and it follows from these four conditions that f has all the desired properties.

For $n = 0$, set $c_0 = 1$, $f_0 = \{(0,0)\}$, and $\equiv_0 = \{(0,0)\}$.

Given f_n, c_n, and $\equiv_n$, we build the next approximations in three steps. It will be convenient to work not with $\equiv_n$ but with its associated partition π of $\{0,\ldots,n\}$. First, extend π to a partition π^+ of the set $\{0,\ldots,n,n+1\}$ by deciding which class to put $n+1$ in. To do this, enumerate enough of the complement of $\equiv$ to rule out all but one class. Then, for $i < c_n$, set $f(n+1,i) = f(k,i)$ for any $k \leq n$ in the same π^+ class as $n+1$.

Second, carry out $n+1$ steps in the enumeration of the complement of $\equiv$. A pair (u,v) of numbers is called *critical with respect to π at stage n* iff both u and v are at most $n+1$, u and v are in the same class of π, and if either (u,v) or (v,u) was enumerated into the complement of $\equiv$ in $n+1$ steps. The idea is that if (u,v) is critical for π^+, then we should arrange that u and v will not be in the same $\equiv_{n+1}$ class. We can define a refinement π^* of π^+ as follows:

Fix a critical pair (u,v). Consider each element w of the π^+ class of u and v, and enumerate enough of the complement of $\equiv$ to decide either $w \not\equiv u$ or $w \not\equiv v$. At least one (and possibly both) of these holds, but we only enumerate until one is decided. Then we put w in the π^* class of v if $w \not\equiv v$ was verified no later than $w \not\equiv u$, and we put w in the π^* class of u otherwise. In this way, we split the π^+ class of critical pairs in finitely many steps.

We carry out this splitting on behalf of all critical pairs. That is, there are only finitely many critical pairs (on behalf of π^+) at stage $n+1$. Enumerate them in a list $(u_1,v_1),\ldots,(u_k,v_k)$. First split on behalf of (u_1,v_1). Once this is done, we have a refinement π_1^+ of π^+. (u_2,v_2) may not be critical with respect to π_1^+. If it is not, set $\pi_2^+ = \pi_1^+$, and proceed to the next pair (u_3,v_3). If it is, split a class of π_1^+ to get π_2^+. Continue on in this way k times so that all of the critical pairs for π^+ are in different classes of π_k^+. Set $\pi_{n+1} = \pi_k^+$, and let $\equiv_{n+1}$ be the equivalence relation associated to π_{n+1}. The construction has been arranged so that (1) holds. This is because all of the equivalence relations constructed are coarser than (the restrictions to their fields) the original relation $\equiv$. Formally, of course, this is shown by induction.

Finally, we extend f_n in such a way that (2) holds. Number the $\equiv_{n+1}$–classes in binary. By adding 0's, we can assume that all of the numbers are of length $M+1$ for some $M \geq 0$. Let $c_{n+1} = c_n + M + 1$, so (4) holds. Extend f_n to f_{n+1} by coding, for each u, the binary number of the class of u into $f_{n+1}(u,c_n),\ldots,f_{n+1}(u,c_n+M)$. That is, set $f_{n+1}(u,c_{n+1})$ to be the $(i+1)$–st binary digit of the class of u. Now (2) holds by the induction hypothesis, by the fact that $u \equiv_{n+1} v$ and $u,v, \leq n$ implies $u \equiv_n v$, and by the way that we extended f_n to get f_{n+1}. Finally (3) holds because if $x \not\equiv y$, the for some n, (x,y) gets enumerated into the complement of $\equiv$ in $n+1$ steps. If $x \equiv_n y$, then at step n, the pair (x,y) is critical with respect to $\equiv_n$, and so by construction $x \not\equiv_{n+1} y$. ⊣

5 Degrees of Unsolvability of Minimal Algebras and Initial and Final Algebra Realizations of the R.E. Degrees

The equivalence of Gödel codings noted in Corollary 2 allows us to define the Turing degree a of minimal algebra. Given two sets X and Y of natural numbers, we write $X \leq_T Y$ (X is *Turing reducible* to Y) if there is a Turing machine with an oracle for Y which computes the characteristic function of X. The oracle answers questions of the form $n \in Y$? to as single steps in the computation of the characteristic function of X. We will not use any deep facts about Turing reduction. In fact, all our uses of this notion are confined to applications of Lemma 6 and Lemma 10 below. The relation $\leq_T$ is reflexive and transitive, and it partitions the family of sets of natural numbers into *Turing degrees*. Following standard notation, we write $X \equiv_T Y$ when $X \leq_T Y$ and $Y \leq_T X$. For more on the notion of Turing reduction, cf., e.g., Rogers [14].

If A is a minimal Σ-algebra to be the Turing degree of the congruence $\equiv_A$ induced on T_Σ by A. In order for this definition to make sense, it should be the case that the Turing degree of $\equiv_A$ be independent of the choice of Gödel coding of T_Σ. Moreover, if A is of the form $R/\equiv$, where R is recursive, then we would also like to know that $\equiv\ \equiv_T\ \equiv_A$. Fortunately, both of these facts are easy consequences of Theorem 1.

Lemma 6 *Let Σ be a finite signature and suppose that A and B are recursive minimal Σ-algebras. Let $\equiv_A$ and $\equiv_B$ be Σ-congruences such that $A/\equiv_A \cong B/\equiv_B$. Then $\equiv_A\ \equiv_T\ \equiv_B$.*

Proof Let $nat_A : A \to A/\equiv_A$ and $nat_B : B \to B/\equiv_B$ be the natural maps. So by initiality, $nat_A \circ \epsilon_A$ is the unique Σ-homomorphism from T_Σ to $A/\equiv_A$. And the same is true for B and $B/\equiv_B$. Now since $A/\equiv_A$ and $B/\equiv_B$ are isomorphic, $nat_A \circ \epsilon_A$ and $nat_B \circ \epsilon_B$ induce the same congruence on T_Σ. This means that that for all $t, t' \in T_\Sigma$, $\epsilon_A(t) \equiv_A \epsilon_A(t')$ iff $\epsilon_B(t) \equiv_B \epsilon_B(t')$.

Here is the computation of the characteristic function of $\equiv_A$ using an oracle for $\equiv_B$: Given (a_1, a_2), find the least terms t_1 and t_2 in T_Σ such that $\epsilon_A(t_i) = a_i$. (We are using the fact that ϵ_A is surjective (by minimality) and recursive. We are also using the recursiveness of T_Σ.) Then compute $\epsilon_B(t_1)$ and $\epsilon_B(t_2)$ and ask the oracle for $\equiv_B$ whether $t_1 \equiv_B t_2$ or not. This shows that $\equiv_A$ is reducible to $\equiv_B$. The converse holds by symmetry, and thus the two are Turing equivalent. ⊣

We define the Turing degree of a minimal algebra A to be the Turing degree of any congruence relation which A induces on a recursive algebra B such that there is a surjective homomorphism from B onto A. By the lemma, the degree is independent of the choice of B. In particular, B need not be initial. We might also note a property of these definitions which will be used at the very end of this paper. Suppose that B is a recursive Σ-algebra and that $\equiv$ is a Σ-congruence of degree $\mathbf{d}$ on B. Suppose we extend Σ to a new signature Σ' (without adding new sorts) and that we extend B to a recursive Σ'-algebra B' in such a way that $\equiv$ is a Σ'-congruence on B'. That is, the interpretations of the new symbols preserved $\equiv$. Then the degree of $B'/\equiv$ is also $\mathbf{d}$. This is precisely the situation that arises when we use the BT Theorem.

We mentioned above that there are final algebras with computable V-behavior that are cosemi-computable but not computable. The proofs that these algebras are not computable actually shows that they are isomorphic to quotients $R/\equiv$, where $\equiv$ is a congruence relation on the recursive algebra R of degree $\mathbf{0}'$ (the complete r.e. degree). It is natural to ask whether for some (or for any) signature Σ and any r.e. degree $\mathbf{d}$ there is an expansion Σ' and a finite set E of Σ'-equations such that the final (Σ', E)-algebra has degree $\mathbf{d}$.

These questions are answered in Theorem 11 below. It should be mentioned that our results hold for BT-finality as well, and the proofs are similar except appeals to Theorem 3 should be replaced by uses of Bergstra and Tucker's characterization of BT-final algebras.

Our work involves an examination of the possible Turing degrees of Σ-congruences on initial Σ-algebras. This leads to a determination of the number of possible Σ-congruences on the initial Σ-algebras. The main result in this area is Theorem 8 which may be of interest for studies wholly unrelated to finality.

Definition Given a sort set S, and a signature Σ over a S, consider a new signature Σ^+ obtained by adding one new variable x_s for each sort $s \in S$. This new symbol is added as a constant symbol to the signature; that is, as a function symbol of empty arity. An **s-pattern** is a term $p \in T_{\Sigma^+}$ of sort s with the following properties:

1. p is not one of the new variables.

2. There is exactly one sort r such that the new variable x_r occurs in p.

3. The new variable x_r which occurs in p occurs exactly once in p.

If p is an s-pattern, the sort of the new variable is called the **domain** of p. An s-pattern is **operative** if its domain is s.

Given an operative s-pattern p, we define a map $h_p : \omega \times (T_\Sigma)_s \to (T_\Sigma)_s$ by recursion on ω:

$$h_p(0,t) = t$$

$$h_p(n+1,t) = p(x_s \leftarrow h_p(n,t))$$

For each $t \in (T_\Sigma)_s$, the map $n \mapsto h_p(n,t)$ is injective. This uses the assumption that p is not merely a new variable.

A term $t \in (T_\Sigma)_s$ is **s-primitive** if t has no proper subterm of sort s. An operative s-pattern is **s-universal** if for all $t \in (T_\Sigma)_s$ there is an $n \in \omega$ and an s-primitive term u such that $t = h_p(n,u)$.

A signature Σ is **productive** if for some sort r such that $(T_\Sigma)_r$ is nonempty and for some sort s, there is an operative r-pattern q and an s-pattern p with domain r such that neither p nor q is a subterm of the other. It is possible that s and r might be the same sort. Note that if Σ is productive, then both $(T_\Sigma)_r$ and $(T_\Sigma)_s$ are infinite. If for all sorts r and s there are no patterns p and q as above, then of course we say that Σ is **non-productive**.

Examples The signature for the successor operation on the natural numbers (sort set $\{s\}$, constant 0 and unary function *succ*) is non-productive. Every s-pattern in this case is of the form $succ^n(x_s)$ for some n. Note also that there is a universal s-pattern, namely $succ(x_s)$. Even if we add a finite number of additional constant symbols, we still get a non-productive signature. On the other hand, if we add a second unary function symbol *succ'*, then the signature becomes productive via $succ(x_s)$ and $succ'(x_s)$. In this case, we no longer have a universal pattern. If we take the signature above (0, *succ* on a sort s) and add a new sort r together with a unary function f of arity s and co-arity r. Then the resulting signature is productive via $succ(x_s)$ and $f(x_s)$. Note that there are no universal r-patterns, and that there are infinitely many r-primitive terms. The signature Σ_0 of Section 1 is productive. Every signature can be extended to a productive signature without adding new sorts, and every extension of a productive signature is productive.

Lemma 7 *Let Σ be a finite non-productive signature over the sort set S. Then for each $s \in S$,*

1. *There are only finitely many s-primitive terms $u_1, \ldots, u_n$.*

2. *Assume that there are infinitely many terms of sort s. Then there exists a unique s-universal s-pattern.*

Proof It is sufficient to prove this for signatures over finite sort sets, and we show by induction on the cardinality of S that for every signature Σ over S the lemma holds. Suppose first that S has only one sort s. Part (1) is trivial in this case, so suppose that T_Σ is infinite. Then the definition of productivity implies either Σ has no constant symbols or else Σ contains at most one non-constant symbol f which must be of arity exactly s. If T_Σ is infinite, then such a symbol f exists, and in this case $f(x_s)$ is a universal s-pattern.

Suppose the lemma true for all S of cardinality at most $N > 1$, and let S have cardinality $N+1$. If the lemma fails, then there is some sort s for which either (1) or (2) fails. Let $S' = S - \{s\}$, and let Σ' be the subset of Σ containing the symbols of arities in $(S')^*$ and co-arities in S'. Note that Σ' is non-productive over S', and thus by induction hypothesis, (1) and (2) hold for it.

Suppose that (1) fails for s. Since Σ is finite, for some function symbol f of arity $w = r_1 \cdots r_n \in (\Sigma')^*$ and co-arity s, there are infinitely many terms of the form $f(t_1, \ldots, t_m)$ for $t_i \in T_{\Sigma'}$. Since w is a finite word, for some fixed i, $(T_{\Sigma'})_{r_i}$ is infinite. By induction hypothesis (2) applied to Σ' and r_i, there is a r_i-universal pattern q (for $T_{\Sigma'}$). In particular, q is r_i-operative, and q has no subterm of sort s. But we also have an s-pattern p with domain r_i, as follows: Fix ground terms t_j of sort r_j for $j \neq i$ and let $p = f(t_1, \ldots, t_{i-1}, x_{r_i}, t_{i+1}, \ldots, t_n)$. Then p has no subterm of sort s, so we get a contradiction because now Σ is productive.

For (2), suppose that there are infinitely many terms of sort s. We first claim that there is an operative s-pattern in T_Σ. Suppose towards a contradiction that this were false. For every r such that there is an s-pattern in T_Σ with domain r, we cannot have $(T_{\Sigma'})_r$ infinite. (For the induction hypothesis would imply that there is an operative r-pattern, and then again Σ would be productive.) In addition, if there is an s-pattern with domain r, then there can be no r-pattern in T_Σ with domain s, or else we would have by composition an operative s-pattern. So if there is an s-pattern with domain r, then $(T_\Sigma)_r = (T_{\Sigma'})_r$. It follows that if there is an s-pattern in T_Σ with domain r, then $(T_\Sigma)_r = (T_{\Sigma'})_r$ is finite. Since there are only finitely many sorts, $(T_\Sigma)_s$ is finite. This contradiction proves the claim of this paragraph that there is an operative s-pattern.

Now let p be a minimal operative s-pattern; i.e., one containing the fewest symbols. We claim that p is s-universal. If not, let $t \in T_\Sigma$ be a minimal term of sort s which is not of the form $h_p(n, u)$ for any n and u. Then t is not s-primitive, so form the operative s-pattern q by replacing some maximal subterm of sort s by the variable x_s, and replacing all of the remaining subterms of sort s (if any) by s-primitive terms. We first claim that $p \neq q$. Suppose towards a contradiciton that they were equal. The first case is when both have only one subterm of sort s. Then t has only one maximal proper subterm t_0 of sort s, so by minimality of t, there is some n and u such that $t_0 = h_p(n, u)$. But $p = q$ is the result of replacing t_0 by x_s in t, so t is the result of replacing x_s by t_0 in p. Therefore $t = h_p(n+1, u)$, and this contradicts the definition of t. The second case is when $p = q$ has more than one subterm of sort s. By the construction of q, neither of these subterms is itself a proper subterm of the other. Then we can form q' by interchanging in q the subterm x_s with one of the other subterms of sort s. Then neither q not q' is a subterm of the other. It follows that Σ is productive, and this is a contradiction. Therefore $p \neq q$. Now by the minimality of p, q cannot be a proper subterm of p. By construction, the only subterms of q of sort s are either s-primitive or the variable x_s. It follows that p is not a proper subterm of q. So Σ is productive, and this once again is a contradiction. It follows that p is s-universal.

Finally, the non-productivity of Σ implies that any two s-universal patterns would have to be subterms of each other. This uniqueness result will not be used in the sequel, and so we omit the proof. ⊣

The next theorem can be regarded as a kind of Cantor–Bendixson theorem for Σ-congruences on initial Σ-algebras since it states that there are either countably many such congruences or that there is a continuum of them. Indeed that much of the theorem can be proved outright as follows: Let Σ be a finite or countable signature. The Gödel coding of Σ establishes an isomorphism of T_Σ with the space of sequences from some finite set. In this way, the set $\mathcal{S}$ of subsets of T_Σ can be regarded as a metrtic space homeomorphic to the Cantor set. In particular, this space is complete. Now in this space, the family of Σ-congruences on T_Σ is a closed set, so by the Cantor–Bendixson theorem, it is either countable or has the cardinality of the continuum. (If the original signature were countably infinite, then this argument works because $\mathcal{S}$ will be isomorphic to the Baire space.) The next result improves this observation by giving the criterion of productivity which separates the cases and by considering the computability properties of the possible congruences.

We need a definition at this point. A congruence $\equiv$ is **finitely generated** if there are finitely many ground instances of it such that $\equiv$ is the least Σ-congruence containing those instances.

Theorem 8 *Let Σ be a finite signature.*

1. *If Σ is non-productive, then there are at most countably many Σ-congruences over the initial Σ-algebra, and each of these is generated by finitely many equations and is recursive.*
2. *If Σ is productive, then there are $2^{\aleph_0}$ Σ-congruences over the initial Σ-algebra, and for every Turing degree* $\mathbf{d}$, *there is a Σ-congruence of degree* $\mathbf{d}$. *Moreover, for every r.e. degree* $\mathbf{d}$, *there are r.e. and co-r.e. Σ-congruences of degree* $\mathbf{d}$.

Proof Let Σ be non-productive. We show first that every Σ-congruence on the initial Σ-algebra is finitely generated, that is, that for each such congruence, there is a finite set of ground equations such that the given congruence is the least Σ-congruence extending the equations. Suppose toward a contradiction that $\equiv$ is a Σ-congruence which is not finitely generated. Then there is a sequence of instances $\langle e_i : i \in \omega\rangle$ such that each e_i is not in the least Σ-congruence extending $\langle e_j : j < i\rangle$. Since S has only finitely many sorts, let s be fixed so that infinitely many of the e_i are of equations between two terms of sort s. Now $(T_\Sigma)_s$ must be infinite, so Lemma 7 tells us that there are finitely many s-primitive terms $u_1, \ldots, u_K$, and that there is a s-universal pattern p.

Let $h(m,n)$ $(m \in \omega, 1 \leq n \leq K)$ be given by $h(m,n) = h_p(m, u_n)$. Set

$$Y_n = \{h(m,n) : m \in \omega\}.$$

Then the Y_n partition the terms of sort s into K classes. Fix two numbers $j, k < K$ such that for infinitely many i, e_i is an equation between an element of Y_j and an element of Y_k. We can assume that all of the e_i are of this form. Suppose e_1 is

$$h(a,p) \equiv h(b,q).$$

Then for some $j > 1$, e_j is of the form

$$h(c,p) \equiv h(d,q)$$

with $c > a$ and $d > b$. The first equation implies by repeated substitutions that

$$h(c,p) \equiv h(b + (a-c), q),$$

and therefore $h(d,q) \equiv h(b + (a-c), q)$. Note that $d \neq b + (a-c)$ lest e_j be a consequence of e_1. Suppose that $d > b + (a-c)$ (the argument is the same in the other case). By this last equation, the congruence generated by $\{e_1, e_j\}$ is such that every element $h(m,q)$ of Y_q is $\equiv$ some $h(e,q)$ with $e \leq d$, namely $h(m - (b + (a-c) \bmod d, q)$. A similar argument shows that this congruence restricted to Y_p also has only finitely many classes. But now we have a contradiction because there cannot be infinitely many independent e_k with $k > j$. That is, let e_k and e_l $(k, l > j)$ be two different equations such that the corresponding terms are in the same classes in the equivalence relation generated by $\{e_1, e_j\}$. Then the equivalence relations generated by $\{e_1, e_j, e_k\}$ and $\{e_1, e_j, e_l\}$ are identical. This contradicts the original hypothesis on the equations $\langle e_i : i \in \omega\rangle$.

Now that we know that all for non-productive Σ, all Σ-congruences are finitely generated, we will outline the argument that all the Σ-congruences are recursive.

It might help at this point to give a simple example; this will motivate the general argument and will also familiarize the reader with some of the notation. Suppose Σ has a single sort **Nat**, *two* constants c and d, and a single successor s. Suppose also that $L = 2$ and that e_1 is $s^4(c) \equiv s^7(d)$ and e_s is $s^6(c) \equiv s^9(d)$. In the notation above, for $i = 1, 2$,

$$Y_1 = \{s^n(c) : n \in \omega\} \text{ and } Y_2 = \{s^n(d) : n \in \omega\}.$$

It is easy to verify that e_2 is a consequence of e_1, and that $s^n(c) \equiv s^m(d)$ iff $m = n + 3$. Of course, $\equiv$ is recursive. The important point is that e_2 is a consequence of e_1. If e_2 had instead been, say, $s^6(c) = s^{10}(d)$, then as in the previous paragraph, we can argue that each element of each Y_i is equivalent under $\equiv$ to an element of a finite set $Z_i \subseteq Y_i$. These equivalences are recursive, and it follows that $\equiv$ is recursive. In fact, adding more equations generates a finer congruence than $\equiv$, and such a congruence is essentially be a relation on $Z_1 \cup Z_2$. Since this set is finite, the finer congruence is also recursive.

The case $K > 2$ is more involved than this example. That is, there are more than two s-primitive terms, and we need to elaborate the argument. Again, we use the represenation above to write e_i as $(h(m_{i0}, n_{i0}), h(m_{i1}, n_{i1}))$ where again $m_{i0}, m_{i1} \in \omega$ and $1 \leq n_{i0}, n_{i1} \leq K$. Consider the relation R on $\{1, \ldots K\}$ defined by $(j, k) \in R$ if one of the equations e_i is an equation between an element of Y_j and an element of Y_k. So R is symmetric and transitive. Now if $(j, k) \in R$ then there is a finite set of equations between an element of Y_j and an element of Y_k such that all equations in $\equiv$ between elements of these two sets are consequences of transitivity and repeated substitutions into these. This is what was shown above. For $(j, k) \in R$, we can fix a finite *independent* set of such equations $E_{j,k}$.

Each Y_j is recursive, and to show that $\equiv$ is recursive is the same as showing that if there are infinitely many terms of sort s, and if $(j, k) \in R$, then $Q = \{(m_1, m_2) : h(m_1, j) \equiv h(m_2, k)\}$ is recursive. If $E_{j,k}$ is a singleton, say $\{h(a, j) \equiv h(b, k)\}$ then $(m_1, m_2) \in Q$ iff $m_1 \geq a$, $m_2 \geq b$, and $m_1 - m_2 = a - b$. If $E_{j,k}$ is not a singleton, then since it is a set of independent equations, there is some b such that every element $h(a, j)$ of Y_j is $\equiv$ to $h(a \bmod b, j)$ (with finitely many exceptions). It follows that in this case Q is very simple indeed and certainly recursive.

For the second assertion of the theorem, we use a general lemma which holds for any signature.

Lemma 9 *Let Σ be productive. Then there is a recursive set $\{t_i : i \in \omega\}$ of ground terms of the same sort such that if $i \neq j$, then t_i is not a subterm of t_j.*

Proof Suppose Σ is productive via patterns p and q of sorts s and r, respectively, and suppose that the domain of each of these is r. To expose the main ideas, we will first argue a special case, when both p and q are compositions of unary functions. Fix a term of sort r, and let $t_i = p(x_r \leftarrow h_q(i + 1, u))$. We can write p as vx_r for some (formal) word v, and q as wx_r for some w. Suppose $i < j$ but $t_i = vw^{i+1}u$ is a subterm of $t_j = vw^{j+1}u$. It follows that either v is a word of w or vice-versa, but this would contradict the definition of productivity.

In the general case, we need some notations and definitions. We will use s (possibly with subscripts, primes, etc.) to denote a sequence of nonzero natural numbers. (This will only occur in the proof of this lemma, and there should be no confusion of our use of 's' to denote a sort.) Then $s^\frown s'$ denotes the concatenation of s with s'. One term sequences are denoted $\langle m \rangle$. We also recall the definition of the subterm t/s of t reached by the sequence s. The definition is by recursion on terms:

$$x/\langle\ \rangle = x \text{ if } x \text{ is a constant symbol}$$

$$f(t_1, \ldots, t_n)/\langle m \rangle^\frown s = t_m/s \text{ if } 1 \leq m \leq n$$

In all other cases t/s is undefined. If t/s is defined, then we say that s is a **path** through t. If s is a path through t, and if t' is any term, then we use $p(s \leftarrow t')$ to denote the result of replacing p/s by

t' in p. In this notation, the following replacement lemma is crucial: If s^* is a path through t_1, and if s^{**} is a path through t_1/s^*, then for all terms t_2,

$$(t_1/s^*)(s^{**} \leftarrow t_2) = (t_1(s^{*\frown}s^{**} \leftarrow t_2))/s^*.$$

We will use often the fact that if s and s' are incompatible paths through t (i.e., neither extends the other), then $p/s' = p(s \leftarrow t')/s'$.

Fix patterns p and q of sorts s and r, respectively, and of domain r which witness the productivity of Σ. There is a unique path s_0 such that $q/s_0 = x_r$, and there is a unique s_1 such that $p/s_1 = x_r$. We write the concatenation of a path s with itself n times as $n \cdot s$. We define terms u_n by recursion on n:

$u_0 = x_r$

$u_{n+1} = u_n(x_r \leftarrow q)$

After proving a series of claims about these terms, we will define the terms t_n as in the statement of the lemma. We will use the fact that $u_{n-m}(x_r \leftarrow u_m) = u_n$ freely in the sequel.

The first claim here is that for all $m < n$, $(n-m) \cdot s_0$ is the only path s through u_n such that $u_n/s = u_m$. We prove this by induction on n_m. If $n - m = 1$, then clearly $u_n/s_0 = u_m$. Any other path would be either a proper subpath of s_0, or a proper extension of s_0, or would be incompatible with this path. If s is a proper subpath of s_0, then we u_n/s_0 is a proper subterm of u_n/s, so the latter cannot equal u_n/s_0. Likewise if s is a proper extenstion of s_0, then u_n/s' is a proper subterm of u_n/s_0, so $u_n/s' \neq u_m$. And if s and s_0 are incompatible paths through u_n, then s is also a path through q, and $u_n/s = q/s$. Now the latter cannot equal u_m because x_r is a subterm of u_m but not of q/s. This completes the case $n - m = 1$. Suppose this claim true for $n - m$, we argue the case for $n + 1 - m$. The existence is immediate — $u_n/(n-m) \cdot s_0 = u_m$. Suppose s is another path through u_n with this property. By the arguments in the case $n - m = 1$, s cannot be compatible with $(n - m + 1) \cdot s_0$, and s cannot be incompatible with s_0. The only alternative is that s is a proper extension of s_0 which is incompatible with $(n - m + 1) \cdot s_0$. Let s' be such that $s_0{}^\frown s' = s$. Then $s' \neq (n-m) \cdot s_0$, but $u_{n-1}/s' = q(x_r \leftarrow u_{n-1})/s_0{}^\frown s' = u_n/s = u_m$. But this contradicts the uniqueness part of the induction hypothesis.

The second claim is that for all $n > 1$, p is not a subterm of u_n and u_n is not a subterm of p. This is true for $n = 1$ by the assumption of this lemma, since $u_1 = q$. So we turn to the induction step. Now u_{n+1} is not a subterm of p as u_n is a subterm of u_{n+1}. Suppose toward a contradiction that p is a subterm of u_{n+1}, and fix s such that $u_{n+1}/s = p$. If s extends s_0, then p would be a subterm of $u_{n+1}/s_0 = u_n$, contradicting the induction hypothesis. We get as well a contradiction if s is a proper subterm of s_0. The only remaining possibility is that s and s_0 are incompatible. But then $u_{n+1}/s = q/s$, and we get a contradiction because now p is a subterm of q.

Our third claim is that for all m and n, $p(x_r \leftarrow u_m)$ is not a subterm of u_n. This is the main step of the proof. Suppose toward a contradiction that $u_n/s = p(x_r \leftarrow u_m)$. Then $u_n/s^\frown s_1 = u_m$, so $s^\frown s_1 = (n-m) \cdot s_0$. We apply the replacement lemma from above with $t_1 = u_n$, $s^* = s$, $s^{**} = s_1$, and $t_2 = x_r$. Then

$$\begin{aligned} p &= (p(x_r \leftarrow u_m))(s_1 \leftarrow x_r) \\ &= (u_n/s)(s_1 \leftarrow x_r) \\ &= (u_n(s^\frown s_1 \leftarrow x_r))/s \\ &= (u_n((n-m) \cdot s_0 \leftarrow x_r))/s \\ &= u_{n-m}/s. \end{aligned}$$

Thus p is a subterm of u_{n-m}, and this is a contradiction.

Next we argue that for $m < n$, $p(x_r \leftarrow u_m)$ is not a subterm of $p(x_r \leftarrow u_n)$. Note that there is only one s' such that $p(x_r \leftarrow u_m)/s' = x_r$, namely $s_1{}^\frown(m \cdot s_0)$. The same hold with '$n$'

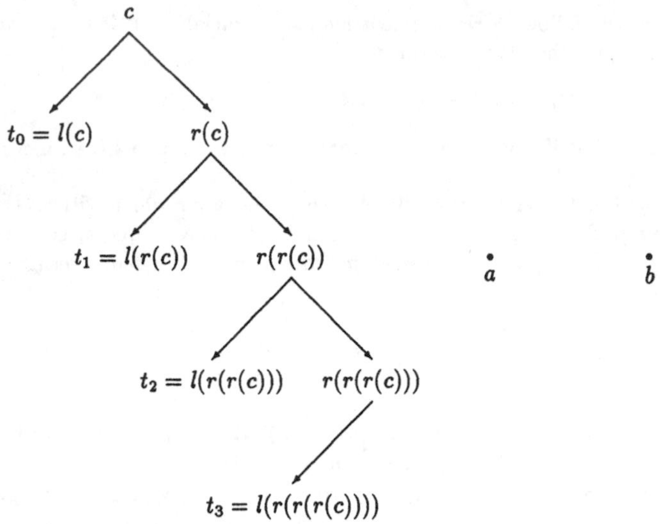

Figure 2: Some terms of T_{Σ_0}.

replacing 'm' of course. Suppose that s is a path such that $p(x_r \leftarrow u_n)/s = p(x_r \leftarrow u_m)$. Then $s^\frown s_1^\frown(m \cdot s_0) = s_1^\frown(n \cdot s_0)$. By the comparability of initial segments, s is either a subsequence of s_1 or an extension. In the former case, $m \geq n$, and in the latter, $p(x_r \leftarrow u_m) = p(x_r \leftarrow u_n)/s$ is a subterm of $p(x_r \leftarrow u_n)/s_1 = u_n$, and this too is a contradiction.

Finally, we can define the terms t_i in the statement of the lemma. Let $\bar{t}$ be any term of sort r. Let $t_i = p(x_r \leftarrow u_i)(x_r \leftarrow \bar{t})$. Suppose that t_i were a subterm of t_j; we will get a contradiction. Let s^* be such that $t_j/s^* = t_i$, and let s^{**} be such that $t_i/s^{**} = \bar{t}$. By the replacement lemma,

$$t_i(s^{**} \leftarrow x_r) = (t_j(s^{*\frown}s^{**} \leftarrow x_r))/s^*.$$

Now $t_i(s^{**} \leftarrow x_r) = p(x_r \leftarrow u_i)$, and $t_j(s^{*\frown}s^{**} \leftarrow x_r) = p(x_r \leftarrow u_j)$. Thus we contradict the result of the last paragraph. ⊣

Lemma 10 *Suppose $F = \{t_i : i \in \omega\}$ is a recursive family of terms of the same sort such that for all $i \neq j$, t_i is not a subterm of t_j. For $X \subseteq \omega$, let $\equiv_X$ be the least Σ-congruence on T_Σ extending $\{(t_{2i}, t_{2i+1}) : i \in X\}$. If $i \notin X$, then (t_{2i}, t_{2i+1}) does not belong to $\equiv_X$. Moreover, $\equiv_X$ and X are Turing equivalent, and if either is (co–)r.e., so is the other.*

Before turning to the proof, we present a picture which showcases the main idea. Consider the signature Σ_0 from Section 1. It has a constants c, a, b, and two unary operations r and l. A portion of the tree of elements of T_Σ is given below in Figure 2 together with some of the members of F. The terms t_i noted in the figure are the ones obtained following Lemma 9 where we take $l(x_s)$ for p, $r(x_s)$ for q, and c for $\bar{t}$.

If $i \in X$, then $\equiv_X$ identifies corresponding elements of the trees below t_{2i} and t_{2i+1}. Now none of the other elements of F lie in either of those trees. This is intuitively the reason that if $(t_{2j}, t_{2j+1}) \in \equiv_X$, then it must be the case that $j \in X$.

Proof of Lemma 10 Fix F and X, define S-sorted relations $\equiv_n$ by recursion on n:

$$\equiv_0 = \{(t_1, t_2) : t_1 = t_2 \in T_\Sigma\} \cup \{(t_{2i}, t_{2i+1}) : i \in X\} \cup \{(t_{2i+1}, t_{2i}) : i \in X\};$$

$$\equiv_{n+1} = \equiv_n \bigcup \{(f(u_1, \ldots, u_n), f(v_1, \ldots, v_n)) : f \in \Sigma \text{ and for all } i, u_i \equiv_n v_i\}.$$

It is clear that each $\equiv_n$ is reflexive and symmetric, and we will show that each $\equiv_n$ is also transitive. Then it follows that $\equiv_X = \bigcup\{\equiv_n : n \in \omega\}$ because the union preserves the action of the function symbols.

We argue by induction on n that each $\equiv_n$ is transitive. For $n = 0$ this is direct, so we prove it for $n+1$ assuming it for n. Suppose that (u_1, u_2) and (u_2, u_3) both belong to $\equiv_{n+1}$. The induction hypothesis applies if both pairs belong to $\equiv_n$. It also applies when neither belongs to $\equiv_0$ because then the outermost symbol of the three u's is identical, and we can peel it away, use our induction hypothesis, and then conclude that $(u_1, u_3) \in \equiv_{n+1}$. So assume that the first pair does not belong to $\equiv_n$ but that the second belongs to $\equiv_0$. We may assume that the second is of the form (t_{2i}, t_{2i+1}) for some $i \in X$ and that the first is of the form $(f(u_1, \ldots, u_n), f(v_1, \ldots, v_n))$. Now by assumption, the first pair does not belong to $\equiv_n$, so it does not belong to $\equiv_0$. Thus some subterm of some v_i is of the form t_{2i^*} or t_{2i^*+1} for some $i^* \in X$. But then either t_{2i^*} or t_{2i^*+1} is a proper subterm of t_{2i+1}, and this contradicts the assumption on F. This completes the induction; the proof of the last two sentences of this lemma is an application of the representation of $\equiv_X$.

Suppose that $i \notin X$. Then by induction on n, the only element of $\equiv_n$ containing t_{2i} as either a first or second component is $(t_{2i}, t_{2i}) \in \equiv_0$. The argument is as in the last paragraph. Therefore, $(t_{2i}, t_{2i+1}) \notin \equiv_X$.

Next we argue that X and $\equiv_X$ are Turing equivalent. Clearly $X \leq_T \equiv_X$, so we show the converse. There are several general observations here. One is that if $Y \subseteq X$, then we can form subequivalence relations $\equiv_n^Y \subseteq \equiv_n$ by starting with Y instead of X. If Y is finite (or even recursive) then the characteristic function of $\equiv_n^Y$ is recursive, uniformly in n and (a code for) Y. There is a recursive function $h(u, v)$ which when given two (codes of) terms, gives a code of the $\subseteq$-minimal set Y such that $(u, v) \in \equiv_Y$ if some such Y exists, and $h(u, v)$ is some error message otherwise. The idea is to look at the trees of subterms of u and v and thereby decide exactly whether there could be equalities between sets of pairs of appropriate elements of F which would imply that $u = v$, and if so what the least set of such pairs is. Formally, the definition of h is by recursion on the pairs of terms. If it is not the case that both u and v have subterms in the recursive set F, then $h(u, v)$ is a code of the empty set if $u = v$; it is some error message if $u \neq v$. (The error message can be any object which is not a natural number.) If u and v are terms in F, say t_i and t_j, then if for some k, either $i = 2k$ and $j = i + 1$, or $j = 2k$ and $i = j + 1$, then $h(u, v) = \{k\}$; if no k exists, give the same error message. Finally, if u and v are $f(u_1, \ldots, u_n)$ and $f(v_1, \ldots, v_m)$ and h is defined on the u's and the v's, then give the error message unless $f = g$. If $f = g$ (and thus $m = n$), set $h(u, v) = \bigcup\{h(u_i, v_i) : i \leq n\}$ (assuming none of these values of h gave error messages). In all cases where there was no error, one checks that

$$(u, v) \in \equiv_X \text{ iff } h(u, v) \text{ is a finite subset of } X.$$

We show by induction on n that if $(u, v) \in \equiv_n$ then $h(u, v)$ is a finite subset of X. And we show by induction on pairs of terms that if $h(u, v) \subset X$, then $(u, v) \in \equiv_X$. The map h is the required Turing reduction showing that $\equiv_X \leq_T X$.

This same relation shows that if X is (co)-r.e, then so is $\equiv_X$, and vice-versa. For example, suppose X is co-r.e., say $i \in X$ iff $(\forall m)\, R(i, m)$ where R is a recursive predicate. Then $(u, v) \in \equiv_X$ iff $(\forall i \in h(u, v))(\forall m)\, R(i, m)$. This concludes the proof of the lemma. ⊣

This concludes the proof of the theorem. The proof also gives a number of equivalences for the non-productivity of a signature.

Corollary 11 *The following are equivalent for a finite signature Σ:*

1. *Σ is non-productive.*
2. *There are only countably many Σ-congruences on the initial algebra.*
3. *Every quotient of T_Σ is computable.*
4. *The conjunction of conditions (1) and (2) of Lemma 7.*
5. *There is no infinite set Q of terms such that if q_1 and q_2 are distinct elements of Q, then q_1 is not a subterm of q_2.*

Finally, we can connect the work we did on finality with the work on congruences of productive signatures to show that there are both initial and final algebras of any r.e. degree.

Theorem 12 *There is a finite signature Σ such that for every r.e. degree* **d***, there is a finite set E of Σ equations such that $T_\Sigma/{\equiv_E}$ is semicomputable and has degree* **d**.

Proof Let Σ^* be the productive signature from above. Fix terms t_i as above as well. These have the property that for any term $t \in T_{\Sigma^*}$, there is at most one i such that t_i is a subterm of t. By Theorem 8, there is an r.e. congruence of degree **d** on T_{Σ^*}. Now there is a way to recursively partition T_{Σ^*} into infinitely many classes in a way that is coarser than each of the $\equiv_X$-classes of T_{Σ^*} produced according to Lemma 9. That is, there is a partition such that for all $X \subseteq \omega$, the $\equiv_X$-classes are finer than the classes of the partition we construct. Let $m : T_{\Sigma^*} \to \omega$ be given by $m(t) = i$ if either t_{2i} or t_{2i+1} is a subterm of t. (Set $m(t) = 0$ if t does not begin with an r.) Then m is well defined, and the congruence $\equiv$ it induces on T_{Σ^*} is coarser than each $\equiv_X$. Fix some such r.e. X of degree **d**. Now Bergstra and Tucker ([3], Theorem 4.2) prove that in this situation there is a finite extension Σ of Σ^*(it's the one containing $ENUM$, $SUCC$, etc.), a recursive Σ'-algebra B, and a set E of conditional Σ-equations with several properties. First, $\equiv_X$ is a Σ'-congruence on B. Second, the reduct of B to Σ^* is exactly T_{Σ^*}. And third, $T_\Sigma/{\equiv_E}$, the initial Σ,E-algebra, is (isomorphic to) $B/{\equiv_X}$. It follows from our remarks on Turing degrees of algebras following Lemma 6 that the degree of $B/{\equiv_X}$ is the same as the degree of $\equiv_X$, and this degree is **d**. So $T_\Sigma/{\equiv_E}$ has this degree as well. ⊣

Theorem 13 *If Σ_0 is the signature from Section 1, and $V = \{s'\}$ is the set of visible sorts, then there is an extension Σ' of Σ_0 such that for every r.e. degree* **d** *there is a finite set E of Σ' equations such that $I_{\Sigma',E,V}$ is computable and $F_{\Sigma',E,V}$ has degree* **d**.

Proof Let X be a co-r.e. set of degree **d**. By the proof of the last theorem, there is a co-r.e. Σ-congruence $\equiv$ of degree **d**. Thus, for $\langle B, p\rangle$ a Gödel coding of T_{Σ_0}, $B/{\equiv}$ is a cosemicomputable algebra of degree **d**. Moreover, it has computable V-behavior because $(B/{\equiv})_{s'}$ contains just the (singletons of) the constants a and b. Indeed, the construction in the previous theorem produced a congruence $\equiv$ whose restriction to the visible carrier is the identity, so that B and $(B/\equiv_{s'})$ are V-behaviorally equivalent. Now we argue exactly as in Theorem 3, taking B for R to get an extension Σ' of Σ_0 with no new sorts and a set E of Σ' equations with the following properties: First, B can be extended to a recursive Σ'-algebra B'. And second, $\equiv$ is a Σ'-congruence on B'. Third, $B'/{\equiv}$ is the final (Σ', E)-algebra. Now by our remarks following Lemma 6, it follows that $F_{\Sigma',E,V} = B'/{\equiv}$ has degree **d**. ⊣

6 References

[1] Bergstra, J.A. and J.-J. Meyer, I/O-Computable data structures. *SIGPLAN Notices*, vol. 16, no. 4 (1981), pp. 27–32.

[2] Bergstra, J.A. and J.V. Tucker, Algebraic specifications of computable and semicomputable data structures. Research Report IW 115, Mathematical Centre, Dept. of Computer Science, Amsterdam, 1979. Revised as University of Leeds CTCS Report 2.86. To appear in *Theoretical Computer Science.*

[3] Bergstra, J.A. and J.V. Tucker, Initial and final algebra semantics for data type specifications: two characterisation theorems. *SIAM J. Comput.*, vol 12 (1983), pp. 366–387.

[4] Bergstra, J.A. and J.V. Tucker, Characterization of computable data types by means of a finite equational specification method. In J. W. de Bakker and J. van Leeuwen (eds.), **Automata, Languages and Programming, Seventh Colloquium,** Noordwijkerhout, Springer Lecture Notes in Computer Science, vol. 81 (1980), pp. 76–90.

[5] Giarrantana, V., F. Gimona, and U. Montanari, Observability concepts in abstract data type specification. **Mathematical Foundations of Computer Science '76**, Springer Lecture Notes in Computer Science, vol. 45, pp. 576-587, 1976.

[6] Goguen, J.A., Realization is universal. *Mathematical System Theory*, vol 6 (1973), pp. 359–374.

[7] Goguen, J.A. and J. Meseguer, Completeness of many-sorted equational logic. *Algebra Universalis* vol. 11 (1985), no. 3, pp. 307-334. Extended abstract appeared in *SIGPLAN Notices,* July 1981, vol. 16, no. 7, pp. 24-37.

[8] Guttag, J.V., *The Specification and Application to Programming of Abstract Data Types*, Ph.D. Thesis, University of Toronto, 1975. Computer Science Department, Report CSRG-59.

[9] Huet, G. and D.C. Oppen, Equations and rewrite rules. In R. Book (ed.), **Formal Language Theory**, Academic Press, 1980, 350:405.

[10] Majster, M.E., Data types, abstract data types and their specification problem. *Theoretical Computer Science*, vol. 8 (1979), pp. 89–127.

[11] Malcev, A.I., Constructive algebras I, *Russian Mathematical Surveys* 16(3), 1961, pp. 77–129.

[12] Meseguer, J. and J.A. Goguen, Initiality, induction, and computability. In M. Nivat and J. Reynolds (eds.), **Algebraic Methods in Semantics**, pp. 459-541, Cambridge University Press, 1985. Also appeared as SRI CSL Tech. Rep. 140, 1983.

[13] Rabin, M., Computable algebra: general theory and theory of computable fields. *Transactions of the American Mathematical Society* 95(1960), pp. 341–360.

[14] Rogers, H., **The Theory of Recursive Functions and Effective Computability**, McGraw-Hill, New York, 1967.

[15] Wand, M., Final algebra semantics and data type extension. *J. Comp. Sys. Sciences*, vol. 19 (1979), pp. 27–44.

Good functors ...

are those preserving philosophy!

Gilles Bernot

Laboratoire de Recherche en Informatique
Bât 490, Université PARIS-SUD
F-91405 ORSAY CEDEX
FRANCE

Abstract :

The aim of this paper is to prevent the abstract data type researcher from an improper, naive use of category theory. We mainly emphasize some unpleasant properties of the *synthesis functor* when dealing with so-called *loose semantics* in a hierarchical approach. All our results and counter-examples are very simple, nevertheless they shed light on many common errors in the abstract specification field.

We also summarize some properties of the category of models "protecting predefined sorts."

Keywords : abstract data types, abstract specifications, category theory, completeness, consistency, initial model, structured specifications.

1. Introduction

In the following pages, we focus our attention on results which seem to be "trivially ensured" in the basic abstract data type framework. We sometimes give proofs ... often counter-examples of such results. In order to get striking counter-examples, we provide very simple ones, if not trivial (mainly based on elementary algebraic properties of natural numbers). Nevertheless, many common errors, or misinterpretations found in the abstract data type litterature result from similar mechanisms. This emphasizes the fact that category theory should be carefully used in the abstract data type field, including for (very) low level concepts.

More provocatively: this paper mainly points out the fact that the synthesis functor F of abstract data types "does not preserve philosophy." However, since about teen years [ADJ 76], it is well known that this functor is crucial for defining a hierarchical, modular approach of abstract specifications!

Some elementary reminders about abstract data types are given in the next section (Section 2). Section 3 discusses about the well known *forgetful* and *synthesis functors*, U and F, associated with a hierarchical approach. Section 4 shows the difficulty of properly defining sufficient completeness and hierarchical consistency with loose semantics. In Section 5, we show what happens when combining enrichments. Lastly, Section 6 discusses about a loose semantics obtained by "protecting" predefined sorts.

The following discussions are mainly centered on pairs [positive fact / proof] (respectively: [negative fact / counter-example]).

2. Elementary reminders

Let us begin with basic definitions and properties [ADJ 76]:
Given a *signature* $\mathbf{\Sigma}$ (i.e. a finite set S of *sorts* and a finite set $\mathbf{\Sigma}$ of *operation*-names with arity in S), a $\mathbf{\Sigma}$-algebra, A, is a heterogeneous set partitioned as $\{ A_s \}_{s \in \mathbf{S}}$, and for each operation-name $op\colon s_1 \cdots s_{n-1} \to s_n$ of $\mathbf{\Sigma}$ there is an operation $op_A\colon A_{s_1} \times \cdots \times A_{s_{n-1}} \to A_{s_n}$. A Σ-*morphism* from A to B is a sort-preserving, operation-preserving application from A to B. This defines a category, denoted by Alg($\mathbf{\Sigma}$) ; it has an initial object: the ground-term algebra $T_{\mathbf{\Sigma}}$.
In the following, a *specification* SPEC will be defined by a signature $\mathbf{\Sigma}$ and a finite set E of *positive conditional equations* of the form:

$$v_1=w_1 \wedge \cdots \wedge v_{n-1}=w_{n-1} \Rightarrow v_n=w_n$$

where v_i and w_i are $\mathbf{\Sigma}$-terms with variables.
Given a specification SPEC, Alg(SPEC) is the full sub-category of Alg($\mathbf{\Sigma}$) whose objects are the $\mathbf{\Sigma}$-algebras which validate each axiom of E. The category Alg(SPEC) has an initial object, denoted by $T_{\mathbf{SPEC}}$ [BPW 82].
Since $T_{\mathbf{SPEC}}$ exists, Gen(SPEC) can be defined as the full sub-category of Alg(SPEC) such that the initial morphism is an epimorphism (i.e. is *surjective*, in our framework). Gen(SPEC) is the category of the *finitely generated algebras*. Our first "fact" will be devoted to the following remark:
It is well known that Gen(SPEC) is a particularly interesting category for the abstract data type computer scientist; nevertheless, this is not exactly due to its large spectrum of morphisms, as reminded below.

Fact 1 : *Morphisms from a finitely generated algebra*

Let Γ be an object of Gen(SPEC) and A an object of Alg(SPEC). The set $Hom_{\mathbf{Alg(SPEC)}}(\Gamma,A)$ contains *at most one element.* Consequently, for all objects X and Y of Gen(SPEC), $Hom_{\mathbf{Gen(SPEC)}}(X,Y)$ contains at most one morphism.

Proof :

By initiality properties, if there exists a morphism μ, then the following triangle commutes:

(For printing facilities, our triangles will become squares!)

$$
\begin{array}{ccc}
\Gamma & -\mu\rightarrow & A \\
\uparrow & & \uparrow \\
\text{initial} & & \text{initial} \\
\text{morphism} & & \text{morphism} \\
| & & | \\
T_{\mathbf{SPEC}} & =\!=\!=\!= & T_{\mathbf{SPEC}}
\end{array}
$$

Thus, the unicity of μ results from the surjectivity of the initial morphism associated with Γ. □

One of the most important aspect of abstract data types is its structured, hierarchical, modular approach. This is obtained by means of *presentations*. A presentation **PRES** over **SPEC** is a new "part of specification" **PRES=<S',Σ',E'>** such that the disjoint union **SPEC+PRES=<S∪S',Σ∪Σ',E∪E'>** is a specification. Sorts and operations of **SPEC** are often called the *predefined* sorts and operations. Relations between the categories Alg(**SPEC**) and Alg(**SPEC+PRES**) are handled by the well known *forgetful functor* and *synthesis functor*:

(U: Alg(**SPEC+PRES**)→Alg(**SPEC**)) and (F: Alg(**SPEC**)→Alg(**SPEC+PRES**)).

The functor F is a left adjoint for the functor U. Consequently, for each **SPEC**-algebra A, there is a particular morphism from A to $U(F(A))$: the morphism deduced from the adjunction unit (or *adjunction morphism*). This morphism is absolutely crucial for the hierarchical approach: it allows to evaluate the modifications performed on A under the action of **PRES**.

Example 0 :

If A is equal to **N** over the signature {*0,succ_*} (without axioms) and if **PRES** adds *pred_* with the axioms [*pred(succ(n))=succ(pred(n))=n*] , then $U(F(\mathbf{N}))$ is isomorphic to **Z**. The unit of adjunction leads to the natural inclusion; and this morphism permits to show that **N** has been modified by adding negative values.

If the axioms were [*pred(succ(n))=n* and *pred(0)=0*], then the unit of adjunction leads to the identity over **N** showing that this second specification of *pred* does not change **N**.

3. Forgetful and synthesis functors

We first present a rather obvious reminder about the *forgetful functor*. Let B be a **SPEC+PRES**-algebra. The forgetful functor removes all subsets B_s where $s\in$ S', and all operations of **Σ'** are forgotten (including those with arity in S only), but *it does not remove any value of predefined sort*: $U(B_s)=B_s$ for each $s\in$ S. For instance, in Example 0, $U(\mathbf{Z})=\mathbf{Z}\neq\mathbf{N}$.

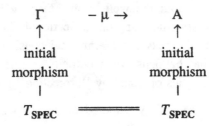

Let us remind the classical definition of the *synthesis functor* (although classical, this definition is the starting point of some misinterpretations!): Let A be a SPEC-algebra and let $T_{\Sigma+\Sigma'(A)}$ be the algebra of $\Sigma+\Sigma'$-terms with variables in A; we denote by *eval*: $U(T_{\Sigma+\Sigma'(A)}) \rightarrow A$ the canonical evaluation morphism. $F(A)$ is the quotient of $T_{\Sigma+\Sigma'(A)}$ by the smallest congruence containing both the fibers of *eval* and the close instanciations of E+E'.

Because E+E' is required in the definition of F (instead of E' alone), $F(A)$ does not only depend on A and **PRES** ; it also depends on **SPEC**.

Fact 2 :

Given a presentation **PRES**, the action of the synthesis functor F over a given, fixed algebra A is highly dependent of the predefined specification.

As outlined in the following example, this fact considerably restricts the possibility of writing "implementation independent" specifications (see for instance [EKMP 80], [SW 82], or [BBC 86a] about abstract implementations).

Example 1 :

Let **SPEC** be a classical specification of *NAT* with operations *0*, *succ_* and *_+_* :

$$x + 0 = x$$
$$x + succ(y) = succ(x+y)$$

Let **SPEC'** be the specification obtained by adding the following axiom to **SPEC** :

$$x + y = x + z \implies y = z$$

The specifications **SPEC** and **SPEC'** have clearly the same initial object: **N** . In fact, they have the same finitely generated algebras (**N** and $\{\frac{\mathbf{Z}}{n\mathbf{Z}}\}$) because the previous axiom can be proved from **SPEC** via structural induction.

Let **PRES** be the presentation adding no sort, adding the operation _×_, and adding the axioms:

$$x \times succ(0) = x \qquad \text{(1 is neutral)}$$
$$x \times succ(y) = x + (x \times y) \qquad \text{(recursive definition)}$$

When **PRES** is shown as a presentation over **SPEC**, $F(\mathbf{N})$ is a model where all terms containing a multiplication by *0* cannot be evaluated. When **PRES** is shown as a presentation over **SPEC'**, $F(\mathbf{N})$ is isomorphic to **N**, because:

$$x + 0 = x = x \times succ(0) = x + (x \times 0)$$

and the simplification axiom of **SPEC'** leads to $0 = x \times 0$.

Notice that, in spite of the fact that **SPEC** and **SPEC'** have the same finitely generated models and the same initial algebra, the presentation **PRES** is not completely specified over the first specification, but is completely specified over the second one.

4. Consistency and completeness

The subject of this section is an examination of some *a priori* possible definitions of the notions of *sufficient completeness* and *hierarchical consistency* with loose semantics. We start with the most loose semantics: the entire category Alg(SPEC). We will show that the simplest definitions are unacceptable for abstract specification purposes.

All the counter-examples provided in this section are based on the following specification+presentation example. Hopefully, we believe that this counter-example cannot be suspected to be too much unusual, complicated or *ad hoc*.

Example 2 :

Let **SPEC** be a specification of natural numbers (for instance the specification given in Example 1) together with a sort *BOOL* and boolean operations *True* and *False*. We consider the presentation **PRES** enriching **SPEC** by an equality predicate *eq?* :

$$\begin{aligned} eq?(0,0) &= True \\ eq?(0,succ(n)) &= False \\ eq?(succ(m),0) &= False \\ eq?(succ(m),succ(n)) &= eq?(m,n) \end{aligned}$$

Looking at this presentation **PRES**, we can affirm that a "good notion" of sufficient completeness (resp. hierarchical consistency) should be satisfied by **PRES**. This example is simply written by taking into account each possible value for the arguments of *eq?*, with respect to the constructors of **SPEC**, moreover there are no axioms between constructors (fair presentation [Bid 82]).

We may of course imagine more sophisticated presentation examples, in particular examples which add new sorts to **SPEC**. But our goal is simply to prevent the abstract data type researcher from using a naive, rather unrealistic definition of sufficient completeness or hierarchical consistency.

4.1. Sufficient completeness

In the initial approach, sufficient completeness is defined as follows [Gau 78].

"*The adjunction morphism associated with the initial algebra is surjective:*"

$$T_{\mathbf{SPEC}} \rightarrow U(F(T_{\mathbf{SPEC}}))$$

This condition exactly means that **PRES** does not add new values to $T_{\mathbf{SPEC}}$. Remind that $F(T_{\mathbf{SPEC}})=T_{\mathbf{SPEC+PRES}}$, due to adjunction properties.

Fact 3 :

The following definition of sufficient completeness is not suitable in the general case: "**PRES** *is sufficiently complete if and only if for all algebras in* Alg(SPEC) *the adjunction morphism is surjective*".

Using Example 2, we convince ourselves of this fact by considering the SPEC-algebra obtained by two copies of N. This algebra, ($N\times\{0,1\}$ and $\{True,False\}$), is not finitely generated, but is an object of Alg(SPEC) by sending the operation-name *0* over the element *(0,0)*, and *succ((n,a))=(succ(n),a)*. Terms of the form *eq?((n,0),(m,1))* cannot be evaluated using the **PRES** axioms of Example 2. Consequently, they add new boolean values, and the adjunction morphism is not surjective.

Fact 4 :

The following two definitions of sufficient completeness are logically equivalent:

1) *the adjunction morphism associated with the initial algebra* $T_{\mathbf{SPEC}}$ *is surjective*

2) *for all algebras in* Gen(SPEC) *the adjunction morphism is surjective.*

Proof :

[2⟹1] is trivial because the initial algebra is finitely generated.

[1⟹2] : let A be a finitely generated SPEC-algebra. By construction of *F*, *F*(*A*) is finitely generated over the signature of SPEC+PRES. Consequently, the image of the initial morphism via the forgetful functor is surjective:

$$U(init_A)\colon U(F(T_{\mathbf{SPEC}}))=U(T_{\mathbf{SPEC+PRES}}) \rightarrow U(F(A))$$

Our conclusion results from the commutativity of the following diagram:

$$\begin{array}{ccc} A & \xrightarrow{\text{adjunction}} & U(F(A)) \\ \uparrow & & \uparrow \\ \textit{surjective} & & \textit{surjective} \\ | & & | \\ T_{\mathbf{SPEC}} & \xrightarrow{\text{surjective}} & U(T_{\mathbf{SPEC+PRES}}) \end{array}$$

Restricting ourselves to finitely generated algebras has several disadvantages. For instance, parameterized presentations require a non finitely generated semantics [ADJ 80].

4.2. Hierarchical consistency

In the initial approach, hierarchical consistency is defined as follows:

"*the adjunction morphism associated with the initial algebra is a monomorphism*" (i.e. is *injective* in our framework).

Fact 5 :

The following definition of hierarchical consistency is not suitable in the general case: "**PRES** *is hierarchically consistent if and only if for all algebras in* Alg(SPEC) *the adjunction morphism is injective*".

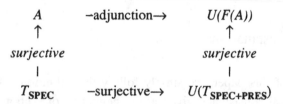

Let us return to Example 2. If we consider the **SPEC**-algebra **Z** (which is a non finitely generated algebra), we get the following inconsistency:

$$True = eq?(0,0) = eq?(0,succ(-1)) = False$$

Restricting hierarchical consistency checks to finitely generated algebras does not yield better results:

Fact 6 :

The following definition of hierarchical consistency is not suitable in the general case: "**PRES** *is hierarchically consistent if and only if for all algebras in* Gen(**SPEC**) *the adjunction morphism is injective*".

Using Example 2 again, we consider a finitely generated algebra of the form $\frac{\mathbf{Z}}{n\mathbf{Z}}$, and we get the following inconsistency:

$$True = eq?(0,0) = eq?(0,n) = eq?(0,succ(n\text{-}1)) = False$$

These facts prove that "defining sufficient completeness on Alg(**SPEC**)" , "defining hierarchical consistency on Alg(**SPEC**)" or "defining hierarchical consistency on Gen(**SPEC**)" are too strong requirements. Extension from the purely initial semantics to a loose semantics must be done more carefully.

5. Combining presentations

In the remainder of this paper, we simply follow the definitions of sufficient completeness and hierarchical consistency given at the beginning of sections 4.1 and 4.2 (i.e. the initial approach). Given a specification **SPEC**, we consider two presentations $\mathbf{PRES}_1$ and $\mathbf{PRES}_2$ *with disjoint signatures.*
Let **PRES** be the union of $\mathbf{PRES}_1$ and $\mathbf{PRES}_2$, we care about the sufficient completeness and hierarchical consistency of **PRES**. In spite of the strong hypothesis described here, we have sometimes to be careful, as detailed in the following two subsections.

5.1. Sufficient completeness

Fact 7 :

If $\mathbf{PRES}_1$ and $\mathbf{PRES}_2$ are both sufficiently complete over **SPEC**, then $\mathbf{PRES}=\mathbf{PRES}_1+\mathbf{PRES}_2$ remain sufficiently complete. Moreover, under the same hypothesis, $\mathbf{PRES}_2$ is sufficiently complete over $\mathbf{SPEC}+\mathbf{PRES}_1$.

Proof : *(using elementary tools)*

$T_{\mathbf{SPEC}+\mathbf{PRES}_1+\mathbf{PRES}_2}$ is the quotient of $T_{\mathbf{\Sigma}+\mathbf{\Sigma}_1+\mathbf{\Sigma}_2}$ by the smallest congruence containing the close instanciations of the SPEC+$PRES_1$+$PRES_2$ axioms [BPW 82]. Consequently, it suffices to prove that each $\Sigma+\Sigma_1+\Sigma_2$-ground-term of sort in S (resp. in $S+S_1$) belongs to the equivalence class of a Σ-term (resp. $\Sigma+\Sigma_1$-term). This can be trivially proved via structural induction. □

Obviously, the converse is false: the sufficient completeness of PRES does not imply the sufficient completeness of $PRES_1$ or $PRES_2$.

5.2. Hierarchical consistency

Fact 8 :

The hierarchical consistency of $PRES_1$ and $PRES_2$ over SPEC *does not imply* the hierarchical consistency of PRES=$PRES_1$+$PRES_2$ over SPEC.

Example 3 :

Let SPEC be a specification of natural numbers. Let $PRES_1$ be the presentation simply containing the following axiom:

$$succ(n) = 0 \quad \Rightarrow \quad n = 0$$

$PRES_1$ is clearly consistent (in fact, the premise cannot be satisfied in the initial object, thus this axiom is never applied). Let $PRES_2$ be the presentation adding the operation *pred_* with [*pred(succ(n))=succ(pred(n))=n*] . $PRES_2$ is clearly hierarchically consistent over natural numbers (even though it is not sufficiently complete). The union PRES=$PRES_1$+$PRES_2$ is not hierarchically consistent because from *succ(pred(0))=0* we get:

$$0=pred(0)\ , \quad \text{which leads to} \quad succ(0)=succ(pred(0))=0$$

Another example of the same fact is the following:

Example 4 :

Let $PRES_1$ be the presentation described in Example 2 (adding equality predicate to natural numbers), and let $PRES_2$ be the same presentation as Example 3 before (adding *pred*). $PRES_1$ and $PRES_2$ are clearly hierarchically consistent over natural numbers, but the union PRES=$PRES_1$+$PRES_2$ is not hierarchically consistent because:

$$True = eq?(0,0) = eq?(0,succ(pred(0))) = False$$

(a similar example was first presented in [EKP 80], for abstract implementation purposes).

Fact 9 :

If PRES=$PRES_1$+$PRES_2$ is hierarchically consistent over SPEC , then $PRES_1$ and $PRES_2$ are hierarchically consistent over SPEC.

Proof :

Assume that $\mathbf{PRES}_1$ is not consistent: the morphism from $T_{\mathbf{SPEC}}$ to $U(T_{\mathbf{SPEC+PRES_1}})$ is not injective. Since the following diagram commutes, the adjunction morphism from $T_{\mathbf{SPEC}}$ to $T_{\mathbf{SPEC+PRES_1+PRES_2}}$ is not injective:

$$\begin{array}{ccc}
U_1(T_{\mathbf{SPEC+PRES_1}}) & \rightarrow & U(T_{\mathbf{SPEC+PRES_1+PRES_2}}) \\
\uparrow & & \uparrow \\
\mathbf{PRES}_1 & & \mathbf{PRES}_1\mathbf{+PRES}_2 \\
\textit{adjunction morphism} & & \textit{adjunction morphism} \\
| & & | \\
T_{\mathbf{SPEC}} & = & T_{\mathbf{SPEC}}
\end{array}$$

(the horizontal arrow is the forgetful of the adjunction morphism for $\mathbf{PRES}_2$ over $\mathbf{SPEC+PRES}_1$).

It results that **PRES** is not hierarchically consistent over **SPEC**. □

Fact 10 :

If $\mathbf{PRES}_1$ and $\mathbf{PRES}_2$ are both hierarchically consistent and sufficiently complete over **SPEC**, then $\mathbf{PRES=PRES_1+PRES_2}$ too. Moreover, under the same hypothesis, $\mathbf{PRES}_2$ is hierarchically consistent and sufficiently complete over $\mathbf{SPEC+PRES}_1$.

(This fact is well known; a demonstration with conditional axioms, including exception handling, can be found in [Ber 86]).

6. Loose semantics with "Protect"

Clearly, abstract specifications do not necessarily directly lead to executable specifications. It is often convenient to specify some operations via "universal properties." For instance the subtraction can be specified via:

$$z - y = x \quad \Leftrightarrow \quad x + y = z$$

Sometimes, such axioms may lead to uncompletely specified presentations, as in the following example.

Example 5 :

Let **SPEC** be an initial specification of integers with operations *0*, *succ_*, *pred_*, *_+_*, *_−_* and *_×_* . Let us specify a presentation **PRES** adding the operation *_div_* as follows:

$$\begin{array}{rcl}
0 \leq (a - (b \times (a \; div \; b))) & = & True \\
(a - (b \times (a \; div \; b))) < b & = & True
\end{array}$$

These axioms characterize *(a div b)* among all integers *finitely generated* with respect to *succ* and *pred*. However, in the initial model $T_{\mathbf{SPEC+PRES}}$, the term

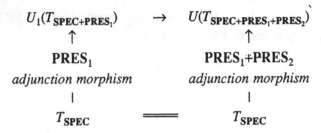

(a div b) is not reached by *succ* and *pred*. Its value is only a unreachable value such that the (unreachable) remainder $(a - (b \times (a\ div\ b)))$ returns the specified boolean values when compared with *0* and *b*.
Consequently, this presentation is uncompletely specified according to the usual definition of sufficient completeness.

In such examples, the only interesting models are those which do not modify the predefined initial model (**Z**). This leads to a (loose) semantics where models are those *protecting* predefined sorts [Kam 80]. Indeed, when writing relatively large specifications, this semantics seems to be highly suitable (ASL [Wir 82] [SW 83], PLUSS [Gau 84], OBJ [FGJM 85], LARCH [GH 83] ...).

Let us define the associated category:

Definition : *The "Protect" category*

Let **SPEC** be a specification and let **PRES** be a presentation over **SPEC**. The category of **PRES**-*models protecting* **SPEC** is the full subcategory of Alg(**SPEC**+**PRES**) whose objects are the **SPEC**+**PRES**-algebras *A* such that *U*(*A*) is isomorphic to the initial predefined algebra $T_{\mathbf{SPEC}}$. We denote this category by Pro(**SPEC**,**PRES**).

Notice that the object class of Pro(**SPEC**,**PRES**) can be empty.

Fact 11 :

If Pro(**SPEC**,**PRES**) is not an empty category, then **PRES** is hierarchically consistent over **SPEC**.

(Here, consistency is defined with respect to the initial algebra $T_{\mathbf{SPEC}}$ only)

Proof :

If $T_{\mathbf{SPEC+PRES}}$ is inconsistent over $T_{\mathbf{SPEC}}$, then a fortiori all **SPEC**+**PRES**-algebras are inconsistent over $T_{\mathbf{SPEC}}$ (because $T_{\mathbf{SPEC+PRES}}$ is minimal). □

Fact 12 :

Even if **PRES** is consistent over **SPEC**, Pro(**SPEC**,**PRES**) may be empty.

Example 6 :

Let **SPEC** be the boolean specification with *True* and *False*. Let **PRES** be a specification of *SET(BOOL)* with ∅, *insert*, ∈ and *choose* :

$$\begin{array}{rcl} True \in \varnothing & = & False \\ False \in \varnothing & = & False \\ b \in insert(b',X) & = & b{=}b' \text{ or } b{\in}X \\ choose(X) \in X & = & True \end{array}$$

This specification is clearly hierarchically consistent (even though it is not sufficiently complete). However, the Protect category is empty, because the term *choose(∅)* can neither be equal to *True* nor to *False* (both choices induce *True=False*).
(Fortunately, this example can be easily specified without inconsistency using abstract data types with exception handling [Bid 84] [GDLE 84] [BBC 86b] [Ber 86], or with partial functions [BW 82].)

Fact 13 :
Even if Pro(SPEC,PRES) contains models, it has not necessarily an initial object.

Example 7 :
Let **SPEC** be the boolean specification with *True* and *False*. Let **PRES** be the presentation adding the constant operation *maybe*, without any axiom. Pro(**SPEC**,**PRES**) contains two models, no one is initial.

Fact 14 :
If **PRES** is sufficiently complete over **SPEC**, then either Pro(**SPEC**,**PRES**) has an initial object, either it is empty.

Proof :
If **PRES** is consistent, then the initial model $T_{\mathbf{SPEC+PRES}}$ belongs to Pro(**SPEC**,**PRES**); it is then necessarily initial in Pro(**SPEC**,**PRES**). If **PRES** is not hierarchically consistent, then Fact 11 implies that Pro(**SPEC**,**PRES**) has no object. □

Fact 15 :
There are presentations **PRES** which are not sufficiently complete over **SPEC**, such that Pro(**SPEC**,**PRES**) is not empty and has an initial object.

It suffices to refer to Example 5, where the axioms characterize *div* by a "universal property among integers." The division is incompletely specified according to classical initial definition of sufficient completeness, but Pro(**SPEC**,**PRES**) only contains one model (**Z**) which is necessarily initial.

7. Conclusion

We have investigated how a hierarchical approach of abstract data types, with the notions of *hierarchical consistency* and *sufficient completeness*, could be defined when dealing with so-called *loose semantics.* The results shown in sections 2 to 5 seem to be somewhat pessimistic:

- The synthesis functor is ''implementation dependent'' with respect to the predefined specification (Fact 2).
- Sufficient completeness cannot be checked on all models (Fact 3).
- Hierarchical consistency cannot be checked on all models (Fact 5).
- Hierarchical consistency cannot be checked on all finitely generated models, a smaller class of models must be investigated (Fact 6).
- Combining hierarchically consistent presentations does not result on a hierarchically consistent presentation (Fact 8).

However, we showed some positive results:

- Checking sufficient completeness on all finitely generated algebras is equivalent to check it on the initial algebra only (Fact 4).
- Combining sufficiently complete presentations results on sufficiently complete presentations (Fact 7); the same occurs for presentations that are both sufficiently complete and hierarchically consistent (Fact 10).

In the last section (Section 6), we defined the category of models *protecting predefined sorts.* We have investigated the relations between the classical notions of completeness/consistency and the elementary properties of this category:

- The category is empty if the presentation is not hierarchically consistent, but the converse is false (Facts 11 & 12).
- The category has not necessarily initial models (Fact 13).
  It has initial models if the presentation is sufficiently complete and hierarchically consistent, but the converse is false (Facts 14 & 15).

In conclusion: From facts 2, 3, 5, 6 and 8, we showed that the synthesis functor of classical abstract data types ''does not always preserve philosophy'' when dealing with loose semantics. Moreover, with a loose semantics based on protection of predefined sorts, the corresponding category has few systematic relations with sufficient completeness or hierarchical consistency (facts 11 to 15).

Acknowledgements :

It is a pleasure to express gratitude to Michel Bidoit, Christine Choppy and Marie-Claude Gaudel for encouragements to write this paper and careful proof readings. The title was suggested by Stephane Kaplan.

This work is partially supported by the ESPRIT Project METEOR.

8. References

[ADJ 76] **Goguen J., Thatcher J., Wagner E.** : "*An initial algebra approach to the specification, correctness, and implementation of abstract data types*", Current Trends in Programming Methodology, Vol.4, Yeh Ed. Prentice Hall, 1978. Also : IBM Report RC 6487, Oct. 1976.

[ADJ 80] **Ehrig H., Kreowski H., Thatcher J., Wagner J., Wright J.** : "*Parameterized data types in algebraic specification langages*", Proc. 7th ICALP, July 1980.

[BBC 86.a] **Bernot G., Bidoit M., Choppy C.** : "*Abstract implementations and correctness proofs*", Proc. 3rd STACS, January 1986, Springer-Verlag LNCS 210, January 1986. Also : LRI Report 250, Orsay, Dec. 1985.

[BBC 86b] **G. Bernot, M. Bidoit, C. Choppy** : "*Abstract data types with exception handling : an initial approach based on a distinction between exceptions and errors*", Theoretical Computer Science, Vol. 46, n° 1, p. 13-45, November 1986.

[Ber 86] **Bernot G.** : "*Une sémantique algébrique pour une spécification différenciée des exceptions et des erreurs : application à l'implémentation et aux primitives de structuration des spécifications formelles*", Thèse de troisième cycle, LRI, Université de Paris-Sud, Orsay, Février 1986.

[Bid 82] **Bidoit M.** : "*Algebraic data types: structured specifications and fair presentations*", Proc. AFCET Symposium on Mathematics for Computer Science, Paris, March 1982.

[Bid 84] **Bidoit M.** : "*Algebraic specification of exception handling by means of declarations and equations*", Proc. 11th ICALP, Springer-Verlag LNCS 172, July 1984.

[BPW 82] **Broy M., Pair C., Wirsing M.** : "*A systematic study of models of abstract data types*", Theoretical Computer Sciences, p. 139-174, vol. 33, October 1984.

[BW 82] **Broy M., Wirsing M.** : "*Partial abstract data types*", Acta Informatica, Vol.18-1, Nov. 1982.

[EKMP 80] **Ehrig H., Kreowski H., Mahr B., Padawitz P.** : "*Algebraic implementation of abstract data types*", Theoretical Computer Science, Oct. 1980.

[EKP 80] **Ehrig H., Kreowski H., Padawitz P.** : "*Algebraic implementation of abstract data types: concept, syntax, semantics and correctness*", Proc. ICALP, Springer-Verlag LNCS 85, 1980.

[FGJM 85] **Futatsugi K., Goguen J., Jouannaud J-P., Meseguer J.** : "*Principles of OBJ2*", Proc. 12th ACM Symp. on Principle of Programming Languages, New Orleans, January 1985.

[Gau 78] **Gaudel M-C.** : "*Spécifications incomplètes mais suffisantes de la représentation des types abstraits*", Laboria Report 320, 1978.

[Gau 84] **Gaudel M-C.** : "*A first introduction to PLUSS*", LRI Report, Orsay, December 1984.

[GDLE 84] **Gogolla M., Drosten K., Lipeck U., Ehrich H.D.** : "*Algebraic and operational semantics of specifications allowing exceptions and errors*", Theoretical Computer Science 34, North Holland, 1984.

[GH 83] **Guttag J.V., Horning J.J.** : "*An introduction to the LARCH shared language*", Proc. IFIP 83, REA Mason ed., North Holland Publishing Company, 1983.

[Kam 80] **Kamin S.** : "*Final data type specifications : a new data type specification method*", Proc. of the 7th POPL Conference, 1980.

[SW 82] **Sannella D., Wirsing M.** : "*Implementation of parameterized specifications*", Report CSR-103-82, Department of Computer Science, University of Edinburgh.

[SW 83] **Sannella D., Wirsing M.** : "*A kernel language for algebraic specification and implementation*", Proc. Intl. Conf. on Foundations of computation Theory, Springer-Verlag, LNCS 158, 1983.

[Wir 82] **Wirsing M.** : "*Structured algebraic specifications*", Proc. of AFCET Symposium on Mathematics for Computer Science, Paris, March 1982.

Viewing Implementations as an Institution

Christoph Beierle
LILOG, IBM Deutschland GmbH
Postfach 80 08 80, D-7000 Stuttgart 80
EARN/BITNET: BEIERLE at DSØLILOG

Angelika Voß
GMD, Forschungsgruppe Expertensysteme
Postfach 12 40, D-5205 St. Augustin 1
USENET: AVOSS%GMDXPS at GMDZI

Abstract:

The theory of institutions as introduced by Goguen and Burstall formalizes the notion of a logical system, and it allows to study many aspects of programming methodology independently of the underlying logical system. In this paper we present an institutional study of implementations. As an illustration we show how implementations of loose algebraic specifications constitute an institution. In particular, the horizontal structuring properties of implementations are investigated and it is shown that parameterization and implementation operations are compatible with each other. Conditions are given under which these results carry over to implementations in arbitrary institutions.

1. Introduction

The concept of institution was first introduced by Burstall and Goguen in [BG 80] as a "language" and was studied in detail in [GB 83]. One of its main purposes is "to support as much computer science as possible independently of the underlying logical system" [GB 86]; for instance, the structuring mechanisms of Clear [BG 80] are applicable to any other language that is based upon a suitable institution.

Whereas in languages like Clear the focus of attention has been on specifications and their combinations and instantiation, we propose to view implementation relations between specifications as another type of object in a specification language. It should be possible to develop implementation relations stepwise and alternately with the development of the involved specifications. Like specifications, implementation relations should have a syntactic part expressing the connection between the implemented specification and the implementing one and a semantic part dealing with the relationship on the level of models. There should be a notion of implementation refinement so that implementation variants, differing e.g. in the efficiency of individual implementing operations, can be obtained.

The observations above led us to study implementations in the framework of institutions. Assuming that the underlying specification method is formalized as an institution we show how implementations between such specifications constitute again an institution. For illustration throughout the paper we use a loose algebraic specification method with sorted signatures and heterogeneous strict algebras, but no assumption is made about the types of sentences which could be equations, Horn clauses, first order predicate formulas, constraints, behavioral abstraction etc. In particular, we investigate the horizontal structuring properties of implementations and show that parameterization and implementation operations are compatible with each other. Finally, we state some conditions under which these results carry over to implementations in an arbitrary institution.

In section 2, we describe a two-dimensional development scenario for algebraic specifications and implementations. As a base to formalize this scenario section 3 defines institutions for both specifications and implementations, and Section 4 contains an example development. In Section 5 the horizontal structuring properties of implementations are studied in detail.

Section 6 carries the development over to arbitrary institutions, and Section 7 contains some concluding remarks.

Acknowledgements: We would like to thank Joseph Goguen, Brian Mayoh and Andrzej Tarlecki for some valuable discussions and suggestions on an earlier version of this paper. Thanks also to the anonymous referees for their helpful comments. Most of this work was carried out while the authors were at the University of Kaiserslautern. It was supported partly by the Bundesministerium für Forschung und Technology (IT 8302363) and partly by the Deutsche Forschungsgemeinschaft (SFB 314).

2. A two-dimensional development scenario

Loose abstract data type (ADT) specifications (e.g. [BG 80], [CIP 85]) have in general many non-isomorphic models as opposed to initial or terminal specifications (e.g. [GTW 78], [EM 85], [Wa 79]) that have only isomorphic models. Thus, in a loose approach the process of stepwise refinement is captured by the successive elaboration of a specification: horizontal structuring activities such as extending its signature, adding new axioms and constraints, instantiating subparts, etc. represent design decisions and correspond to a restriction on the level of models or solutions. The specifications thus developed are related via specification morphisms, which we will also call refinements.

However, specification refinements are not sufficiently flexible since they do not allow for a change of data structures [Hoa 72]. Such a change of data structures can be expressed by an implementation in the fixed case as demonstrated in e.g. [EKMP 82]. Hence, the specification development scenario should be extended to include not just specification refinements but also implementation relations in order to describe the transition from more abstract to more concrete specifications.

For fixed ADT specifications there is a well-elaborated notion of implementation ([EKMP 82], see also e.g. [GTW 78], [Ehc 82], [Ga 83]). In [SW 82] Sannella and Wirsing generalize this concept to Clear-like loose specifications. In both cases, the essential components of an implementation are (1) an abstract specification to be implemented, (2) a concrete specification implementing the abstract one, (3) a signature translation from abstract to concrete names, and (4) on the level of models an abstraction function from a concrete algebra to an abstract one. The difference lies in point (4): For implementations between fixed specifications there is only one pair of models with an abstraction function in between, whereas in the loose case many different pairs of models must be taken into account.

Sannella and Wirsing require that for every concrete model there should be some abstract model and an abstraction function connecting them. If such a complete set of triples exists, the concrete specification is said to implement the abstract one, otherwise it does not. In this approach the syntactic level of specifications and the semantical level of models are not clearly distinguished, and it gives no room for a notion of refinement between implementations since there is no way to characterize and restrict the set of triples any further - e.g. by constraints on the concrete or abstract models. Furthermore, the requirement that every concrete model must implement an abstract one seems to be rather restrictive.

Due to these observations and analogous to the idea of loose specifications to consider at first an arbitrary large set of models and to restrict this set stepwise by refining the specification, we think the adequate idea of implementations between loose specifications is to accept all meaningful combinations of an abstract model, a concrete model, and an abstraction function and to restrict them stepwise by refining the implementation.

A simple implementation consisting of an abstract specification, a concrete one, and a signature translation between them denotes the set of all triples consisting of an abstract model, a concrete one, and an abstraction function from the concrete to the abstract model.

Such a tripel is called an implementation model. As in the fixed case, the abstraction function may be partially defined and it must be surjective and homomorphic.

Now we extend these simple implementations to a concept incorporating a notion of refinement between implementations. Such a refinement should restrict the set of implementation models which can be done componentwise by restricting the set of abstract models, restricting the set of concrete models, and restricting the set of abstraction functions.

In the framework of loose specifications the set of models - like the abstract and the concrete ones - is restricted by adding sentences to the respective specification.
Since the abstraction functions operate on both concrete and abstract carriers we propose to view them as algebra operations from concrete to abstract sorts. These operations can be restricted as usually by adding sentences over both the concrete and the abstract signatures extended by the abstraction operation names. These sentences will be called implementation sentences.

Summarizing we propose an implementation specification to be

- a simple implementation
- together with a set of implementation sentences and
- denoting all implementation models of the simple implementation which satisfy the implementation sentences.

Analogously to specifications which consist of a signature in the simplest case, a simple implementation will also be called an implementation signature.

We already claimed that an implementation should be refinable by adding more implementation sentences to it and thus reducing the class of implementation models. This idea is extended analogously to loose ADT specifications by admitting a change of signature: There, a specification morphism is a signature morphism such that the translated sentences of the refined specification hold in the refining specification. Since an implementation contains two specifications, an implementation morphism should consist of two specification morphisms, an abstract one between the abstract specifications and a concrete one between the concrete specifications. With these two morphisms, the sentences of the refined implementation can be translated into sentences over the refining implementation by mapping the sorts and operations according to the two specification morphisms and by mapping the abstraction operation names to the corresponding abstraction operation names in the refining implementation. Thus a refinement between two implementations is given by an abstract and a concrete specification morphism such that the translated sentences of the refined implementation hold in the refining one.

Refinements between implementations may be distinguished by the components being modified: Refinements where only the abstract or concrete specification is changed describe specification development steps, while refinements where the implementation specification is changed without affecting its component specifications describe proper implementation development steps. Both types may be combined into steps where all components are changed simultaneously. All three types occur when switching attention from the development of individual specifications to the entire process of elaborating more abstract specifications and more concrete specifications as implementations thereof. Depicting this process graphically we obtain a two dimensional structure similar as suggested in the CAT paper [GB 80]:

horizontal direction of specification morphisms, expressing the elaboration of specifications, e.g. by extending their signatures and adding sentences

vertical direction of implementations, expressing the transition from more abstract specifications to more concrete ones including changes of data structures

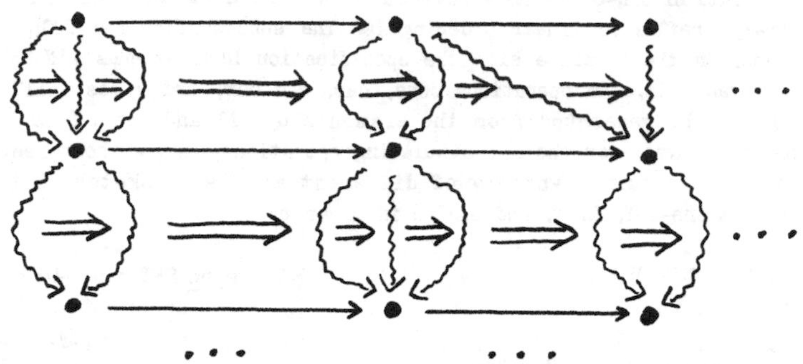

In this figure, • stands for a specification, → for a specification refinement, ~~> for an implemenetation specification, and => for an implementation refinement. In the following, we will formalize these notions and show their horizontal structuring properties; in [BV 85b, 87a] we additionally define a vertical composition operation for implementations so that CAT's double law is satisfied.

3. The institutions of algebraic specifications and implementations

3.1 Introduction

We formalize the above scenario in the framework of institutions ([GB 83]) in the following two subsections. This formalization is rather technical; for a more elaborated version containing further examples we refer to [BV 85a,b].

3.2 The institution of loose specifications

We assume that the loose specifications have equational signatures with error constants and denote strict algebras; no assumptions are made about the types of sentences. Formally, specifications are defined as the theories of the following institution

SPEC-institution := <SIG, EAlg, ESen, $|\hat{=}$ >

where:

- SIG is a category of equational signatures with an error constant error-s for each sort s.
- EAlg is a model functor mapping a signature Σ to all strict Σ-algebras, which have cpos as carriers, strict operations, and the error constants denoting the bottom element.
- ESen is a sentence functor mapping a signature Σ to a set of Σ-sentences.
- $|\hat{=}$ is the strict satisfaction relation.

SPEC denotes the category of theories in the SPEC-institution which will be called (loose) specifications, and Sig: SPEC → SIG is the functor forgetting specifications to their signatures.

We should note that the assumptions about the signature category could be weakened to include predicates as well, we only have to ensure that it is still cocomplete (for a proof that SIG is cocomplete see e.g. [GB 84]); likewise the model functor could be generalized. However, for this paper we stick to the simpler versions for notational convenience.

Throughout this paper we will use various set-by-list implementations for illustration purposes. Here, we start with some of the involved specifications:

On the abstract side we have the specification SET of sets with the empty set as constant, and operations to insert an element, to determine or remove the minimum element in a set, and to test for the empty set or for the membership of an element. Beside standard sets, bags or unreachable elements of sort set need not be excluded. The set elements are described in the specification LIN-ORD which introduces a sort elem with an equality operation eq and an arbitrary reflexive linear ordering le. The subspecification BOOL of LIN-ORD specifies the booleans. On the concrete side the specification LIST extends LIN-ORD to standard lists with the constant nil, the operations cons, car, and cdr, and a test nil? for the empty list. All lists must be generated from the elements by nil and cons. LIST is extended to LIST-S by introducing names for the set simulating operations, but without restricting these operations in order to obtain a variety of different models. A sketch of the presentations of the specifications SET, LIST, and LIST-S is given by:

```
spec LIST = LIN-ORD u
    sorts list
    ops   nil: → list
          cons: elem list → list
          car: list → elem
          cdr: list → list
          nil?: list → bool
    sentences ... < specifying standard
                  lists over elem generated
                  by nil and cons >

spec LIST-S = LIST u
    ops   l-insert: elem list → list
          l-min: list → list
          l-remove-min: list → list
          l-in?: elem list → bool

spec SET = LIN-ORD u
    sorts set
    ops   empty: → set
          insert: elem set → set
          min: set → elem
          remove-min: set → set
          empty?: set → bool
          in?: elem set → bool
    sentences  ... < specifying the set
                   operations with their
                   usual meaning, but not
                   necessarily excluding
                   non-standard sets >
```

In our examples we assume that the error constants are implicitly declared. As sentences we use first order formulas where the quantified variables are not interpreted as bottom elements. Besides, some constraint mechanism is needed to exclude unreachable elements (e.g. initial [HKR 80], data [BG 80], hierarchy [SW 82], or algorithmic constraints [BV 87b]).

3.3 The institution of implementation specifications

We now formalize the notions of implementation signatures and models (Section 3.3.1), and implementation sentences and their satisfaction (Section 3.3.2). In Section 3.3.3 the new institution is summarized and compared with the SPEC-institution.

3.3.1 Implementation signatures and models

According to Section 2, a simple implementation or i-signature IΣ is a triple IΣ = <SPa, σ, SPc> and an i-signature morphism between two i-signatures IΣ_j = <SPa$_j$, σ_j, SPc$_j$> for $j \in \{1,2\}$ is a pair

$$\tau = \langle \rho a, \rho c \rangle : I\Sigma_1 \to I\Sigma_2$$

consisting of an abstract specification morphism ρa: SPa$_1$ → SPa$_2$ and a concrete specification morphism ρc: SPc$_1$ → SPc$_2$.

However, the following compatibility requirement should also be satisfied. Assume we have an i-signature from sets over arbitrary elements to extended lists over arbitrary elements, and another i-signature from sets over natural numbers to extended lists over natural numbers. Then it should not matter whether we first represent sets over arbitrary elements by lists over arbitrary elements and then refine to lists over natural numbers, or if we first refine the sets over arbitrary elements to sets over natural numbers and then represent them as lists over natural numbers. In general that means that the following diagram commutes when viewing ρa and ρc as signature morphisms:

$$\begin{array}{ccc} SPa_1 & \xrightarrow{\sigma_1} & SPc_1 \\ {\scriptstyle \rho a}\downarrow & & \downarrow{\scriptstyle \rho c} \\ SPa_2 & \xrightarrow{\sigma_2} & SPc_2 \end{array}$$

These notions of i-signatures and i-signature morphisms constitute a category. In fact, it is the comma category induced by the functor Sig forgetting specifications to their signatures.

Definition 3.1 [ISIG, i-signature]
Given the forgetful functor Sig: SPEC → Sig, the comma category

ISIG = (Sig↓Sig)

is the category of implementation signatures (i-signatures).

For an i-signature $I\Sigma$ = <SPa,σ,SPc> ε ISIG, SPa is called the abstract specification of $I\Sigma$, SPc the concrete specification, and σ the translation. For an i-signature morphism τ = <ρa,ρc>: $I\Sigma_1 \rightarrow I\Sigma_2$, ρa is called the abstract specification morphism and ρc the concrete one.

Continuing our set-by-list example we can give a first simple i-specification implementing sets by lists

```
ispec I:S/LS =
      isig σS/LS: SET → LIST-S
```

which just consists of the signature morphism

```
σS/LS: Sig(SET) → SIG(LIST-S)
       set        → list,        empty  → nil,
       empty?     → nil?,        insert → l-insert,
       in?        → l-in?,       min    → l-min,
       remove-min → l-remove-min
       x          → x            for x ε Sig(LIN-ORD)
```

translating sort set to list and translating the set operations to their simulating list operations without renaming the signatures of the common subspecifications LIN-ORD and BOOL.

When putting together i-signatures we will be interested in computing their colimits. Conditions for a comma category (F↓G) to be cocomplete when one of the functors F or G is an identity functor are given in e.g. [McL 71] and [GB 84]. These conditions can be generalized to arbitrary functors so that colimits in (F↓G) may be computed from the colimits in the source categories of F and G [BV 85b].

Since the category SIG is cocomplete, SPEC is cocomplete as well and the functor Sig: SPEC → SIG preserves all colimits (c.f. [GB 83]). Therefore, ISIG is cocomplete, too, by the general property of comma categories mentioned above.

Fact 3.2 [colimits] ISIG is cocomplete.

An $I\Sigma$-implementation model, or just $I\Sigma$-i-model MA is a tripel MA = <Ac,α, Aa> consisting of on an abstract algebra Aa, a concrete algebra Ac, and a partial, homomorphic, and surjective abstraction function α from EAlg(σ)(Ac), the forgetful image of Ac along σ, to Aa.

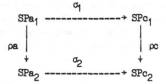

Proceeding analogously to i-signature morphisms, we obtain a notion of i-model morphism as a connection between two i-models. Given another $I\Sigma$-i-model MB = <Bc, β, Ba> an i-model morphism from MA to MB should be a pair <hc,ha> consisting of an SPa-homomorphism ha: Aa → Ba and an SPc-homomorphism hc: Ac → Bc. Analogously to i-signature morphisms the compatibility condition for i-model morphisms should express that it does not matter whether we first abstract Ac-elements with α to Aa-elements and then map them with hc to Bc-elements, or wether we first map the Ac-elements with hc to Bc-elements and then abstract them with β to Ba. This condition may be expressed graphically by requiring that the square

$$\begin{array}{ccc} EAlg(\sigma)(Ac) & \xrightarrow{\alpha} & Aa \\ {\scriptstyle EAlg(\sigma)(hc)}\downarrow & & \downarrow{\scriptstyle ha} \\ EAlg(\sigma)(Bc) & \xrightarrow{\beta} & Ba \end{array}$$

commutes. Note that we must forget hc along σ because we also forget its source and target along σ. We now formalize this description.

<u>Definition 3.3</u> [Σ-p-homomorphism]

Let A, B ε EAlg(Σ) with Σ = <S,Op> ε SIG. An S-sorted family of functions

$$h = \{h_s: A_s \to B_s \mid s \in S\}$$

is a partially-homomorphic Σ-homomorphism (or just Σ-p-homomorphism) iff

$$\forall\, op: s_1 \ldots s_n \to s \in \Sigma \,.$$
$$\forall\, x_1 \in A_{s1} \,.\, \ldots\, \forall\, x_n \in A_{sn} \,.$$
$$h_{s1}(x_1) \neq error\text{-}s_{1B} \;\&\; \ldots \;\&\; h_{sn}(x_n) \neq error\text{-}s_{nB}$$
$$\Rightarrow h_s(op_A(x_1,\ldots,x_n)) = op_B(h_{s1}(x_1),\ldots,h_{sn}(x_n))$$

Note that Definition 3.3 could also be taken as a definition of homomorphism between partial algebras by forgetting the error elements and viewing the algebra operations as partial functions.

In our set-by-list example, consider a LIST-S algebra A_{LS} of standard lists and a SET algebra A_S of standard sets. If the set operations in A_{LS} simulate the standard set operations on sorted lists, a p-homomorhpism from (the forgetful image along $\sigma_{S/LS}$ of) A_{LS} to A_S would map every sorted list to the set of its elements and every unsorted list to error-set.

Since we assume signatures with error constants, every Σ-p-homomorphism is a family of strict functions and it is therefore easy to show that Σ-p-homomorphisms are closed under composition. Thus, together with the strict Σ-algebras they form a category.

<u>Definition 3.4</u> [IMod(IΣ), i-model]

For every IΣ ε ISIG, IMod(IΣ) is the category of IΣ-implementation models (or just IΣ-i-models) with:

<u>objects</u>: <Ac, α, Aa>
where: Ac ε EAlg(SPc), Aa ε EAlg(SPa), and α: EAlg(σ)(Ac) → Aa is a surjective Σa-p-homomorphism

<u>morphisms</u>: <ha, ha>: <Ac, α, Aa> → <Bc, β, Ba>
such that $\beta \circ EAlg(\sigma)(hc) = ha \circ \alpha$

In Section 6 we will give an alternative definition for IMod(IΣ) using a comma category construction similar to the definition of ISIG in Definition 3.1.

Similar to ordinary signatures, every i-signature morphism induces a forgetful functor between the respective model categories in the reverse direction. It is defined componentwise.

<u>Fact 3.5</u> [IMod(τ)]

Let τ = <ρa,ρc>: IΣ_1 → IΣ_2 ε ISIG. Defining IMod(τ) on objects by

IMod(τ)(<Ac,α,Aa>) := <EAlg(ρc)(Ac), EAlg(ρa)(α), EAlg(ρa)(Aa)>

and on morphisms by

IMod(τ)(<hc,ha>) := <EAlg(ρc)(hc),EAlg(ρa)(ha)>

yields a forgetful functor

$IMod(\tau): IMod(I\Sigma_2) \to IMod(I\Sigma_1)$.

Proof (idea)

In order to show that $IMod(\tau)$ is well-defined on objects let $I\Sigma_j = \langle SPa_j, \sigma_j, SPc_j \rangle$ for $j \in \{1,2\}$ and consider $(Ac_2, \alpha_2, Aa_2) \in IMod(I\Sigma_2)$. We have $EAlg(\rho c)(Ac_2) \in EAlg(SPc_1)$ and $EAlg(\rho a)(Aa_2) \in EAlg(SPa_1)$ as required since $\rho c: SPc_1 \to SPc_2$ and $\rho a: SPa_1 \to SPa_2$ are morphisms in SPEC. $Ealg(\rho a)(\alpha)$ has the correct arity because $\rho c \circ \sigma_1 = \sigma_2 \circ \rho a$ due to the properties of τ as a comma category morphism, and α is surjective because $EAlg(\rho a)$ respects surjective morphisms. An analogous argument shows that $Imod(\tau)$ is well-defined on morphisms as well.

Definition 3.6 [IMod]

$IMod: ISIG \to CAT^{op}$ is the modelling functor for implementation signatures.

3.3.2 Implementation sentences and their satisfaction

According to Section 2, implementation sentences over an i-signature $I\Sigma$ shall be expressed over the abstract signature Σa, the concrete signature Σc, and so called abstraction operations to be interpreted as abstraction functions. Thus, we can define the set of $I\Sigma$-implementation sentences to be the set of all ordinary $\psi(I\Sigma)$-sentences, where $\psi(I\Sigma)$ is a suitable equational signature combining Σa, Σc, and the abstraction operations.

We must define $\psi(I\Sigma)$ such that the sentences which are to restrict the abstract algebras do not affect the concrete ones and vice versa. Therefore, taking the set theoretic union of signatures is not suitable, since the abstract and the concrete signatures need not be disjoint. Thus we take the disjoint union (or coproduct). Moreover, for reasons of convenience we will use standard names for the abstraction operations:

Fact 3.7 [ψ]

Let $I\Sigma = \langle\langle \Sigma a, Ea\rangle, \sigma, \langle \Sigma c, Ec\rangle\rangle \in ISIG$ and $\tau = \langle \rho a, \rho c\rangle: I\Sigma \to I\Sigma' \in /ISIG/$.

Defining ψ on objects by

$$\psi(I\Sigma) := \Sigma a \uplus \Sigma c \uplus \{abs\text{-}s_{I\Sigma}: \sigma(s) \to s \mid s \in \Sigma a\}$$

and on morphisms by

$$\psi(\tau) := \rho a \uplus \rho c \uplus \{(abs\text{-}s_{I\Sigma}, abs\text{-}\rho a(s)_{I\Sigma'}) \mid s \in \Sigma a\}.$$

yields a functor

$\psi: ISIG \to SIG$.

Proof (idea)

$\psi(I\Sigma)$ is a well defined signature since $(I\Sigma)$ are operation symbols over $\Sigma a \uplus \Sigma c$. To show that ψ is well defined on morphisms as well, we observe that $\psi(\tau)$ fulfills the properties of a signature morphism for all sorts and operations coming from the abstract and the concrete specification since both ρa and ρc are in particular signature morphisms. For an abstract sort $s \in \Sigma a$ we have

$$\begin{aligned} &\psi(\tau)(abs\text{-}s_{I\Sigma}: \sigma(s) \to s) \\ = {} & abs\text{-}\rho a(s)_{I\Sigma'}: \sigma'(\rho a(s)) \to \rho a(s) \\ = {} & abs\text{-}\rho a(s)_{I\Sigma'}: \rho c(\sigma(s)) \to \rho a(s) \\ = {} & \psi(\tau)(abs\text{-}s_{I\Sigma}): \psi(\tau)(\sigma(s)) \to \psi(\tau)(s) \end{aligned}$$

where the middle equation holds because of the i-signature morphism property $\sigma' \circ \rho a = \rho c \circ \sigma$ of τ, implying the signature morphism property of $\psi(\tau)$ for the abstraction operations.

In the set-by-list example, the implementation sentence

le(e2,e1) = true & eq(e1,e2) = false =>
    abs-set(cons(e1,cons(e2,x))) = error-set

would require that the abstraction function maps every unsorted list to error-set.

Having defined an $I\Sigma$-implementation sentence to be an ordinary $\psi(I\Sigma)$-sentence p we must determine whether an $I\Sigma$-i-model $MA = \langle Ac, \alpha, Aa\rangle$ satisfies p. Since the abstract symbols in $\psi(I\Sigma)$

shall be interpreted by the abstract algebra Aa, the concrete symbols by the concrete algebra Ac, and the abstraction operations by the abstraction function α, we can take the disjoint union of Aa, Ac, and α to obtain a $\psi(I\Sigma)$-algebra interpreting $\psi(I\Sigma)$.

Definition 3.8 [$join_{I\Sigma}(MA)$]
For an i-signature $I\Sigma$ = <SPa, σ,SPc> and an $I\Sigma$-i-model MA = <Ac, α,Aa>

$join_{I\Sigma}(MA) := Aa \,\dot{\cup}\, Ac \,\dot{\cup}\, \alpha$

is the $\psi(I\Sigma)$-algebra A obtained by taking the union of Aa and Ac and by interpreting the abstraction operation names abs-s by the abstraction functions α_s.

All operations coming from Aa and Ac are strict. Since the abstraction functions α_s are p-homomorphic w.r.t. the error constants, they are strict as well, making the algebra $join_{I\Sigma}(MA)$ a strict algebra. (In fact the join operation constitutes a natural transformation join: IMod ==> EAlg $\circ$ ψ ([BV 85b])). Thus the question whether an $I\Sigma$-i-model MA satisfies the implementation sentence p has been reduced to the question whether $join_{I\Sigma}(MA)$ satisfies p in the framework of the SPEC-institution.

Definition 3.9 [ISen]
The implementation sentence functor is given by ISen := Sen $\circ$ ψ: ISIG $\rightarrow$ SET.

Definition 3.10 [$\models^{i}$]
Let $I\Sigma \in$ ISIG, MA $\in$ IMod($I\Sigma$) and p $\in$ ISen($I\Sigma$). MA satisfies p, written MA $\models^{i}_{I\Sigma}$ p iff $join_{I\Sigma}(MA) \models^{e}_{\psi(I\Sigma)}$ p.

Fact 3.11 [satisfaction condition]

$\forall\ \tau: I\Sigma_1 \rightarrow I\Sigma_2 \in ISIG\ .$
$\quad \forall\ MA \in IMod(I\Sigma_2)\ .$
$\quad\quad \forall\ p \in ISen(I\Sigma_1)\ .$
$\quad\quad\quad MA \models^{i}_{I\Sigma 2} ISen(\tau)(p) \iff IMod(\tau)(MA) \models^{i}_{I\Sigma 1} p.$

Proof
We have

$MA \models^{i}_{I\Sigma 2} ISen(\tau)(\rho) \iff join_{I\Sigma}(MA) \models^{e}_{\psi(I\Sigma 2)} ESen(\psi(\tau))(\rho)$

by the definitions of join and $\models^{i}$. Now applying the satisfaction condition of the SPEC-institution we get

$EAlg(\psi(\tau))(join_{I\Sigma 2}(MA)) \models^{e}_{\psi(I\Sigma 1)} p$

which in turn is equivalent to

$join_{I\Sigma 1}(IMod(\tau)(MA)) \models^{e}_{\psi(I\Sigma 1)} p$

since join is a natural transformation. Applying again the definition of $\models^{i}$ yields the desired result

$IMod(\tau)(MA) \models^{i}_{I\Sigma 1} p.$

3.3.3 The institution

Since the satisfaction condition is satisfied, the notions defined above constitute an institution whose theories constitute our implementation specifications.

Definition 3.12 [IMP-institution, IMP]

IMP-institution := <ISIG, ISen, IMod, $\models^{i}$ >

is the institution of implementation specifications. Its theory category is denoted by IMP and is called the category of implementation specifications.

Thus an implementation specification or just i-specification ISP is a pair ISP = <$I\Sigma$,IE> consisting of an i-signature $I\Sigma$ and a set of $I\Sigma$-i-sentences IE, and an i-specification morphism is an i-signature morphism respecting the i-sentences. If $I\Sigma$ = <SPa, σ,SPc> we use the notation

SPa ~~~ISP~~~> SPc

in order to indicate the abstract and the concrete specification of ISP.

Continuing our set-by-list example, we give an i-specification

SET ~~~IS:S/LS~~~> LIST-S

where the abstraction operation for the sort set is restricted in such a way that only sorted lists may represent sets:

```
ispec IS:S/LS = I:S/LS u
     isentences
        (∀ e, e1, e2:elem . ∀ x: list .
              abs-set(cons(e,nil)) = insert(abs-elem(e),empty)                    &
              le(e1,e2) = true & eq(e1,e2) = false =>
                abs-set(cons(e1,cons(e2,x))) = insert(abs-elem(e1),abs-set(cons(e2,x))) &
              le(e2,e1) = true & eq(e1,e2) = false =>
                abs-set(cons(e1,cons(e2,x))) = error-set                                 )
```

From general institution properties we know that the forgetful functor from theories to signatures reflects colimits [GB 83]. Therefore, since ISIG is cocomplete, IMP is cocomplete as well.

Fact 3.13 [colimits] IMP is cocomplete.

Note that we now have a very general condition for IMP being cocomplete: We only require that the signature category of the underlying institution of ADT specifications is cocomplete. In our assumptions of Section 3.2 this is satisfied because the category SIG of equational signatures is cocomplete.

Summarizing the development above the following table shows the corresponding notions for specifications and implementations:

	SPEC-institution	IMP-institution
signature	Σ = <S,Op>	$I\Sigma$ = <SPa,σ,SPc>
model	strict Σ-algebra A	$I\Sigma$-i-model MA = <Ac,α,Aa>
sentence	Σ-sentence p, e.g. 1st order formula over Σ or Σ-constraint	$\psi(I\Sigma)$-sentence p
satisfaction relation	strict satisfaction $A \models^{s}_{\Sigma} p$	$MA \models^{i}_{I\Sigma} p$ defined by $join(MA) \models^{s}_{\psi(I\Sigma)} p$
theory	loose ADT specification SP = <Σ,E>	i-specification ISP = <$I\Sigma$,IE>
refinement	specification morphism $\rho: SP_1 \to SP_2$	i-specification morphism $\tau = \langle \rho a, \rho c\rangle: ISP_1 \to ISP_2$

Having formalized our concept of implementation specifications we will now compare it to some other approaches.

Among the approaches for fixed specifications, the one in [EKMP 80, 82] is most elaborated. It introduces implementations between initial specifications where the syntactic level of specifications and the semantic level of algebras and functors are clearly distinguished. Syntactic conditions and semantic requirements are elaborated which guarantee that the concrete algebra correctly implements the abstract one. The underlying abstraction functions are not defined explicitly but result implicitly from the so-called sorts-implementing operations and equations.

In [Hup 81] Hupbach proposes an implementation concept for the loose canon specifications of [HKR 80] which allows to extend the concrete specification as part of the implementation. If this feature is not exploited, the translation σ from the abstract to the concrete specification must be a specification morphism such that every abstract algebra is the forgetful image under σ of a concrete algebra. Hence the ideas of abstraction functions and changes of data structures are not incorporated.

As mentioned in Section 2, an implementation concept for loose specifications including a change of data structures via abstraction functions has been proposed by Sannella and Wirsing ([SW 82, 82a]). An implementation in their sense is - in our terminology - an i-signature with the condition that for every abstract algebra there is a concrete one with an abstraction function in between. Therefore, we could call an i-specification ISP = <IΣ,IE> with IΣ = <SPa,σ,SPc> an SW-implementation if IE is empty and if for every SPc-model Ac there is an IΣ-i-model <Ac,α,Aa>.

Like the SW-property, the consistency of i-specifications is another model theoretically defined notion for which sufficient criteria should be established. For example, the problem of consistency, which asks whether there is any model at all, arises for i-specifications in direct analogy as for loose specifications and should be solved by the same techniques, e.g. by suitably restricting the class of admissible sentences or by providing a constructively defined model. Such a model may be the program obtained by stepwse refinement in the software development process ([BOV 86]).

The frameworks underlying the approaches of [Sch 82], [GM 82], and [Li 83] are quite different from our own, since the former two consider modules (with internal states) rather than specifications, and the latter considers the semantical level of algebras independent of a syntactical level of specifications. The implementation concept for the kernel language ASL ([SW 83], [Wir 83]) merely requires that the abstract specification is included in the concrete one. It can be so simple since ASL has very powerful specification building operations, but for an arbitrary specification method as we assume it in our approach this concept would be too restrictive. In the recent paper [ST 87] a notion of implementation is presented which as in [SW 82] requires that for every concrete model there exists an implemented abstract one; the correspondence is given essentially by a so-called constructor function.

4. Examples

Our approach to implementations supports a very flexible development technique where both specifications and implementations are elaborated stepwise and alternately. In this section we demonstrate this development technique by continuing our set-by-list example and elaborating some implementations of sets by lists which are summarized in Figure 4.1.

The i-specification SET ~~~I:S/LS~~~> LIST-S was already given in Section 3.3.1. Since I:S/LS contains no i-sentences, its i-models comprise all possible implementations of sets by lists.

I:S/LS can be refined in various ways by adding i-sentences restricting the abstraction operations of sort set, such that

- all lists represent sets (IA:S/LS),
- only lists with unique entries may represent sets (IU:S/LS),
- only sorted lists may represent sets (IS:S/LS), or
- only sorted lists with unique entries may represent sets (ISU:S/LS).

Each of these refinements of I:S/LS represents a proper implementation development step since the abstract and the concrete specifications are left unchanged. The last i-specification refines not only I:S/LS, but also IU:S/LS and IS:S/LS. As an example the i-specification IS:S/LS was given in Section 3.3.3.

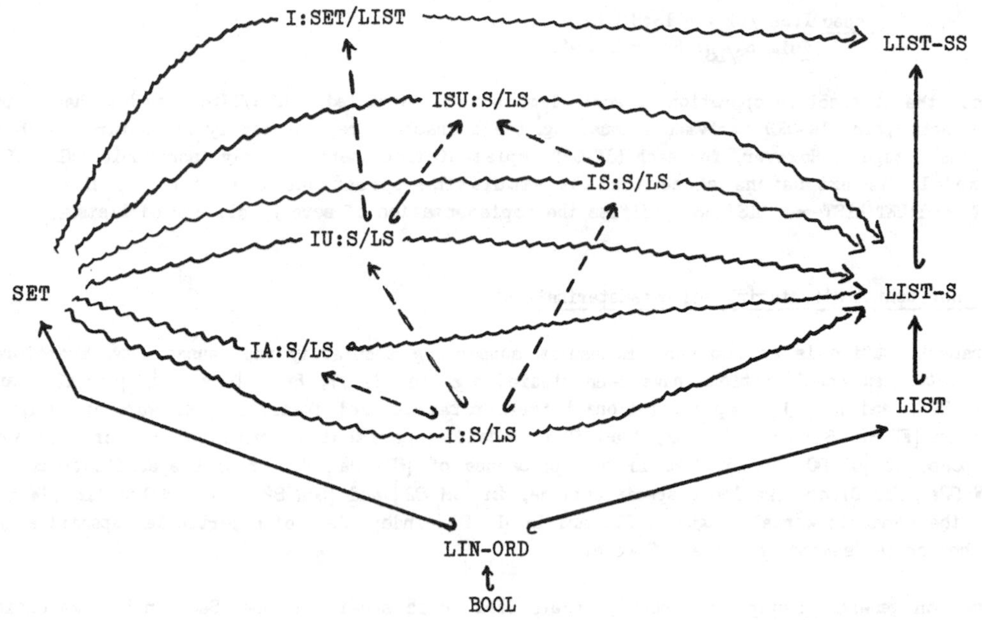

<u>Figure 4.1:</u> The relation between the specifications and i-specifications

By restricting the abstraction operations these alternative i-specifications constrain their i-models not only w.r.t. the abstraction function of sort set, but also w.r.t. the set simulating list operations. Correspondingly, we can specify four refinements of the concrete LIST-S specification by adding sentences fixing the set simulating operations, such that they generate (and operate upon) exactly

(1) - <u>all</u> lists ,
(2) - all lists with <u>u</u>nique entries ,
(3) - all <u>s</u>orted lists , or
(4) - all <u>s</u>orted lists with <u>u</u>nique entries.

To give an example we elaborate the specification LIST-SS for alternative (3):

```
spec LIST-SS = LIST-S u
     sentences
          l-min(nil) = error-elem
          l-remove-min(nil) = nil
          (∀ e: elem . l-insert(e,nil) = cons(e,nil))
          (∀ x: list . ∀ e: elem .
               l-min(cons(e,x)) = e                                     &
               l-remove-min(cons(e,cons(e,x))) = l-remove-min(cons(e,x)) )
          (∀ x: list . ∀ e1, e2: elem .
               le(e1,e2) = true =>
                    l-insert(e1,cons(e2,x)) = cons(e1,cons(e2,x))              &
                    l-in?(e1,cons(e2,x)) = eq(e1,e2))                          &
                    (eq(e1,e2) = false =>
                         l-insert(e2,cons(e1,x)) = cons(e1,l-insert(e2,x)) &
                         l-in?(e2,cons(e1,x)) = l-in?(e2,x)                &
                         l-remove-min(cons(e1,cons(e2,x))) = cons(e2,x)      ) )
```

Now we can in turn refine the i-specification IS:S/LS by replacing the concrete specification LIST-S by its refinement LIST-SS yielding the i-specification

<u>ispec</u> I:SET/LIST = IS:S/LS u
<u>isig</u> $\sigma_{S/LS}$: SET → LIST-SS

Since the abstraction operation of sort elem is not restricted, I:SET/LIST i-models have many non-isomorphic LIN-ORD implementations, e.g. the characters represented by the natural numbers or the integers. However, for each LIN-ORD implementation there are only isomorphic I:SET/LIST i-models as extensions since the set simulating operations are fixed by now. Thus SET ~~~I:SET/LIST~~~> LIST-SS specifies the implementation of sets by all sorted lists.

5. Horizontal structuring and parameterization

Parameterization is an important mechanism supporting modularity and reusability. Therefore, parameterized specifications have been studied e.g. in [BG 77, 80], [TWW 82], [Ehc 82], and [Ehg 81], and [Ga 83]. Implementations between parameterized specifications were investigatd e.g. in [EK 82, 82a] and [SW 82, 82a]. The so-called horizontal composition operation was proposed in [GB 80] and studied in the approaches of [EK 82a] for initial specifications, in [SW 82a] for Clear-like loose specifications, in [GM 82] and [Sch 82] for modules, in [SW 83] for the semantic kernel language ASL, and in [Li 83] independent of a particular specification method on the semantical level of models.

Based on general assumptions about parameterized loose specifications (Section 5.1) we define implementations between them (Section 5.2) and obtain a horizontal composition operation as a generalization thereof (Section 5.3). By a further generalization we arrive at the new concept of parameterized implementations (Section 5.4). We show that the instantiation of implementation specifications is associative and compatible with the instantiation of the involved specifications.

The results are based on the following general theorem stating that the computation of colimits of i-specifications is compatible with the computation of the colimits of the abstract and the concrete specifications. For this theorem, let ISig: IMP → ISIG be the forgetful functor from i-specifications to their i-signatures, and Pa, Pc: ISIG → SPEC be the projection functors yielding the abstract resp. concrete specification of an i-signature.

<u>Fact 5.1</u> [colimits in IMP and SPEC]

For any diagram D: G → IMP the colimit object of D is an i-specification implementing the colimit object of the abstract part of G by the colimit object of the concrete part of G, i.e.

colimit-object(Pa ∘ ISig ∘ D) ~~~colimit-object(D)~~~> colimit-object(Pc ∘ ISig ∘ D)

<u>Proof</u>

ISig ∘ D is a diagram in the comma category ISIG. Since colimits in comma categories can be computed from the colimits in the component categories (c.f. [McL 71] or [GB 84]) we have

colimit-object(ISig ∘ D) =

<colimit-object(Pa ∘ ISig ∘ D), σ, colimit-object(Pc ∘ ISig ∘ D)>

where σ is uniquely determined. Since the functor ISig reflects colimits ([GB 83]), we have furthermore

colimit-object(ISig ∘ D) = ISig(colimit-object(D))

which implies the assertion.

In the following, we will demonstrate how our propositions about parameterization of specifications and implementations and about horizontal composition can be reduced to special cases of the situation given in Fact 5.1.

5.1 Parameterized specifications

We assume a parameterization concept allowing to actualize in a specification SP (the parameterized one) a subspecification SPf (the formal parameter) by another specification SPa (the actual parameter) via a specification morphism δ: SPf $\to$ SPa (the fittig morphism). The result of such an instantiation is given by a pushout construction such that the actual specification is again a subspecification of the pushout object defining the new instance. We will denote it by SP(δ):

```
SP  ---------------→ SP(δ)
↑                p.o. ↑
|                     |
|          δ          |
SPf ---------------→ SPa
```

Note that these assumptions are sufficiently general in order to subsume the parameterization concepts for loose specifications given e.g. in [BG 77, 80] for Clear, in [ZLT 82] for Look, or the dynamic parameterization concept suggested in [BV 85b]. In principle, there is no need to distinguish formal parameter specifications from ordinary specifications when dealing with a loose approach as opposed to a fixed approach, where formal parameters must be interpreted loosely while the other specifications have an initial or final semantics (e.g. [TWW 82], [Ga 83]).

5.2 Implementations of parameterized specifications

We consider an i-specification ISP with abstract specification SP_1 and concrete specification SP_2 where both SP_1 and SP_2 are parameterized specifications. Following the approaches cited before, it seems natural to require that SP_1 and SP_2 have the same formal parameter, say SPf:

```
SP1 ~~~ISP~~~> SP2
   ↖         ↗
       SPf
```

We call ISP an i-specification of parameterized SPEC specifications, or just SPEC-parameterized i-specification. Now given a fitting morphism δ: SPf $\to$ SPa we would like to get an induced implementation ISP′ between the instantiations of SP_1 and SP_2 defined by replacing SPf via δ by SPa:

```
SP1 ~~~ISP~~~> SP2          SP1(δ) ~~~ISP'~~~> SP2(δ)
   ↖         ↗                    ↖          ↗
       SPf  ---------------δ---------------→ SPa
```

In order to determine ISP′ we first observe that requiring SP_1 and SP_2 to have the same formal parameter means that SPf is left unchanged in the implementation. Thus, we may say that SPf is implemented identically by itself in a sub-i-specification ISPf of ISP as shown below where the signature translation of ISPf is the identity on Sig(SPf):

```
SP1 ~~~~~~~~ISP~~~~~~~> SP2
 ↑           ↑           ↑
 |           |           |
SPf ~~~~~~~ISPf~~~~~~~> SPf
```

Furthermore, we observe that both $SP_1(\delta)$ and $SP_2(\delta)$ have SPa as a common subspecification which, analogously to the formal parameter SPf, should be implemented identically by itself. Thus, let

SPa ~~~~ISPa~~~~> SPa

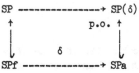

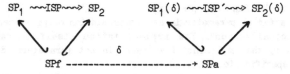

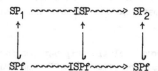

be an i-specification where the signature translation is the identity on Sig(SPa). Now if

$<\delta, \delta>$: ISPf $\rightarrow$ ISPa

is an i-specification morphism in IMP we can construct the following pushout situation in IMP:

```
ISP ----------------→ ISP(<δ,δ>)
 ↑              p.o. ↑
 |                   |
 |        <δ,δ>      |
ISPf ---------------→ ISPa
```

Thus, Fact 5.1 gives us immediately the compatibility of i-specifications with specification instantiation: the pushout object ISP($<\delta, \delta>$) in the diagram above is the induced implementation ISP'.

For illustration we use again our set-by-list example and take the i-specification

SET ~~~ I:SET/LIST~~~> LIST-SS

LIN-ORD

from the end of Section 4 viewing LIN-ORD as formal parameter. As actual parameter we may take a specification NAT of the natural numbers. With $\rho_{LO/N}$: LIN-ORD $\rightarrow$ NAT as the obvious specification morphism we can instantiate I:SET/LIST to an i-specification implementing sets over natural numbers by lists over natural numbers:

SET($\rho_{LO/N}$) ~~~I:SET/LIST($<\rho_{LO/N}, \rho_{LO/N}>$)~~~> LIST-SS($\rho_{LO/N}$)

NAT

or more suggestively, replacing a fitting morphism by its target:

SET(NAT) ~~~I:SET/LIST(NAT)~~~> LIST-SS(NAT)

NAT

Similar situations are discussed in e.g. [EK 82a] and [SW 82a]:
In [EK 82a] the process of actualizing a formal parameter in an implementation of parameterized specifications is called "inner actualization". The implementation itself is called "parameterized implementation"; however, the implementation itself is not parameterized but rather the abstract and the concrete specification.
In [SW 82a] such a situation is called "implementation of parameterized theories". In addition to the setting discussed so far, the formal parameter of the abstract and the concrete specification (called "theory" in [SW 82a]) need not be identical, but may be different specifications which are connected by a specification morphism μ from the concrete to the abstract parameter. The idea is that on the level of models μ induces a forgetful functor which takes every model of the abstract parameter specification to a model of the concrete parameter specification. In [EK 82a] the identity of the parameter specification induces the identity on the level of models.

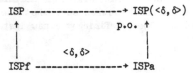

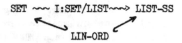

In our approach of i-specifications we allow for a more general relationship between formal and actual parameters. Instead of considering just the individual specifications SPf and SPa in the situation above we take into account the sub-i-specifications ISPf and ISPa. Although both ISPf and ISPa have identical abstract and concrete sides, we do not make any more specific assumptions about these i-specifications except that the signature translation is the identity. Thus, on the level of models a change of data representation may take place, and the relationship between a concrete parameter model and an abstract parameter model may be a nontrivial abstraction function. Moreover, ISPf and ISPa may be restricted by adding i-sentences, the only additional requirement to be fulfilled is that $<\delta,\delta>$ must respect these i-sentences. Therefore, setting e.g.

ISPf := I:loose(SPf)

ISPa := I:loose(SPa)

where I:loose(SPx) := $<<$SPx, $id_{Sig(SPx)}$, SPx$>$, $\emptyset>$ is an unrestricted implementation containing no sentences, allows for arbitrary changes between formal and actual parameter models. On the other hand, setting

ISPf := I:fix(SPf)

ISPa := I:fix(SPa)

with I:fix(SPx) := $<<$SPx, $id_{Sig(SPx)}$, SPx$>$, IEx$>$ excludes any such changes if IEx contains i-sentences enforcing an isomorphism between abstract and concrete parameter models ([BV 85b]).

5.3 Horizontal composition

The horizontal composition operation takes two, in our sense SPEC-parameterized implementations and actualizes the formal parameters of the abstract and the concrete specification of the first implementation by the abstract and the concrete specification of the second implementation, respectively. The horizontal composition property holds if this induces an implementation between the respective instantiations. Thus, horizontal composition is a generalization of the situation discussed before:

We start again with the SPEC-parameterized i-specification ISP having a common formal parameter SPf of the abstract and the concrete specification. We assume another i-specification ISPa whose abstract and concrete sides SPa_1 and SPa_2 need not be identical now

SPa_1 ~~~~ISPa~~~~> SPa_2

and further we assume two fitting morphisms

δ_1: SPf → SPa_1

δ_2: SPf → SPa_2

from the SPEC parameter SPf of the first implementation into the abstract and into the concrete specification of the second implementation. We would like to get an induced implementation ISP(δ_1, δ_2) as indicated in

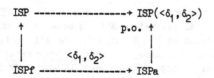

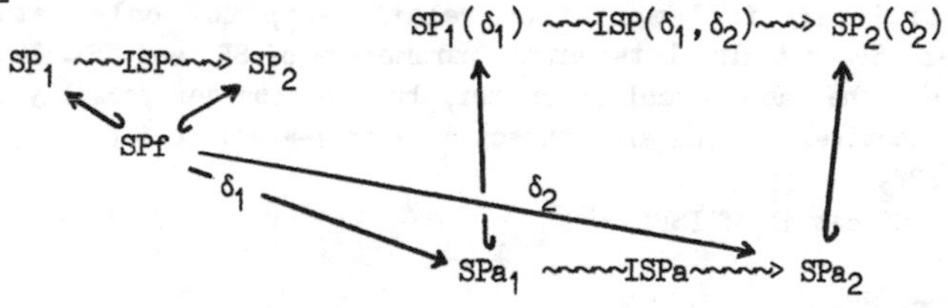

As in the situation discussed before, we conclude that ISP must have a sub-i-specification ISPf implementing the common formal parameter SPf by itself. The condition to be fulfilled by the fitting morphisms δ_1 and δ_2 is again that $<\delta_1,\delta_2>$: ISPf → ISPa respects the i-sentences of ISPf which holds trivially e.g. for ISPf = I:loose(SPf). Now the pushout diagram in IMP

```
ISP  ----------------→ ISP(<δ1,δ2>)
 ↑                p.o. ↑
 |                     |
 |       <δ1,δ2>       |
ISPf ----------------→ ISPa
```

together with Fact 5.1 ensures the horizontal composition property: the pushout object ISP($<\delta_1,\delta_2>$) is the composed implementation

$SP_1(\delta_1)$ ~~~ISP($<\delta_1,\delta_2>$)~~~> $SP_2(\delta_2)$

between the respective specification instantiations SP1(δ1) and SP2(δ2). - As an example, we reconsider the i-specification

SET ~~~I:SET/LIST~~~> LIST-SS

with LIN-ORD as formal parameter specification. Given an arbitrary implementation

LIN-ORD ~~~I:LO/N~~~> NAT

of linear orderings by the natural numbers, we can horizontally compose I:SET/LIST with I:LO/N to obtain an implementation of sets over arbitrary linear orderings by lists over natural numbers. As fitting morphisms we use the identity on LIN-ORD for the abstract side and $\rho_{LO/N}$ for the concrete side:

SET ~~~I:SET/LIST(<$id_{LIN\text{-}ORD}$, $\rho_{LO/N}$>)~~~> LIST-SS($\rho_{LO/N}$)

or, more suggestively:

SET ~~~I:SET/LIST(I:LO/N)~~~> LIST-SS(NAT)

LIN-ORD ~~~I:LO/N~~~> NAT

In addition to "inner actualization" Ehrig and Kreowski ([EK 82a]) also discuss the problem of what they call "outer parameterization": Given a parameterized specification and an implementation, the formal parameter of the specification is actualized once by the abstract and once by the concrete specification of the implementation. If this yields again an implementation, inner actualization and outer parameterization are used to define the process of horizontal composition.

For the horizontal composition of parameterized theories as investigated in [SW 82a], again a specification morphism μ between the formal parameter specifications is taken into account.

5.4 Parameterized implementation specifications

In the previous section we considered an i-specification ISP with abstract specification SP_1 and concrete specification SP_2 where both SP_1 and SP_2 are parameterized specifications with the same formal parameter SPf. Whereas in [SW 82] Sannella and Wirsing already suggest that abstract and concrete formal parameters may not be identical but related to each other by a specification morphism, we will now show how the situation can be generalized in our approach.

The idea is to allow an arbitrary implementation relationship not only between the parameterized parts of SP_1 and SP_2 but also between the parameters of SP_1 and SP_2. Thus, we do not require SP_1 and SP_2 to have the same formal parameter, but we consider possibly different parameters SPf_1 and SPf_2, respectively, which are connected by an i-specification

SPf_1 ~~~ISPf~~~> SPf_2

such that ISPf is a sub-i-specification of ISP:

SP_1 ~~~~~~ISP~~~~~~> SP_2

SPf_1 ~~~~~~ISPf~~~~~~> SPf_2

In accordance to the parameterization concept for specifications sketched in Section 5.1, ISP may be viewed as a parameterized i-specification with parameter ISPf. In order to distinguish this more general setting from the situation discussed in Section 5.2 we will call ISP a parameterized i-specification of parameterized specifications, or just <u>IMP-parameterized</u> i-specification. As far as we know, no generalization comparable to such IMP-parameterized i-specifications has been suggested in the literature.

Now proceeding analogously as in the case of the horizontal composition in the previous section, we can compose ISP with another i-specification ISPa

SPa_1 ~~~ISPa~~~> SPa2

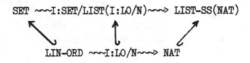

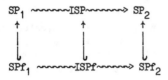

by supplying appropriate fitting morphisms δ_1: $SPf_1 \to SPa_1$ and δ_2: $SPf_2 \to SPa_2$ such that

$\tau := \langle\delta_1, \delta_2\rangle$: ISPf $\to$ ISPa

is an i-specification morphism. This should give us a new i-specification ISP(τ) as indicated in:

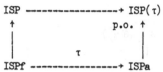

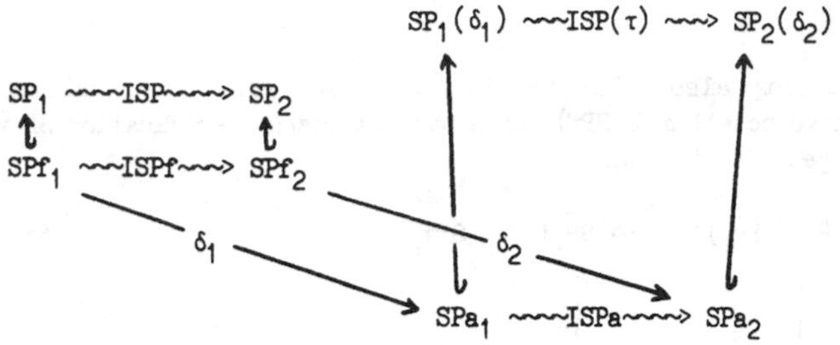

This generalized horizontal composition of ISP and ISPa along τ is defined again by a pushout diagram in IMP

```
ISP ----------------→ ISP(τ)
 ↑               p.o. ↑
 |                    |
 |         τ          |
ISPf ---------------→ ISPa
```

which always exists since IMP is cocomplete. From Fact 5.1 we conclude the general compatibility condition of parameterized implementations: the pushout object ISP(τ) is the induced implementation

$SP_1(\delta_1)$ ~~~ISP(τ)~~~> $SP_2(\delta_2)$

between the respective specification instantiations $SP_1(\delta1)$ and $SP_2(\delta2)$. Thus, the horizontal composition property is still satisfied by the generalized horizontal composition or, in other terms, instantiation of IMP-parameterized i-specifications is compatible with the instantiation of the parameterized specifications involved. - As an example of implementation instantiation, we reconsider the i-specification

```
SET ~~~I:SET/LIST(I:LO/N) ~~~> LIST-SS(NAT)
   ↖             ↑                ↑
    LIN-ORD ~~~I:LO/N~~~~~~~~~~~> NAT
```

which was obtained by a horizontal composition in the previous section. We assume a specification CHAR of the characters and a fitting morphism $\rho_{LO/C}$: LIN-ORD $\to$ CHAR. Now every i-specification

CHAR ~~~I:C/N~~~> NAT

that refines the i-specification I:LO/N via $\langle\rho_{LO/C}, id_{NAT}\rangle$: I:LO/N $\to$ I:C/N may be used as an actual parameter for I:LO/N to instantiate I:SET/LIST(I:LO/N) to an implementation of sets over characters by lists over characters implementing natural numbers:

```
SET(ρLO/C) ~~~I:SET/LIST(I:LO/N)(<ρLO/C,idNAT>)~~~> LIST-SS(NAT)
   ↖                         ↑                           ↑
    CHAR ~~~~~~~~~~~~~~~~I:C/N~~~~~~~~~~~~~~~~~~~~~~~> NAT
```

or more suggestively:

```
SET(CHAR) ~~~I:SET/LIST(I:LO/N)(I:C/N)~~~> LIST-SS(NAT)
   ↖                 ↑                          ↑
    CHAR ~~~~~~~~I:C/N~~~~~~~~~~~~~~~~~~~~~~~> NAT
```

Note that so far we have not considered the actual parameter implementation ISPa to be a SPEC- or IMP-parameterized i-specification. Since every SPEC-parameterized i-specification is also IMP-parameterized, we assume the latter: Let ISPa have ISPx as formal parameter

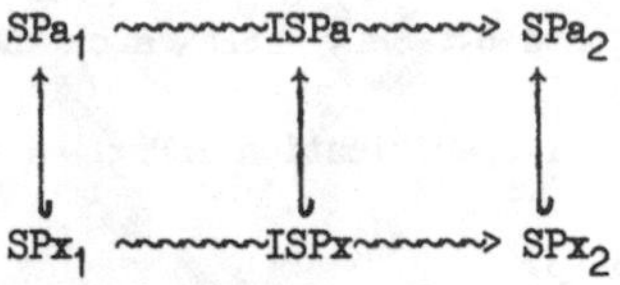

The horizontal composition property also holds in this case. In order to visualize the situation we use the more suggestive notation SP(SPf) for a parameterized specification SP with parameter SPf and neglect the i-specification names:

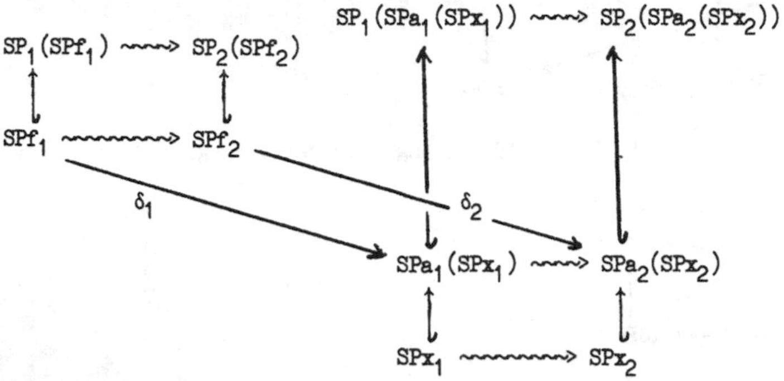

The resulting i-specification

$$SP_1(SPa_1(SPx_1)) \rightsquigarrow ISP(ISPa(ISPx)) \rightsquigarrow SP_2(SPa_2(SPx_2))$$

may in turn be composed horizontally with an i-specification

$$SPb_1(SPy_1) \rightsquigarrow ISPb(ISPy) \rightsquigarrow SPb_2(SPy_2)$$

by replacing ISPx by ISPb(ISPy) along a suitable fitting morphism, yielding

$$(1) \quad SP_1(SPa_1(SPb_1(SPy_1))) \rightsquigarrow ISP(ISPa(ISPb(ISPy))) \rightsquigarrow SP_2(SPa_2(SPb_2(SPy_2))).$$

On the other hand, we could first actualize the formal parameter ISPx of ISPa by ISPb(ISPy) and use the resulting i-specification

$$SPa_1(SPb_1(SPy_1)) \rightsquigarrow ISPa(ISPb(ISPy)) \rightsquigarrow SPa_2(SPb_2(SPy_2))$$

as actual parameter for ISPf in ISP, yielding the same i-specification as in (1). These observations demonstrate the associativity of the generalized horizontal composition which holds essentially since this operation is determined by a pushout construction. Thus, we have:

<u>Fact 5.2</u> Horizontal composition and instantiation of i-specifications are associative.

Finally let us note that, although we only considered the one-parameter case in this section, the generalization to more parameters is straightforward: The general colimit construction of Fact 5.1 covers arbitrary parameter sets and is also applicable in the case of dynamic parameterization where all formal parameter declarations are omitted but where any subspecification may be regarded dynamically as a formal parameter and be replaced by some actual parameter ([ZLT 82], [BV 85b]).

6. Implementations in arbitrary institutions

In the previous sections we used the algebraic SPEC-institution as the underlying ADT specification method to illustrate our implementation concept. While it already abstracts from the types of sentences, we now show how implementation specifications can be built over a completely arbitrary institution of specifications. We assume that

$INST = \langle SIGN, Mod, Sent, \models \rangle$

is an arbitrary institution; let $SPEC_{INST}$ denote its category of theories and Sign: $SPEC_{INST} \to SIGN$ the forgetful functor. We will construct the institution

$IMP(INST) = \langle ISIGN, IModels, ISent, \models^{I} \rangle$

of implementation specifications over INST:

As in the construction in Section 3, the comma category ISIGN = $(Sign \downarrow Sign)$ is the category of implementation signatures.

In order to define the implementation model functor we need a notion of implementation relations between models of the underlying INST institution. For $\Sigma \in SIGN$ let $IRel(\Sigma)$ denote the class of all implementation relations between Σ-models. Thus, an implementation model for $I\Sigma = \langle SPa, \sigma, SPc \rangle \in ISign$ is a tripel

$MA = \langle Ac, \alpha, Aa \rangle$

with $Ac \in Mod(SPc)$, $Aa \in Mod(SPa)$, and α: $Mod(\sigma)(Ac) \to Aa \in IRel(\Sigma a)$ where $\Sigma a = Sign(SPa)$.

Following the argumentation for the algebraic case in Section 3, an implementaion model morphism $\langle hc, ha \rangle$ from MA to $MB = \langle Bc, \beta, Ba \rangle$ connects the abstract and the concrete models where a natural compatability condition is that the square

$$\begin{array}{ccc} Mod(\sigma)(Ac) & \xrightarrow{\alpha} & Aa \\ {\scriptstyle Mod(\sigma)(hc)}\downarrow & & \downarrow{\scriptstyle ha} \\ Mod(\sigma)(Bc) & \xrightarrow{\beta} & Ba \end{array}$$

commutes. However, this requires that Σa-model morphisms and Σa-implementation relations can be composed and compared to each other. Thus we assume that each model category $Mod(\Sigma)$ can be extended to a category $ModIRel(\Sigma)$ that also contains all Σ-implementation relations (and possibly others as well), i.e.

$MI: Mod(\Sigma) \to ModIRel(\Sigma)$ and $IRel(\Sigma) \subseteq /ModIRel(\Sigma)/$

where MI is the inclusion functor. Thus, the compatibility condition for implementation model morphisms says that the square above commutes in $ModIRel(\Sigma a)$. - In Section 3, $ModIRel(\Sigma)$ corresponds to the category of strict Σ-algebras with Σ-p-homomorphisms and $IRel(\Sigma)$ is the class of all surjective Σ-p-homomorphisms.

With the definitions above it is easy to show that implementation models and morphisms constitute a category $IModels(I\Sigma)$. Note that we can define this category more concisely by using comma categories: $IModels(I\Sigma)$ is obtained exactly from the comma category

$(MI_{\Sigma a} \circ Mod(\sigma)|_{Mod(SPc)} \downarrow MI_{\Sigma a}|_{Mod(SPa)})$

by taking the subcategory generated by all objects $\langle Ac, \alpha, Aa \rangle$ with $\alpha \in IRel(\Sigma a)$.

In our example in Section 3 we could have defined $IMod(I\Sigma)$ by such a comma category construction where $MI_{\Sigma a}$ would denote the inclusion from $EAlg(\Sigma a)$ into $EAlg(\Sigma a)$ extended by all Σ-p-homomorphisms and where the subcategory construction restricts to objects with surjective α. This example also illustrates why we allow for a subcategory step in the definition of $IModels(I\Sigma)$.

For defining a forgetful functor $IModels(\tau)$ componentwise directly analogously to the construction in Fact 3.5 we need one further assumption: for each signature morphism σ: $\Sigma \to \Sigma'$ there must be a forgetful construction from $IRel(\Sigma')$ to $IRel(\Sigma)$. If $IRel(\sigma)$ denotes this forgetful step, the componentwise definition of Fact 3.5 (with IRel(ρa) applied to α and using Mod instead of EAlg in all other cases) yields the implementation model functor

IModels: ISIGN → CAT^{op}.

For the implementation sentences we assume that IRelSen: SIGN → SET is a functor giving all sentences over implementation relations and that $\models^{r}$ is a satisfaction relation saying when an implemenation relation α satsfies p ε IRelSen(Σ). Generalizing the construction of ψ in Section 3.3.3 the implementation sentence functor

ISent: ISIGN → SET .

is defined by the coproduct

Sent(Σa) × IRelSen(Σa) × Sent(Σc).

Likewise, the implementation satisfaction relation $\models^{I}$ is defined componentwise in the obvious way.

It is interesting to see which of the properties given in Sections 3 - 5 carry over to the general case as constructed above.

Fact 6.1 [cocompleteness of ISIGN]

If the category of ordinary signatures SIGN is cocomplete then the category of implementation signatures ISIGN is cocomplete as well.

Proof

If SIGN is cocomplete the category $SPEC_{INST}$ is cocomplete as well; therefore, the functor Sign respects all colimits (c.f. [GB 83]) and hence the comma category ISIGN = (Sign↓Sign) has all colimits, too.

Fact 6.2 [cocompleteness of $IMP_{IMP(INST)}$]

If SIGN is cocomplete the category of implementation specifications $IMP_{IMP(INST)}$ is cocomplete as well.

Proof

Because of general institution properties since ISIGN is cocomplete (c.f. [GB 83]).

Fact 6.3 [colimits in $SPEC_{INST}$ and $IMP_{IMP(INST)}$]

If SIGN is cocomplete then for any diagram D: G → $IMP_{IMP(INST)}$ the colimit object of D is an implementation specification implementing the colimit object of the abstract part of G by the colimit of the concrete part of G.

Proof

A straightforward generalization of Fact 5.1.

Fact 6.3 ensures that all horizontal structuring properties and the compatibility of parameterization and implementation as elaborated in Section 5 carry over to the general case; the only condition to be satisfied is that the category of the underlying signatures is cocomplete.

Finally, we would like to mention a further generalization. Whereas so far we have considered only implementations between specifications of one institution we could also have two institutions where the first one defines the abstract or implemented objects and the second one the concrete or implementing objects. On the syntactical level we would assume again a translation between signatures of the two institutions, and on the level of models we would assume implementation relations going in the other direction. For instance we could implement first order specifications by specifications with algorithmic sentences, etc. The results given above carry over to this further generalization as well.

7. Conclusions

We presented an institutional study of implementations. It introduces the notions of implementation signatures, - models, and - specifications and it formalizes the transition from a more abstract to a more concrete specification. By providing the notion of refinement between implementations it supports to develop specifications and implementations hand in hand. The concept is illustrated by applying it to algebraic specifications and implementations.

Since implementation specifications form a category, parameterized implementation specifications could be introduced by a pushout construction. This new notion of parameterized implementation specifications generalizes the concepts of implementations between parameterized specifications and the horizontal composition of implementations. It exhibits such desirable properties as associativity and compatibility with the instantiation of the involved specifications.

In this paper we did not address the topic of vertical composition of implementations, which we study in [BV 85b] and [BV 87a]. There, we elaborate the implementation concept proposed here for a particular institution of ADT specifications. An effective procedure is given to convert an implementation specification to a normal form which essentially consists of an ordinary ADT specification. Referring to the normal forms associative vertical composition operations for implementations are defined compatibly on the syntactical and semantical levels. It is shown that horizontal composition and instantiation of paramterized implementations are compatible with vertical composition, allowing to combine implementation specifications interchangeably in different directions with the same result, as required in the double law of [GB 80] and sketched in the figure given in Section 2.

There are several topics in our approach we would like to study in more detail, for instance the treatment of model-theoretic attributes like the SW-property mentioned in Section 4 or the consistency of implementation specifications. For algebraic implementations as given in this paper we think that it will be possible to elaborate sufficient criteria for these attributes as well as for their preservation by the instantiation and composition operations on the basis of [SW 82, 82a], [EKMP 82], and [Urb 85].

References

[BG 77] Burstall, R.M., Goguen, J.A.: Putting Theories together to Make Specifications. Proc. 5th IJCAI, 1977, pp. 1045-1058.

[BG 80] Burstall, R.M., Goguen, J.A.: The semantics of Clear, a specification language. Proc. of Advanced Course on Abstract Software Specifications, Copenhagen. LNCS Vol.86, pp. 292-332.

[BOV 86] Beierle, C., Olthoff, W., Voß, A.: Towards a formalization of the software development process. Proc. Software Engineering '86, Southampton, UK, 1986.

[BV 85a] Beierle, C., Voß, A.: Implementation specifications. In: H.-J. Kreowski (ed): Recent Trends in Data Type Specifications, Informatik Fachberichte 116, Springer, 1985.

[BV 85b] Beierle, C., Voß, A.: Algebraic Specifications and Implementations in an Integrated Software Development and Verification System, Memo SEKI-85-12, FB Informatik, Univ. Kaiserslautern, (joint SEKI-Memo containing the Ph.D. thesis by C. Beierle and the Ph.D. thesis by A. Voß), Dec. 1985.

[BV 87a] Beierle, C., Voß, A.: On implementations of loose abstract data type specifications and their vertical composition. Proc. 4th STACS, LNCS, Vol. 247, 1987.

[BV 87b] Beierle, C., Voß, A.: Theory and practice of canonical term functors in abstract data type specifications. Proc. TAFSOFT'87, Pisa. LNCS, Vol. 250, 1987.

[CIP 85] CIP Language Group: The Munich Project CIP. Vol. I: The Wide Spectrum Language CIP-L. LNCS, VOL. 183, 1985.

[Ehc 82] Ehrich, H.-D.: On the theory of specification, Implementation and Parametrization of Abstract Data Types. JACM Vol. 29, No. 1, Jan. 1982, pp. 206-227.

[Ehg 81] Ehrig, H.: Algebraic Theory of Parameterized Specifications with Requirements, Proc. 6th Colloquium on Trees in Algebra and Programming (E. Astesiano, C. Böhm, eds.), LNCS 112, pp. 1-24, 1981.

[EK 82] Ehrig, H., Kreowski, H.-J.: Parameter Passing Commutes with Implementation of Parameterized Data Types, PROC. 9th ICALP, LNCS 140, pp. 197-211, 1982.

[EK 82a] Ehrig, H., Kreowski, H.-J.: Compatibility of parameter passing and implementation of parameterized data types. TU Berlin, FB Informatik, 1982.

[EKMP 80] Ehrig, H., Kreowski, H.-J., Mahr, B., Padawitz, P.: Compound Algebraic Implementations: an Approach to Stepwise Refinement of Software Systems. Proc. 9th MFCS (P. Dembinski, ed.), LNCS 88, Springer-Verlag, Berlin 1980, pp. 231-245.

[EKMP 82] Ehrig, H., Kreowski, H.-J., Mahr, B., Padawitz, P.: Algebraic Implementation of Abstract Data Types. Theoretical Computer Science Vol. 20, 1982, pp. 209-254, (also:) Bericht Nr. 80-32, Fachbereich Informatik, Techn. Univ. Berlin 1980.

[EM 85] Ehrig, H., Mahr, B.: fundamentals of Algebraic Specificiations 1 - Equations and Initial Semantics, Springer Verlag, 1985.

[Ga 83] Ganzinger, H.: Parameterized Specifications: Parameter Passing and Implementation with respect to Observability. ACM TOPLAS Vol. 5, No.3, July 1983, pp. 318-354.

[GB 80] Goguen, J.A., Burstall, R.M.: CAT, a system for the structured elaboration of correct programs from struc-tured specifications. SRI International, Technical Report CSL-118, Oct. 1980.

[GB 83] Goguen, J.A., Burstall, R.M.: Institutions: Abstract Model Theory for Program Specification. SRI International and University of Edinburgh, January 1983, revised 1985. (see also Proc. Logics of Programming Workshop, LNCS, Vol. 164, 1984.

[GB 84] Goguen, J.A., Burstall, R.M.: Some Fundamental Algebraic Tools for the Semantics of Computation. TCS, Vol 31, No. (1,2) pp. 175 - 209 (Part 1) and No. (3), pp. 263 - 295 (Part 2), 1984.

[GB 86] Goguen, J.A., Burstall, R.M.: A Study in the Foundations of Programming Methodology: Specifications, Institutions, Charters and Parchments. In: D. Pitt, A. Poigne, D. Rydeheard (eds): Category Theory and Computer Programming. LNCS, Vol. 240, 1986.

[GM 82] Goguen, J.A., Meseguer, J.: Universal Realization, Persistent Interconnection and Implementation of Abstract Modules. Proc. 9th ICALP (M. Nielsen/E.M. Schmidt, eds.), LNCS 140, Springer-Verlag, Berlin 1982, pp. 265-281.

[GTW 78] Goguen, J.A., Thatcher, J.W., Wagner, E.G.: An initial algebra approach to the specification, correctness, and implementation of abstract data types, in: Current Trends in Programming Methodology, Vol.4, Data Structuring (ed. R. Yeh), Prentice-Hall, 1978, pp. 80-144. also: IBM Research Report RC 6487, 1976.

[HKR 80] Hupbach, U.L., Kaphengst, H., Reichel, H.: Initial algebraic specifications of data types, parameterized data types, and algorithms. VEB Robotron, Zentrum für Forschung und Technik, Dresden, 1980.

[Hoa 72] Hoare, C.A.R.: Proof of Correctness of Data Representations. Acta Informatica 1 (4), 1972.

[Hup 81] Hupbach, U.L.: Abstract implementation and parameter substitution. Proc. 3rd Hungarian Computer Science Conference, Budapest, 1981.

[Li 83] Lipeck, U.: Ein algebraischer Kalkül für einen strukturierten Entwurf von Datenabstraktionen. Dissertation. Bericht Nr. 148, Universität Dortmund, 1983.

[Sch 82] Schoett, O.: A theory of program modules, their specification and implementation. Draft report, CSR-155-83, Univ. of Edinburgh.

[ST 84] Sannella, D.T., Tarlecki, A.: Building specifications in an arbitrary institution. Symp. Semantics of Data Types, LNCS, Vol. 173, 1984.

[ST 87] Sannella, D.T., Tarlecki, A.: Toward formal development of programs from algebraic specifications: Implementations revisited. Proc. TAPSOFT'87, LNCS, Vol. 249, 1987.

[SW 82] Sannella, D.T., Wirsing, M.: Implementation of parameterized specifications, Proc. 9th ICALP 1982, LNCS Vol. 140, pp 473 - 488.

[SW 82a] Sannella, D.T., Wirsing, M.: Implementation of parameterized specifications, Report CSR-103-82, Univ. of Edinburgh, 1982.

[SW 83] Sannella, D., Wirsing, M.: A kernel language for algebraic specification and implementation. Proc. Intl. Conf. on Foundations of Computing Theory, LNCS 158, 1983

[TWW 82] Thatcher, J.W., Wagner, E.G., Wright, J.B.: Data Type Specification: Parameterization and the Power of Specification Techniques. ACM TOPLAS Vol. 4, No. 4, Oct. 1982, pp. 711-732.

[Urb 85] Urbassek, C.: Ein Implementierungskonzept für ASPIK-Spezifikationen und Korrektheitskriterien. Diplomarbeit, Univ. Kaiserslautern, FB Informatik, 1985.

[Wa 79] Wand, M.: Final algebra semantics and data type extensions. J. Comp. Syst. Sci. 19, 1979.

[Wir 83] Wirsing, M.: Structured algebraic specifications: a kernel language. Habilitation, TU München, 1983.

[ZLT 82] Zilles, S.N., Lucas, P., Thatcher, J.W.: A Look at Algebraic Specifications. RJ 3568 (41985), IBM Research Division Yorktown Heights, New York, 1982.

An interval model for second order lambda calculus

Simone Martini
Dipartimento di Informatica
Università di Pisa
Corso Italia, 40
I-56100 Pisa, Italy.

1. Introduction

In the past few years there has been a lot of interest in the semantics of second order lambda calculus. After the first model constructions (Closure Model: [Scott 80]; Finitary Retractions Model: [McCracken 84]), in [Bruce, Meyer & Mitchell 85] a general notion of model is given, together with a proof of soundness and, under the assumption that every type is not empty, completeness. Shortly afterwards, other models were discovered: the Finitary Projection Model [Amadio, Bruce & Longo 86], which is based on early ideas by Scott; the PER (or "modest") Model by Moggi (see forthcoming papers by Moggi and Hyland, or [Longo 87] for a discussion); the Qualitative Domains Model, in [Girard 85]. Although these models differ in important ways, they all agree in being *extensional* models: the (η) rule (that is: $\lambda x{:}\alpha.Mx = M$ and $\Lambda t.Mt = M$, for x and t not free in M) is sound in all of them. By this, in each type, every function has only one *representative*.

In the first order type-free calculus, on the other hand, no one ever doubted that (η) was a very strong requirement, and much effort was devoted to build non-extensional models, that is structures in which only a *weak extensionality* property (expressed by the (ξ) rule) holds. Models like $P\omega$, D_A, T^ω, are all non-extensional λ–models (see [Barendregt 84] for definitions and details).

As for the (first order) simply typed λ-calculus, a categorical characterization of non-extensional models is given in [Hayashi 85] (semi (or weak) cartesian closed categories); in [Martini 86] it is shown how to obtain a wCCC from the Ideal Model by [MacQueen&Plotkin&Sethi 86], thus obtaining a "concrete" non-extensional model of the simply typed calculus.

The main aim of this paper, thus, is to determine a non-extensional, second order lambda model, by "generalizing", in a sense, the Ideal Model. Let us review that construction and discuss the extension we want to perform.

Let D be a lambda model, and $I,J \subseteq D$ be ideals: in the Ideal Model, function types are interpreted by means of the so-called *simple semantics*:

$$I \to_s J = \{d \in D \mid \forall i \in I \; di \in J\}.$$

As for polymorphic products, they are interpreted by *intersection*: For f:Ideals$\to$Ideals,

$$\forall(f) = \cap\{f(t) \mid t \in \text{Ideals}\}.$$

The meaning to second order terms is then given by means of their *erasure*: to each typed term is associated the element of the domain which is the interpretation of the untyped term obtained erasing all type information and abstractions. As well known, however, this interpretation is not satisfactory for (at least) two reasons: first of all, by interpreting terms with their erasures, it does not seem to be an adequate model of the *explicit* polymorphism; secondly, it is not even weakly extensional (see [Bruce, Meyer & Mitchell 85]).

Suppose now that the set of types could be (isomorphically) embedded in D: we can then imagine a different, more expressive interpretation for the product. Indeed, for $T \subseteq D$ isomorphic to the set of types, set

$$\forall(f) = \{d \in D \mid \forall t \in T \; dt \in f(t)\}.$$

With this interpretation we can maintain the type abstraction in the interpretation of typed terms, thus solving the first problem listed for the ideal model.

As for weak extensionality, the process for interpreting a term M of a function type ($\alpha \rightarrow \beta$, say), is first to obtain form M a function from α to β (and not from the whole D into D, as for the erasure), then to *extend* it to a function from D to D in such a way that different terms which agree on α give raise to the same function, and finally to take its representative in D. The same procedure applies to type abstraction (that is terms M with type $\forall t.\alpha$), from which we obtain a function from T into D that can then be extended to the whole D.

Unfortunately, the simple interpretation above, which makes use of ideals only, is not sufficient. We need to embed it into a richer structure (the *intervals*, [Cartwight 84]), in which all the type formation operations will become continuous functions.

In the next section a short presentation of weak cartesian closed categories (wCCC) is given, together with the construction of a wCCC from the ideals over a λ–model. Section 3 presents the model construction for the second order calculus, while the proof of the main theorems (theorems 3.6 and 3.7) is deferred to the Appendix (Section 5). Section 4 contains a discussion of the properties of the model as well as the sketch of some extensions and a final comment. Some acquaintance with [Bruce, Meyer & Mitchell 86], [MacQueen, Plotkin & Sethi 86], and the syntax of the second order λ-calculus, is needed.

2. An ideal model for $\lambda\beta^t$

We introduce in this section the notion of weak Cartesian Closed Category (wCCC, for short) [Hayashi 85] and show how to build a wCCC from the ideals over a lambda-model.

2.1 Definition A category **C** is a *wCCC* iff

(i) it is cartesian (i.e. it has the terminal object ($1 \in Ob_C$), and the finite products)

(ii) for all $a,b,c \in Ob_C$ there exist

$$b^a \in Ob_C,$$

$$eval_{b,a} \in \mathbf{C}[b^a \times a, b]$$

and $\Lambda_c: \mathbf{C}[c \times a, b] \to \mathbf{C}[c, b^a]$ (in **Set**),

which satisfy:

$$(\beta_{cat}) \qquad eval_{b,a} \circ (\Lambda_c(f) \times id_a) = f$$

$$(\text{nat}-\Lambda) \qquad \Lambda_c(g \circ (h \times id)) = \Lambda_c(g) \circ h$$

for *f*, *g*, *h* in the due types.

(In the sequel we will omit the subscript in *eval* and Λ, whenever possible.)

One could now define, as for CCC's,

$$\Lambda^{-1} = eval \circ \_ \times id.$$

(β_{cat}) than says that $\Lambda^{-1}(\Lambda(f)) = f$; however the lack of the equation

$$(\eta_{cat}) \qquad \Lambda(\text{eval} \circ (\text{h} \times \text{id})) = \text{h}$$

prevents us from proving $\Lambda(\Lambda^{-1}(g)) = g$, that is we have just a retraction between the homsets (in CCC we had isomorphism):

$$\mathbf{C}[c \times a, b] < \mathbf{C}[c, b^a].$$

Remembering that $a \cong 1 \times a$ (where 1 is the terminal object), one has then $\mathbf{C}[a,b] < \mathbf{C}[1,b^a]$, that is, thinking to $\mathbf{C}[1,a]$ as the "points", or "elements", of *a*, b^a is no longer an object exactly representing $\mathbf{C}[a,b]$: there are many points which represent the same morphism. Similarly, $eval \in \mathbf{C}[b^a \times a, b]$ is not (interpreted as) the "functional application".

Remarks (i) Hayashi defines a *semi CCC* as a category with the terminal object, *weak* products and weak exponents.

(ii) For $F,G{:}\mathbf{C} \to \mathbf{D}$ functors, say that there exists a *natural retraction* between *F* and *G* via (ψ,ϕ) (write: $F <_n G$ via (ψ,ϕ)) if *F* is a rectract of *G* in the category of functors and natural trasformations. Then, for **C** a cartesian category, **C** is a wCCC iff for every $a,b \in Ob_C$ there exist an object b^a and a natural retraction

$$\mathbf{C}[\_ \times a, b] <_n \mathbf{C}[\_, b^a] \quad (\Lambda, \Lambda^{-1}).$$

(iii) wCCC's can be neatly characterized by the notion of *semi-adjunction* (or *adjunction of semifunctors)*: a cartesian category **C** is a wCCC iff for each $a \in Ob_C$, the functor $\_ \times a : \mathbf{C} \to \mathbf{C}$ has a right semi-adjoint. (See [Hayashi 85] for definitions and results, or [Wiweger 84] (*pre-adjuction*), or also [Kainen 71] for a different notion of weak adjunction).

As an example of wCCC we introduce **Ide$_D$**, the category of the ideals over a complete algebraic lattice D, which satisfies $(D\to D) < D$ and $(D\times D) < D$, the latter via an ideal preserving retraction.
Remember that a subset X of D is an *ideal* iff it is a non-empty downward and directed closed subset (i.e. it is closed w.r.t. the Scott topology on D). Note first that any ideal X in D, with the induced structure, is a consistently complete algebraic c.p.o. (a domain, for short). Then, for **Dom** the category of domains with continuous functions:

$Ob_{Ide} = \{X \mid X \text{ is an ideal of } D\}$
$\mathbf{Ide}[X,Y] = \mathbf{Dom}[X,Y]$

(i) $\{\bot_D\}$ is the terminal object
(ii) product between ideals is obtained via the ideal preserving retraction $(D\times D) < D$. (If, for instance, we take $D=P\omega$, the pairing

$$<x,y> = \{<0,n> \mid n\in X\} \cup \{<1,n> \mid n\in Y\}, \text{ for } x,y\in P\omega$$

works fine).
(iii) As for exponents, for $X,Y\in Ob_{Ide}$, write $Y^X = \{a\in D \mid Y \supseteq aX\} = X\to_s Y$ (which is an ideal).
Note that any element a of Y^X defines a continuous function (say f_a) in $\mathbf{Dom}[D,D]$; many elements define the same function and for a given $f\in \mathbf{Ide}[X,Y]$, there are obviously many different f_a, in the notation above, which *extend* it, all agreeing on X. The closed structure allows us to choose among them in a canonical and continuous way.

2.2 Proposition Let X be an ideal, and let X^0 be its algebraic elements.
The function $Ext_X : \mathbf{Ide}[X\to Y]\to \mathbf{Ide}[D\to D]$ defined by

$$Ext_X(f) = \lambda x{:}D.\ sup\ \{ f(e_0) \mid e_0\leq x \wedge e_0\in X^0\}$$

is continuous. Moreover $\forall x\in X\ \ Ext_X(f)(x) = f(x)$.

Proof *a.* $Ext_X(f)$ is a continuous function: Let $\mathfrak{I}\subseteq D$ directed; then

$$\begin{aligned} Ext_X(f)(\sup \mathfrak{I}) &= \sup \{ f(e_0) \mid e_0\leq(sup\ \mathfrak{I}) \wedge e_0\in X^0\} \\ &= sup\ \{ f(e_0) \mid \exists d\in \mathfrak{I}\ e_0\leq d \wedge e_0\in X^0\} \\ &= sup\ \{ sup\ \{ f(e_0) \mid e_0\leq d \wedge e_0\in X^0\} \mid d\in \mathfrak{I}\} \\ &= sup\ \{Ext_X(f)(d) \mid d\in \mathfrak{I}\}. \end{aligned}$$

b. For $x\in I$ $Ext_X(f)(x) = f(x)$: X is downward closed.

c. Ext_X is continuous: Let $\mathfrak{R}\subseteq[X\to Y]$ be directed;

$$\begin{aligned} Ext_X(sup\ \mathfrak{R})(x) &= sup\ \{ (sup\ \mathfrak{R})(e_0) \mid e_0\leq x \wedge e_0\in X^0\} \\ &= sup\ \{ sup\ \{f(e_0) \mid f\in \mathfrak{R}\} \mid e_0\leq x \wedge e_0\in X^0\} \end{aligned}$$

$$= sup\ \{\ sup\ \{f(e_0) \mid e_0 \le x \wedge e_0 \in X^0\} \mid f \in \Re\}$$
$$= sup\ \{\ Ext_X\ (f) \mid f \in \Re\}. \qquad \Diamond$$

Finally, for $f \in$ **Ide**$[Z \times X, Y]$, set

$$\Lambda_Z(f) = \lambda z{:}Z.\underline{\lambda}(\lambda d{:}D.\ Ext_X\ (f)\ (z,d)$$
$$eval = \lambda z.(fst\ z)\ (snd\ z)$$

where $\underline{\lambda} \in$ **Dom**[**Dom**$[D,D],D]$ is given by $(D \to D) \lhd D$

By these definitions **Ide**$_D$ is a wCCC: $\Lambda_Z(f) \in$ **Ide**$_D\ [Z, Y^X]$ by the continuity of Ext_X; (β_{cat}) follows from $Ext_X\ (f)$ extending f; (nat-Λ) is proved from the definition of Ext_X and the closed structure:

$$\begin{aligned} \Lambda_Z(f) \circ g\ (x) &= \underline{\lambda}(\ \lambda d{:}D.\ Ext_X\ (f)\ (g(x),d)) \\ &= \underline{\lambda}(\ \lambda d{:}D.\ sup\{f(g(x),y_0) \mid y_0 < y\ \wedge\ y_0 \in X^0\ \}) \\ &= \underline{\lambda}(\ \lambda d{:}D.\ Ext_X\ (f \circ (g \times id))\ (x,d) \\ &= \Lambda_Z(f \circ (g \times id)). \end{aligned}$$

As mentioned in the introduction, the main fact about wCCC's is that they are exactly the models of the first order typed λ-calculus without the (η) rule. By this **Ide**$_D \models \lambda\beta^t$. The next section will extend and generalize this model to the second order calculus.

3. The interval model

Information about the syntax of the second order λ-calculus can be obtained from one of the references. We stress, however, that the equality relation between well typed terms, it is defined here by the α–conversion rule, and the following axiom schemes and inference rules:

(β) $\qquad (\lambda x{:}\alpha.M)N = [N/x]M$

$\qquad\quad (\Lambda t.M)\alpha = [\alpha/t]M$

$$(\xi) \qquad \frac{M=N}{\lambda x{:}\alpha.M = \lambda x{:}\alpha.N} \qquad \frac{M=N}{\Lambda t.M = \Lambda t.N}$$

together the standard rules that make = a congruence.

Note that the equality relation is usually defined by adding to the above axioms, the schemes:

(η) $\qquad \lambda x{:}\alpha.Mx = M \qquad$ for x not free in M

$\qquad\quad \Lambda t.Mt = M \qquad$ for t not free in M

However, the model we will describe is not sound for (η): actually the construction of a model of the only ($\alpha\beta$)-equality and (ξ), was one of the main motivations for this work (see the introduction).

The *intervals* has been introduced in [Cartwight 84] (from which we borrow the following definitions and results) as a generalization of the intervals, with the intended aim to avoid the pathologies caused by the non-monotonic type constructors. Indeed, the set of intervals contains the ideals as a subset, and all the primitive type formation operations on the intervals are extensions of the corresponding ones on ideals.

Let D be a consistently complete, algebraic c.p.o. (a *domain*, for short). For $X \subseteq D$, write $X^0 = \{x \in X \mid x \text{ algebraic}\}$; $\boldsymbol{Ide_D}$ is the set of the ideals on D (i.e. non empty closed sets with respect to Scott topology, [MacQueen, Plotkin & Sethi 86]).

3.1 Definition Let $a, A \subseteq D$ be ideals such that $a \subseteq A$. An *interval* $|a,A|$ on D is the set $\{I \in \boldsymbol{Ide_D} \mid a \subseteq I \subseteq A\}$. The set of intervals over a domain D is denoted $\boldsymbol{Type_D}$.

3.2 Proposition (i) For every domain D, $<\boldsymbol{Type_D}, \leq>$ is a domain, where $|a,A| \leq |b,B|$ iff $|a,A| \supseteq |b,B|$ as sets.

(ii) The maximal elements of $\boldsymbol{Type_D}$ are intervals of the form $|A,A|$ containing a single ideal A.

(iii) $|a,A| \in \boldsymbol{Type_D}$ is algebraic iff there exist two finite sets $I,J \subseteq D^0$ such that

$$a = \downarrow\{w \in D^0 | w \in I\}$$
$$A = D \setminus \uparrow\{w \in D^0 | w \in J\};$$

that is a is algebraic in $\mathbf{Ide_D}$ and A is the complement of an algebraic open set.

Remark $|a,A| \leq |b,B|$ iff $a \subseteq b$ & $A \supseteq B$.

The intuition underlyining the idea of an interval $|a,A|$ is that a type should be characterised not only by the elements that belong to it (the ideal a), but also by a (partial) information on that elements that do not belong (the open $D \setminus A$). An interval $|a,A|$ ($a \neq A$) can then be viewed as a "partial" type, meaning that for $d \in D$ when $d \in A \setminus a$ we actually cannot say if it belongs or not to the type.

Notation Let $\alpha = |a,A|$. $\alpha^+ = a$ and $\alpha^- = A$. (*Warning*: in Cartwright's notation $\alpha^- = D \setminus A$.).

3.3 Definition Let $|a,A|$ and $|b,B|$ intervals. The *function type constructor* over $\boldsymbol{Type_D}$ is defined by $|a,A| \Rightarrow |b,B| = |A \rightarrow_s b, a \rightarrow_s B|$, where "$\rightarrow_s$" is the simple semantics arrow.

3.4 Proposition The function type constructor $\Rightarrow$: $\mathit{Type}_D \rightarrow (\mathit{Type}_D \rightarrow \mathit{Type}_D)$ is continuous and preserves maximal elements (if I and J are maximal intervals (i.e. they are ideals), then $I \Rightarrow J$ is a maximal element.)

Let now $\langle D, \cdot, \lambda \rangle$ be a complete algebraic lattice yielding a λ-model and satisfying the following property:
(*universality*)Every T_0-space X with a countable basis can be embedded in D by a continuous function.

Remark $P\omega$ satisfies these requirements. Engeler's models D_A are complete algebraic lattice: Some of them give also universal domains.

Notation In the sequel we will write ***Type*** and ***Ide*** for Type_D and Ide_D .
$[D \rightarrow D]$ are the continuous function from D into D; e_0 is the generic finite (algebraic, compact) element in D. Using (universality) above, we will identify ***Type*** with the subset of D isomorphic to it; Type^{max} will be the maximal elements of ***Type.***

The following definition introduces the main type constructor for our model and it crucially differs from Cartwright's one, which is essentially an intersection.

3.5 Definition (*product type constructor*)

$\forall$: $[\mathit{Type} \rightarrow \mathit{Type}] \rightarrow \mathit{Type}$

$\forall(f) = |\{d \in D \mid \forall t \in \mathit{Type}^{max}\ dt \in f(t)^+\},\ \{d \in D \mid \forall t \in \mathit{Type}^{max}\ dt \in f(t)^-\ \}|$

Remarks (i) It is easy to see that this definition makes sense ($\forall(f)$ is actually an interval) and that $\forall(f)$ is a maximal element when f preserves maximal elements of ***Type*** (i.e, $\forall t \in \mathit{Type}^{max}\ f(t) \in \mathit{Type}^{max} \Rightarrow \forall(f) \in \mathit{Type}^{max}$).
(ii) In particular, $\forall(f)^+$ (and thus $\forall(f)^-$) is never empty: for any f, $\perp_D \in \forall(f)^+$.
(iii) This definition is the core of our interpretation. Its aim is to generalize at the second order the simple semantics for the arrow. The quantification on Type^{max}, instead that on the whole of ***Type***, is the key for the proof of the following theorem. Note that if f and g are two different functions which agree on Type^{max}, then $\forall(f) = \forall(g)$.
(iv) It is an open question to clarify the relations beetwen this $\forall$ and a categorical product.

3.6 Theorem $\forall$ is continuous over $[\mathit{Type} \rightarrow \mathit{Type}]$.
Proof The proof of this fact requires a deeper insight into the structure of ***Type*** and Type^{max} and it is contained in the Appendix (Section 5). ◊

3.7 Theorem The function

$\bigcap : [\mathbf{Type} \to D] \to [\mathbf{Type} \to D]$

$\bigcap(f)(u) = \inf \{f(t) \mid t \in T^{max} \wedge u \leq t\}$

is well defined (i.e. $\bigcap(f) \in [\mathbf{Type} \to D]$) and it is continuous over $[T \to E]$.

Proof See the Appendix. ◊

Note that, as in the case of the $\forall$ constructor, if f and g, $f \neq g$, agree on $\mathbf{Type}^{max}$, then $\bigcap(f) = \bigcap(g)$.

3.8 Proposition The function $Ext : \mathbf{Ide} \times [D \to D] \to [D \to D]$ defined by

$$Ext(I,f) = \lambda x{:}D.\ \sup \{ f(e_0) \mid e_0 \leq x \ \&\ e_0 \in I^0 \}$$

is continuous in both argument. Moreover $\forall x \in I\ Ext(I,f)(x) = f(x)$.

Proof After Proposition 2.2 we have only to prove that Ext is continuous in the first argument. Let $\mathfrak{I} \subseteq \mathbf{Ide}$ be directed; then

$$
\begin{aligned}
Ext(\sup \mathfrak{I}, f)(x) &= \sup \{ f(e_0) \mid e_0 \leq x \ \&\ e_0 \in \sup \mathfrak{I} \} \\
&= \sup \{ f(e_0) \mid e_0 \leq x \ \&\ e_0 \in \cup_{I \in \mathfrak{I}} I^0 \} \\
&= \sup \{ \sup \{f(e_0) \mid e_0 \leq x \ \&\ e_0 \in I^0\} \mid I \in \mathfrak{I} \} \\
&= \sup \{ Ext(I,f)(x) \mid I \in \mathfrak{I} \}.
\end{aligned}
$$

◊

3.9 Proposition The function $Ext_T : [\mathbf{Type} \to D] \to [D \to D]$ defined by

$$Ext_T(f) = \lambda x{:}D.\ \sup \{ f(e_0) \mid e_0 \leq x \ \&\ e_0 \in \mathbf{Type}^0 \}$$

is continuous. Moreover $\forall t \in \mathbf{Type}\ Ext_T(f)(t) = f(t)$.

Proof By a similar argument to the previous proposition. ◊

We can finally give the interpretation for type expressions and second-order λ-terms. Let $TEnv = \{\tau \mid \tau{:}TVar \to \mathbf{Type}\}$ and define the (rough) interpretation function

$$^{R}[\![.]\!]_{.} : TExp \to TEnv \to \mathbf{Type}:$$

$^{R}[\![t]\!]_\tau = \tau(t)$

$^{R}[\![\alpha \to \beta]\!]_\tau = {}^{R}[\![\alpha]\!]_\tau \Rightarrow {}^{R}[\![\beta]\!]_\tau$

$^{R}[\![\forall t.\alpha]\!]_\tau = \forall(\lambda e{:}\mathbf{Type}.{}^{R}[\![\alpha]\!]_{\tau[e/t]})$

That $\lambda e{:}\mathbf{Type}.{}^{R}[\![\alpha]\!]_{\tau[e/t]}$ is a continuous function follows by a routine induction on α.

For $t \in \mathbf{Type}$, let $Set(t) = t^+$ and define, for $VEnv = \{\rho \mid \rho{:}Var \to D\}$, the interpretation $^{R}[\![.]\!]_{.} : \Lambda 2\text{-}terms \to Env \to D$, where $Env = TEnv \times VEnv$, as follows.

Let $\underline{\lambda}: [D \to D] \to D$, and $\cdot : D \to [D \to D]$ be given by $\langle D, \cdot, \underline{\lambda} \rangle$ as a λ-model.

$^R[x]_{<\tau,\rho>} = \rho(x)$

$^R[MN]_{<\tau,\rho>} = {}^R[M]_{<\tau,\rho>} \cdot {}^R[N]_{<\tau,\rho>}$

$^R[M\beta]_{<\tau,\rho>} = {}^R[M]_{<\tau,\rho>} \cdot {}^R[\beta]_\tau$

$^R[\lambda x{:}\alpha.M]_{<\tau,\rho>} = \underline{\lambda}\,(Ext(\ Set\,(^R[\alpha]_\tau)\,,\,\lambda d{:}D.\,{}^R[M]_{<\tau,\rho[d/x]>}))$

$^R[\Lambda t.M]_{<\tau,\rho>} = \underline{\lambda}\,(Ext_T\,(\bigcap\,(\lambda e{:}Type.\,{}^R[M]_{<\tau[e/t],\rho>})))$

A routine (double) induction shows that this definition makes sense, for the functions involved are continuous.

Now everything seems ready for a soundness theorem; however the interpretation given above is not sound if (open) types are interpreted on generics intervals, for the same reasons that make the soundness fail for intervals and first order simply typed λ-calculus. Indeed, as noted after proposition 3.2, there is not a unique way of defining a membership relation for intervals. If we take the view that, for $d \in D$, d is a member of $\alpha = |a,A|$ iff $d \in a$, then (the interpretation of) a term like $\lambda x{:}\alpha.M$, fails to be a member of $\alpha \to \beta$ for some β type of M. If, on the other hand, we assume that d is a member of $\alpha = |a,A|$ iff $d \in A$, it is the application to fail. Soundness is achieved only if we interprete types over *maximal* intervals (i.e. on ideals). In this way, as described in Section 2, at the first order we obtain a weak Cartesian Closed Category (and the intervals are completely useless), that is a model of the non-extensional typed calculus. The model we obtain for second order is non-extensional as well.

We refer to [Bruce&Meyer 84] (or to [Bruce, Meyer & Mitchell 85]) for the definition of *second order model*. Briefly, and for fixing the notation, we recall that a model is given by showing first a *type structure*

$$<T,\ \Rightarrow{:}T\times T\to T,\ \Delta{:}[T\to T]\to T>,$$

where $[T\to T]$ is some subset of the functions from T into T (write: $(T\to T)$),

and a *second order frame*

$$<Types, Dom,$$
$$\{\Phi_{a,b}{:}Set(a\to b)\to[Set(a)\to Set(b)]\mid a,b\in T\},$$
$$\{\Psi_{a,b}{:}[Set(a)\to Set(b)]\to Set(a\to b)\mid a,b\in T\},$$
$$\{\Phi_f{:}Set(\forall(f))\to[\Pi_{a\in Type^{max}}\,Set(f(a))]\mid f\in[T\to T]\},$$
$$\{\Psi_f{:}[\Pi_{a\in Type^{max}}\,Set(f(a))]\to Set(\forall(f))\mid f\in[T\to T]\}>,$$

where
- *Types* is a type structure,
- *Dom*= $\{Set(a)\mid a\in T\}$ is a collection of sets indexed by elements in T,
- $[Set(a)\to Set(b)]\subseteq(Set(a)\to Set(b))$,
- $[\Pi_{a\in Type^{max}}\,Set(f(a))]\subseteq(\Pi_{a\in Type^{max}}\,Set(f(a)))$
- the (indexed) maps Φ and Ψ give the retractions

$[Set(a) \rightarrow Set(b)] < Set(a \rightarrow b)$ and

$[\Pi_{a \in Type^{max}} Set(f(a))] < Set(\forall(f))$.

A second order frame is a *model* iff, given a syntactic type assignment B, the interpretation for types ($[\alpha]_\tau$) and terms ($[M]_{<\tau,\rho>}$) is defined for all $\alpha \in TExp$, all $M \in \Lambda^2$-*terms*, and all environments $\tau : TVar \rightarrow T$ and $\rho : TVar \rightarrow \cup_{t \in T} Set(t)$, such that $<\tau,\rho> \models B$. (say $<\tau,\rho> \models B$ iff $\rho(x) \in Set([B(x)]_\tau)$, for all $x \in Var$ such that B is defined on x). The critical cases in the definition of the interpretation are, as one would expect, the ones concerned with the product types ($\forall t.\alpha$) and the two abstractions ($\lambda x{:}\alpha.M$, $\Lambda t.M$), where we have to check that the functions involved actually belong to the sets $[T \rightarrow T]$, $[Set(a) \rightarrow Set(b)]$ and $[\Pi_{a \in Type^{max}} Set(f(a))]$, respectively.

The following definitions and theorems show how to obtain a second order model from the intervals, interpreting the product type with the $\forall$ constructor. Weak extensionality is achieved by using the *Ext* functions; $\bigcap$ guarantees that only the behaviour on the maximal elements will be taken into account when the extension is performed.

3.10 Definition (type structure)

$Types = <\mathbf{Type}^{max},$

$\Rightarrow : \mathbf{Type}^{max} \times \mathbf{Type}^{max} \rightarrow \mathbf{Type}^{max},$

$\Delta : [\mathbf{Type}^{max} \rightarrow \mathbf{Type}^{max}] \rightarrow \mathbf{Type}^{max} >$

where $\Rightarrow : \mathbf{Type}^{max} \times \mathbf{Type}^{max} \rightarrow \mathbf{Type}^{max}$ is just the restriction of the function space constructor for intervals to the maximal ones;

$[\mathbf{Type}^{max} \rightarrow \mathbf{Type}^{max}] = \{f{:}\mathbf{Type}^{max} \rightarrow \mathbf{Type}^{max} \mid$

$\exists f^* \in [\mathbf{Type} \rightarrow \mathbf{Type}] \forall t \in \mathbf{Type}^{max} f(t) = f^*(t)\}$;

$\Delta = \forall \circ Ch$

$Ch: [\mathbf{Type}^{max} \rightarrow \mathbf{Type}^{max}] \rightarrow [\mathbf{Type} \rightarrow \mathbf{Type}]$

$Ch(f) = f^*$, in the notation above.

(Remember that $[\mathbf{Type} \rightarrow \mathbf{Type}]$ are the continuous endofunctions on $\mathbf{Type}$).

3.11 Proposition *Types* is a type structure and the interpretation $[\alpha]_\tau$ of a type expression α in the environment τ:TVar$\rightarrow \mathbf{Type}^{max}$ is well defined. Moreover, ${}^R[\alpha]_\tau = [\alpha]_\tau$.

(Note that environments range only on maximal types).

Proof The critical case is in the interpretation of the product:

$$[\forall t.\alpha]_\tau = \Delta(\lambda e{:}\mathbf{Type}^{max}.[\alpha]_{\tau[e/t]})$$

where we must show that $f = \lambda e{:}\mathbf{Type}^{max}.[\alpha]_{\tau[e/t]} \in [\mathbf{Type}^{max} \rightarrow \mathbf{Type}^{max}]$.

An easy induction shows that the function $\lambda e{:}\mathbf{Type}.\ {}^R[\alpha]_{\tau[e/t]}$ is continuous and it extends f, for $\Rightarrow$, $\forall$ preserve maximal elements. ◊

3.12 Definition (second order frame)

(i) $[Set(a)\rightarrow Set(b)] = \{f: Set(a)\rightarrow Set(b) \mid \exists d\in D\ \forall x\in Set(a)\ dx=f(x)\}$

(ii) $\Phi_{a,b}(d) = \lambda x{:}Set(a).\ dx$ for $d\in Set(a\rightarrow b)$

$\Psi_{a,b}(f) = \underline{\lambda}(Ext(Set(a),f))$ for $f\in [Set(a)\rightarrow Set(b)]$

(iii) $[\Pi_{a\in Type^{max}} Set(f(a))] = \{f: \Pi_{a\in Type^{max}} Set(f(a)) \mid \exists d\in D\ \forall a\in \boldsymbol{Type}^{max}\ \ da=f(a)\}$

note that the d above defines a continuous function from ***Type*** into D, and then we have a map

$Ch_f: [\Pi_{a\in Type^{max}} Set(f(a))] \rightarrow [\boldsymbol{Type}\rightarrow D]$

(iv) $\Phi_f(d) = \lambda a{:}\boldsymbol{Type}^{max}\ .\ da$ for $d\in Set(\forall(f))$

$\Psi_f(g) = \underline{\lambda}(Ext_T(\bigcap(Ch_f(g))))$ for $g\in[\Pi_{a\in Type^{max}} Set(f(a))]$

(v) $Int = <\ Types,$

$\{Set(t) \mid t\in \boldsymbol{Type}^{max}\},$

$\{\Phi_{a,b} \mid a, b\in T\}, \{\Psi_{a,b} \mid a, b\in T\},$

$\{\Phi_f \mid f\in [T\rightarrow T]\}, \{\Psi_f \mid f\in [T\rightarrow T]\}\ >$

3.13 Theorem *Int* is a second order model.

Proof That the Φ and Ψ give the due retractions is a straightforward check. As for the interpretation of terms, we give the proof only for second order abstraction.

$[\Lambda t.M]_{<\tau,\rho>} = \underline{\lambda}\ (Ext_T\ (\bigcap\ (Ch_f(\lambda e{:}\ \boldsymbol{Type}^{max}.\ [M]_{<\tau[e/t],\rho>}))))$

where we must show that

$$f = \lambda t{:}\ \boldsymbol{Type}^{max}.\ [M]_{<\tau[e/t],\rho>} \in [\Pi_{a\ \in Type^{max}} Set(f(a))].$$

By induction on M we prove that

$$\lambda t{:}\boldsymbol{Type}.\ {}^{R}[M]_{<\tau[e/t],\rho>}$$

extends f. The only interesting case is when $M = \Lambda s.M'$. Let $e\in \boldsymbol{Type}^{max}$:

$[\Lambda s.M']_{<\tau[e/t],\rho>} = \underline{\lambda}\ (Ext_T\ (\bigcap\ (Ch_f(\lambda d{:}\ \boldsymbol{Type}^{max}.\ [M']_{<\tau[e/t,\ d/s\ ],\rho>}))))$

${}^{R}[\Lambda s.M']_{<\tau[e/t],\rho>} = \underline{\lambda}\ (Ext_T\ (\bigcap\ (\lambda e{:}\boldsymbol{Type}.\ {}^{R}[M']_{<\tau[e/t,\ d/s\ ],\rho>})))$

Then

$\bigcap(Ch_f(\lambda d{:}\ \boldsymbol{Type}^{max}.\ [M']_{<\tau[e/t,\ d/s\ ],\rho>})) = \bigcap\ (\lambda e{:}\boldsymbol{Type}.\ {}^{R}[M']_{<\tau[e/t,\ d/s\ ],\rho>}),$

since $\bigcap$ identifies functions agreeing on ***Type***^{max}, which is the case for the two inner functions, by induction hypotesis. ◊

4. Some properties of the model

We will discuss in this section some facts helping to shed some light on the structure of the model.

The first, key property is that all types are non-empty: this is a trivial consequence of the definition of interval. (For the relevance of having all inhabited types for completeness, one should look at [Meyer, Mitchell, Moggi & Statman 87]).

4.1 Theorem (*inhabited types*)

For all $t \in Type^{max}$, $Set(t)$ is not empty.

Proof Immediate: for all $t \in Type^{max}$, $\bot_D \in Set(t)$, for $Set(t)$ is a non empty closed subset of D (note that both $\Rightarrow$ and $\forall$ preserve non-emptyness). ◊

The second property we list, is the ability to model a *subtype* notion. First and second order type structures with subtyping have been discussed in the literature for modeling the concept of *inheritance* of object oriented languages (see, for instance, [Cardelli 85], [Cardelli&Wegner 85]). A key property of that concept is the contravariance of the function type constructor (in its first argument) with respect to subtyping; that is

$$\alpha' \textbf{ sub } \alpha \text{ and } \beta \textbf{ sub } \beta' \text{ implies } \alpha \Rightarrow \beta \textbf{ sub } \alpha' \Rightarrow \beta'.$$

If type constructors are interpreted with continuous functions, there is no hope to have such a property when one define α' **sub** α iff $\alpha' \leq \alpha$, where $\leq$ is the ordering of the model. In fact every retraction-based model is not a model of inheritance. In our case, the simplicity of the approach suggests the definition of the subtype relation as simple set inclusion:

4.2 Definition An interval $|a,A|$ is a *subtype* of $|b,B|$ (write $|a,A|$ **sub** $|b,B|$) iff $a \subseteq b$ & $A \subseteq B$.

4.3 Proposition Let α, α', β, β' be intervals. If α' **sub** α and β **sub** β' then $\alpha \Rightarrow \beta$ **sub** $\alpha' \Rightarrow \beta'$.

Proof By definition of $\Rightarrow$. ◊

Hence *Int* is a "model of inheritance" in the sense of [Cardelli 85].

In [Cardelli & Wegner 85] a second order calculus with subtyping and *bounded quantification*is proposed, while leaving unanswered the problem of finding a model for it. We claim that a bounded universal quantification can be defined in our model by setting:

$$\forall^*: Type \times [Type \to Type] \to Type$$

$$\forall(\alpha, f) = | \{d \in D \mid \forall t \in Type^{max}\ t^- \subseteq \alpha^- \Rightarrow dt \in f(t)^+\},$$
$$\{d \in D \mid \forall t \in Type^{max}\ \ t^+ \subseteq \alpha^+ \Rightarrow dt \in f(t)^-\} |$$

$\forall^*$ is continuous and $\forall^*(|D,D|, f) = \forall(f)$

One important extension to the calculus described in Section 2 is the introduction of *existential types* (or *sums*) (see [Bruce, Meyer & Mitchell 85] for definitions and syntax). Existential types are interpretable in the interval model, by viewing them as dependent pairs. More precisely, take D such that $D\times D<D$ via ideal preserving pairing $<,>:D\to D\times D$ and projections $p_1, p_2:D\times D\to D$ and set

$$\exists : [\mathit{Type}\to\mathit{Type}]\to\mathit{Type}$$

$$\exists(f) = |\{<t,d>\in D \mid t\in \mathit{Type}^{max} \wedge d\in f(t)^+\},\ \{<t,d>\in D \mid t\in \mathit{Type}^{max} \wedge d\in f(t)^-\}|$$

The function $\exists$ can be easily proved continuous and the projections allow the interpretation of (strong) sums.

Comment Looking at the above construction, one may wonder about the necessity of considering first the intervals: indeed the actual interpretation is given only on *maximal* intervals, that is on ideals. The fact of the matter is that without the richer structure of the intervals, it does not seem possible to

- embed (in a sensible way) the collection of the ideals into the domain
- prove the continuity of the *Ext* function, which seems essential in order to have a well defined interpretation for second-order terms. (*Ext*, in its turn, is necessary for weak extensionality.)

5. Appendix

We will prove here the main theorems of Section 3. In order to do this we must have a closer look at the structure of ***Type***.

5.1 Definition Let D be a domain, D^0 be its basis, and, for $d\in D$, set $\uparrow d = \{x\in D \mid d\leq x\}$. The topology over D generated by the subbasis

$$\{\uparrow d \mid d\in D^0\} \cup \{D\setminus \uparrow d \mid d\in D^0\}$$

is called the *Lawson topology* on D.

5.2 Proposition Type^{max} is compact in ***Type*** with respect to the Lawson topology.

Proof claimed in [Cartwright 84] ◊

In the following we will write T for ***Type*** (and T^{max} for Type^{max}); many results hold for a generic domain T for which T^{max} is compact in T wrt the Lawson topology on T.

Let W be an infinite rooted complete binary tree and let $e:\omega\to T^0$ be an enumeration of the basis of T. For a *path* in W we will always mean a path starting from the root. At level n, label every *left* branch with e_n and every *right* branch with $\neg e_n$. Every element $t\in T$ can then be associated to an infinite path on W: at any level choose left if $e_n\leq t$; choose right otherwise.

5.3 Definition Let Π be a (finite or infinite) path in W and let $S\subseteq T$.

(i) Π is *consistent with S* iff there exists $t\in S$ such that for all n

$$e_n \in \Pi \Rightarrow e_n \leq t$$

$$\neg e_n \in \Pi \Rightarrow \neg (e_n \leq t)$$

(ii) Π is *maximally consistent* iff it is consistent with T_{max}

(iii) For Π consistent with some set S, the *meaning of* Π (write: $\cup\Pi$) is the element

$$\cup\Pi = sup \{ e_n \mid e_n \in \Pi \}$$

(iv) cons(Π) = $\{ t \in T_{max} \mid \Pi$ is consistent with $\{t\} \}$

(v) $\pi < \Pi$ iff the finite path π is an initial segment of the path Π.

5.4 Lemma Let Π be an infinite path not maximally consistent. Then there exists a *finite* initial segment of Π which is not maximally consistent.

Proof by compactness. ◊

5.5 Lemma Let Π be an infinite maximally consistent path: $\cup\Pi$ is maximal.

Proof Suppose $\cup\Pi$ not maximal. For every $t \in T_{max}$ there exist e_p finite such that

$$e_p \leq t \ \& \neg(e_p \leq \cup\Pi).$$

But then for all such e_p, $\neg e_p \in \Pi$ and Π is not maximally consistent. ◊

Remark We can summarize the two lemmas above in the following property:

Let z be a non maximal element of T and denote with Π_z the infinite path in W associated with it. Then there exists a *finite* initial segment of Π_z which is not maximally consistent.

This property seems weaker then the compactness of T_{max} and for the intervals it can be proved directly [Di Gianantonio 87].

In the following E will range over domains. For $f_0 = sup\{s_i \to z_i \mid s_i \to z_i$ step function, $i=1,n\}$ finite function from T into E and π path in W, set $\underline{f_0} = \{s_i \to z_i \mid s_i \to z_i$ step function, $i=1,n\}$ and $S_{\pi,f_0} = \{ s_i \mid s_i \to z_i \in \underline{f_0} \wedge \exists t \in cons(\pi)\ s_i \leq t \}$.

5.6 Lemma Let Π be an infinite consistent path in W and $f_0 = sup\{s_i \to z_i \mid s_i \to z_i$ step function, $i=1,n\}$ be a finite function from T into T. Then for every $\pi < \Pi$:

(i) there exists $\pi' < \Pi$ such that $\forall\, \pi < \Pi\ (\pi' < \pi \Rightarrow S_{\pi,f_0} = S_{\pi',f_0})$.

(ii) $\forall\, s_i \in S_{\pi',f_0}\ s_i \leq \cup\Pi$

Proof (i) is immediate, since $\underline{f_0}$ is a finite set. As for (ii), suppose there exists $s_k \in S_{\pi',f_0}$ such that $\neg(s_k \leq \cup\Pi)$. Then at level k in Π we have an arc with label $\neg s_k$. Certainly π' is a proper initial segment of Π restricted up to the level $k+1$ (call it $\Pi\lceil_{k+1}$), since otherwise $\neg(s_k \leq t)$ for all $t \in cons(\pi')$, and $s_k \notin S_{\pi',f_0}$. But then for all $t \in cons(\Pi\lceil_{k+1})$, $\neg(s_k \leq t)$, which is a contradiction, since $S_{\Pi\lceil_{k+1},f_0} = S_{\pi',f_0}$. ◊

5.7 Definition

(i) A binary tree is *uniform* iff it is finite and every internal node has two sons.

(ii) Let $g\in[T\to E]$, $f_0\in[T\to E]$ finite function and $u\in T$.

V is a *u-g witness tree* for f_0 iff it is a uniform initial subtree of W such that, for every path π from the root to a leaf, either

(1) π is inconsistent with $\{t\in T_{max}\,|u\leq t\wedge(\exists i\leq n\ s_i\leq t)\}$ or

(2) $\cup\pi\geq sup\ S_{\pi,f_0}$ & $f_0(\cup\pi)\leq g(\cup\pi)$.

The above definition generalizes the notion of witness tree given in [Cartwright 84] and it is the major technical tool in the proofs of the following theorems.

5.8 Theorem Let $f_0=sup\{s_i\to z_i\mid s_i\to z_i$ step function, $i=1,n\}$ be a finite function from T into E and $g\in[T\to E]$, $u\in T$, such that

$\forall t\in T_{max}\,(u\leq t\Rightarrow f_0(t)\leq g(t))$.

Then there exists a *u-g* witness tree for f_0.

Proof Assume that a *u-g* witness tree for f_0 does not exist.

Let V denote the tree obtained by deleting from the complete labeled tree W all nodes below (the root stays at the top) any path inconsistent with $Z(u)=\{t\in T_{max}\mid u\leq t\wedge(\exists i\leq n\ s_i\leq t)\}$ or which does not satisfy (2) above. Note that all paths inconsistent with $Z(u)$ have some finite initial inconsistent segment, for $Z(u)$ is closed (and hence compact) in T_{max}.

V is certainly infinite, since otherwise it is a witness tree. By König's Lemma, there exists in V an infinite consistent path Π. By the previous lemma, there exists $\pi'<\Pi$ such that $\forall\,\pi<\Pi\,(\pi'<\pi\Rightarrow S_{\pi,f_0}=S_{\pi',f_0})$ & $\forall\,s_i\in S_{\pi',f_0}\ s_i\leq\cup\Pi$: thus $\forall\,\pi<\Pi\quad\pi'<\pi\Rightarrow\neg(f_0(\cup\pi)\leq g(\cup\pi))$. Note now that $\cup\Pi$ is maximal (by 5.5 above) belongs to $Z(u)$ (by an easy argument). Then

$$g(\cup\Pi)=sup\,\{g(\cup\pi)\mid\pi'<\pi<\Pi\}\qquad\text{by } g \text{ continuous}$$

$$\geq f_0(\cup\Pi)\qquad\text{by hypothesis.}$$

Since $f_0(\cup\Pi)$ is finite, there exists $\pi''<\Pi$ such that $\pi'<\pi''<\Pi$ and $g(\cup\pi'')\geq f_0(\cup\Pi)\geq f_0(\cup\pi'')$, which is absurd. ◊

We are eventually in the position to prove the following

5.9 Theorem $\forall:[\textit{Type}\to\textit{Type}]\to\textit{Type}$ is continuous over $[\textit{Type}\to\textit{Type}]$.

Proof $\forall$ is clearly monotonic. Let $\mathfrak{R}\subseteq[\textit{Type}\to\textit{Type}]$ directed and set

$$|J^+,J^-|=\forall(sup\,\mathfrak{R})$$

$$=|\{d\in D\mid\forall t\in \textit{Type}^{max}\ dt\in(sup\,\mathfrak{R})(t)^+\},\ \{d\in D\mid\forall t\in\textit{Type}^{max}\ dt\in(sup\,\mathfrak{R})(t)^-\}|$$

$$|J^+, J^-| = \sup_{f\in\mathfrak{R}} \forall(f) = |\sup_{f\in\mathfrak{R}} \{d\in D \mid \forall t\in \mathbf{Type}^{max}\ dt\in f(t)^+\},\qquad \inf_{f\in\mathfrak{R}} \{d\in D \mid \forall t\in \mathbf{Type}^{max}\ dt\in f(t)^-\}|$$

1. $I^- = J^-$: easy, since $(\sup \mathfrak{R})(t)^- = \bigcap\{ f(t)^- \mid f\in \mathfrak{R}\}$ (remember that we have reversed ordering on the negative side.)

2. Since I^+ and J^+ are ideals, it will be enough to show they share the same finite elements.

2a. $J^+ \subseteq I^+$: $e_0\in D^0 \cap J^+ \Rightarrow \exists f\in \mathfrak{R}\, \forall t\in \mathbf{Type}^{max}\ e_0t\in f(t)^+ \subseteq (\sup \mathfrak{R})(t)^+$

2b. $I^+ \subseteq J^+$: this is the difficult case, since we must show that, for any finite element e_0 bringing $\mathbf{Type}^{max}$ into $(\sup \mathfrak{R})(t)^+$, there is actually an $h\in \mathfrak{R}$, for which e_0 sends $\mathbf{Type}^{max}$ into $h(t)^+$.

Let $e_0\in I^+$ be a finite element and consider the finite function $f_0 : \mathbf{Type}\to\mathbf{Type}$ defined as

$$f_0(t) = |\downarrow e_0t, D|$$

where, for $d\in D$, $\downarrow d = \{x\in D \mid x\le d\}$. We have

(1) $\qquad \forall t\in \mathbf{Type}\ \ e_0t \in f_0(t)^+$

(2) $\qquad \forall t\in \mathbf{Type}^{max}\ f_0(t) \le (\sup \mathfrak{R})(t)$

by $f_0(t)^+ = \downarrow e_0t \subseteq (\sup \mathfrak{R})(t)^+$ (since $e_0t \in (\sup \mathfrak{R})(t)^+$) and $f(t)^- = D \supseteq (\sup \mathfrak{R})(t)^-$.

Claim There exists a *finite* function f such that

(i) $\qquad f \le \sup \mathfrak{R}$

(ii) $\qquad \forall t\in \mathbf{Type}^{max}\ f_0(t) \le f(t)$.

Thus there exists $h\in \mathfrak{R}$ such that $f\le h$ & $\forall t\in \mathbf{Type}^{max}\ f_0(t) \le h(t)$, by the finiteness of f and (i), (ii) above. Hence $\exists\, h\in \mathfrak{R}\ \forall t\in \mathbf{Type}^{max}\ e_0t \in f_0(t)^+ \subseteq h(t)^+$, that is

$$e_0\in \sup_{f\in\mathfrak{R}} \{d\in D \mid \forall t\in \mathbf{Type}^{max}\ dt\in f(t)^+\}$$

Since I^+ and J^+ are both ideals, this establishes the thesis provided we can prove the claim.

Let V be a $\perp_T$-$(\sup \mathfrak{R})$ witness tree for f_0 (by 5.8) and set

$$f = \sup \{ \cup\pi \to f_0(\cup\pi) \mid \pi \text{ consistent path in } V\}$$

For $t\in \mathbf{Type}$, we have

$$\begin{aligned}(\sup \mathfrak{R})(t) &\ge \sup \{(\sup \mathfrak{R})(\cup\pi) \mid \cup\pi \le t\ \&\ \pi \text{ in } V\}\\ &\ge \sup \{ f_0(\cup\pi) \mid \cup\pi \le t\ \&\ \pi \text{ in } V\} &&\text{by } V\ f_0\text{-}(\sup \mathfrak{R}) \text{ witness tree}\\ &= f(t).\end{aligned}$$

Remember now that, for (2) in the definition of witness tree,

$$\cup\pi \ge \sup S_{\pi, f_0} = \sup \{ s_i \mid s_i \to z_i \in \underline{f_0} \wedge \exists t\in cons(\pi)\ s_i \le t\},$$

that is $f_0(\cup\pi) = f_0(t)$ for every $t\in cons(\pi)$; by this, for $t\in \mathbf{Type}^{max}$

$f(t) = sup\ \{ f_0(\cup\pi) \mid \cup\pi \leq t\ \&\ \pi \text{ in } V\} = f_0(t).$

The following theorem is a streghtening of a result in [Di Gianantonio 87]. E will range over domains.

5.10 Theorem Let $f \in [T \to E]$. The function

$$f^* : T \to E$$

$$f^*(u) = inf\ \{f(t) \mid t \in T_{max} \wedge u \leq t\}$$

is continuous.

Proof The intersection exists, for E is an algebraic c.p.o.; f^* is obviously monotonic.

Let $\mathfrak{I} \subseteq T$ be a directed set. We have

$$f^*(sup\ \mathfrak{I}) = inf\ \{f(t) \mid t \in T_{max} \wedge sup\ \mathfrak{I} \leq t\}$$

$$sup_{u\in\mathfrak{I}} f^*(u) = sup_{u\in\mathfrak{I}}\ inf\ \{f(t) \mid t \in T_{max} \wedge u \leq t\}$$

By monotonicity $f^*(sup\ \mathfrak{I}) \geq sup_{u\in\mathfrak{I}} f^*(u)$.

Let now $e_0 \in T^0$, $e_0 \leq f^*(sup\ \mathfrak{I})$.

Claim There exists a finite element $d_0 \in T^0$ such that

(i) $\quad d_0 \leq sup\ \mathfrak{I}$

(ii) $\quad \forall t \in T_{max}\ (d_0 \leq t \Rightarrow e_0 \leq f(t))$.

Thus there exists $u \in \mathfrak{I}$ such that $d_0 \leq u \leq sup\ \mathfrak{I}$, by (i) and d_0 finite. Moreover, from (ii)

$$\begin{aligned} e_0 &\leq inf\ \{f(t) \mid t \in T_{max} \wedge d_0 \leq t\} \\ &\leq inf\ \{f(t) \mid t \in T_{max} \wedge u \leq t\} \\ &\leq sup_{u\in\mathfrak{I}}\ inf\ \{f(t) \mid t \in T_{max} \wedge a \leq t\} \\ &= sup_{u\in\mathfrak{I}} f^*(u). \end{aligned}$$

We finally prove the claim.

Let V be a $(sup\ \mathfrak{I})$-f witness tree for the finite function $\omega_{e_0} = \{\bot_T \to e_0\}$. Since V is finite, the set

$$H = \{\cup\pi \mid \pi \text{ in } V \text{ consistent with } Z(sup\ \mathfrak{I})\}$$

where $Z(u) = \{t \in T_{max} \mid u \leq t\}$, is finite as well, and

$\forall \cup\pi \in H \quad e_0 \leq f(\cup\pi)$ $\qquad$ by V witness tree

$\forall t \in Z(sup\ \mathfrak{I})\ \exists \cup\pi \in H\ \ \cup\pi \leq t,$

since every finite segment of the path in W associated with such a t is obviously consistent with $Z(u)$.

Let now $d_0 = |d^+, d^-|$, where

$$d^+ = \downarrow\{w \in D^0 \mid w \in (sup\ \mathfrak{I})^+ \wedge \exists \cup\pi \in H\ w \text{ maximal in } \cup\pi^+\}$$

$$d^- = D \setminus \uparrow\{w \in D^0 \mid w \notin (sup\ \mathfrak{I})^- \wedge \exists \cup\pi \in H\ w \text{ minimal in } D \setminus \cup\pi^-\}.$$

d_0 is well defined and it is finite, for every element in H is finite and H is a finite set (remember that an interval $|a, A|$ is finite iff there exist two finite sets $I, J \subseteq D^0$ such that

$$a = \downarrow\{w \in D^0 \mid w \in I\}$$

$$A = D \setminus \uparrow\{w \in D^0 \mid w \in J\};$$

that is a is finite in $\mathbf{Ide}_D$ and A is the complement of a finite open set).

Certainly $d_0 \leq sup\, \mathfrak{I}$; as for (ii) of the claim it is obtained by noting that

$$\forall t \in Z(d_0)\ \exists \cup\pi \in H \quad \cup\pi^+ \subseteq (sup\, \mathfrak{I})^+ \cup (t^+ \cap (sup\, \mathfrak{I})^-) \subseteq \cup\pi^-$$

and showing from this that

$$\forall t \in Z(d_0)\ \exists \cup\pi \in H \quad \cup\pi \leq t.$$

Hence $e_0 \leq f(\cup\pi) \leq f(t)$ and the theorem is proved. ◊

5.10 Theorem The function

$$\bigcap : [T \to E] \to [T \to E]$$

$$\bigcap(f) = f^*$$

is continuous over $[T \to E]$.

Proof Let $\mathfrak{R} \subseteq [T \to E]$ be directed. We will show that, for any $u \in T$,

$$e_0 \leq \bigcap(sup\mathfrak{R})(\mathrm{u}) \ \Rightarrow\ e_0 \leq sup_{f \in \mathfrak{R}} [\bigcap(f)(\mathrm{u})].$$

This is obtained by taking a $\perp_T$-$(sup\mathfrak{R})$ witness tree V for the finite function $\omega_{e_0} = \{\perp_T \to e_0\}$, building the (finite) function

$$h_0 = sup\ \{ \cup\pi \to e_0 \mid \pi \text{ path in } V \text{ consistent with } T^{max} \},$$

and showing that

$$\exists h \in \mathfrak{R}\ \ \forall t \in T^{max}\ \ e_0 = h_0(t) \leq h(t)\,.$$ ◊

Acknowledgments I am endebted to Giuseppe Longo, Pietro Di Gianantonio and Andrea Asperti for helpful discussions and hints.

7. References

Amadio, R., Bruce, K., Longo, G. [1986] "The Finitary Projection Model and the solution of higher order domain equations", Proceedings LICS 86, IEEE (pp.122-130).

Barendregt, H. [1984], *The lambda calculus; its syntax and semantics*, Revised edition, North Holland.

Bruce K., Meyer A. [1984] "The semantics of second order polymorphic lambda-calculus", *Symposium on Semantics of Data Types* (Kahn, MacQueen, Plotkin eds.), LNCS 173, Springer-Verlag (pp. 131-144).

Bruce, K., Meyer, A., Mitchell, J. [1985] "The semantics of second order lambda-calculus", *Information and Control* (to appear).

Cardelli, L. [1985] "Semantics of multiple inheritance", *Information and Control* (to appear).

Cardelli, L., Wegner, P. [1985] "On understanding types, data abstraction and polymorphism", *Computing Surveys*, vol 17(4) (pp. 471-522).

Cartwright R. [1984] "Types as Intervals", Proc. Popl 84, ACM.

Di Gianantonio, P. [1987] "The subtype relation in the interval model" (in italian), Master Thesis, Dip. di Informatica, Univ. di Pisa.

Girard, J.-Y. [1985] "The system *F* of variable types, fifteen years later", *Theor. Comp. Sc.* (to appear).

Hayashi, S. [1985] "Adjunction of semifunctors: categorical structures in no-extensional lambda calculus", *Theor. Comp. Sc.*,vol 41(1) (pp. 95-104).

Kainen, P.C. [1971] "Weak adjoint functors", *Math. Z.* 122 (pp.1-9).

Longo, G. [1987] "On Church's formal theory of (computable) functions", Church Thesis Confererence Proceedings, *Annals Pure App. Logic* (to appear).

MacQueen D., Plotkin G., Sethi R. [1986] "An ideal model for recursive polymorphic types",*Information and Control*, vol 71(1-2) (pp.95-130).

Martini, S. [1986] "Categorical models for typed and type-free non-extensional lambda-calculus", preprint, Pisa.

McCracken N. [1984] "A finitary retract model for the polymorphic lambda-calculus," *Information and Control* (to appear).

Meyer, A., Mitchell J.C., Moggi E., Statman R. [1987] "Empty types in polymorphic lambda calculus", (ACM Conference on) *POPL '87*, Münich.

Scott, D. [1980] "A space of retracts", manuscript, Bremen.

Wiweger A. [1984] "Pre-adjunction and λ-algebraic theories", in *Colloq. Math.* 48(2), 153-165.

Logical Aspects of Denotational Semantics

E. ROBINSON

Department of Pure Mathematics and Mathematical Statistics
16 Mill Lane, Cambridge CB2 1SB, England

Introduction

This paper concerns two approaches to programming language semantics, the denotational and the axiomatic. In the denotational philosophy inspired by Strachey the program, or program fragment, is first given a semantics as an element of some abstract mathematical object, generally a partially ordered set, the semantics of the program being a function of the semantics of its constituent parts; properties of the program are then deduced from a study of the mathematical object in which the semantics lives. The axiomatic approach associated originally with Floyd and Hoare, on the other hand, is concerned directly with assertions about the program, but in a syntax-directed way, so that properties of a program can be deduced from assertions about its constituent parts. In the end we are reduced to considering the properties of some collection of "atomic" programs, such as assignment statements, and these are assumed to be given in advance. Of course, work on denotational semantics tends to have a rather mathematical flavour, while work on axiomatic tends to be more "logical" (or proof-theoretic). The mathematics of denotational semantics has been well-developed—if in a somewhat ad hoc fashion—and provides coherent ways of looking at collections of programs. Conventionally, these semantics have been valued in some one or other of various categories of partially-ordered sets (it is not of course a coincidence that many of the abstract treatments of categories of partial maps use Po-enriched categories). None of these is uniformly satisfactory, but the best at the moment seems to be SFP, a category of profinite posets. (Our use of this category in the latter stages of the paper does not however represent any ideological commitment to it—just the fact that for us it is easy to work with.) If the strengths of domain-theoretic semantics are the relative stability of the theory, and its power as a tool for arguing about the collection of all programs, its weakness is the relative difficulty of treating a particular program in a particular semantics. This is the gap filled by the axiomatic semantics—it is on the whole easier to provide uniform techniques for reasoning about syntax than about some collection of abstract mathematical models, or even (presentations of) partially-ordered sets. Even if the rules for the development of a denotational semantics are clear and well-understood, the means of reasoning in detail about a particular semantics remain rather ad hoc, whereas although there are few general tools for the development of axiomatic semantics (in particular it is hard to find a good class of basic assertions), once one has been obtained it is susceptible to study using established and powerful proof-theoretic techniques.

The main purpose of this paper then is to show how we can use the methods of development used in denotational semantics to obtain an axiomatic semantics. The technical tools are a new description of the power-domains in "modal" terms, and a description of the Scott open subsets of a solution of a recursive domain equation. The axiomatic semantics is then obtained by taking as propositions the Scott open subsets of the domain which supports the denotational semantics.

These results are not supposed to be particularly surprising, and certainly should not shock anyone who has had anything to do with category theory. In particular they are prefigured in the work of Scott on Information Systems (Scott [1982]), which can be read as addressing precisely this problem for the restricted family of bounded complete cpo's. In fact Scott's information systems are schizophrenic objects—they are not really sure whether they are presentations of cpo's or collections of (Horn logic) propositions. We on the other hand shall work with locales, which can also be seen as schizophrenic objects, unsure whether they are intuitionistic propositional theories, or cpo's with their Scott topology. It is this dual role that will enable us to argue that associated with each proof-theoretic semantics there is another which is undeniably denotational, and, more concretely, to show in the latter part of the paper how, given a denotational semantics by means of a recursive domain equation and presentation of the semantic maps, it is possible to extract a class of assertions and a complete set of proof rules for an axiomatic semantics.

For the reader unfamiliar with locales, the first sections of the paper contain a brief primer. I hope this wets the reader's appetite (rather than compels him in desperation) to find out more from the major reference in the field, Peter Johnstone's "Stone Spaces" (Johnstone [1982]), now a major paperback. For a brief introduction to the original topological motivation for studying locales see Johnstone [1983].

I said above that the results of this paper are not supposed to be particularly surprising. In particular they are not going to surprise the members of the Cambridge theory of computation seminar. I would like to thank them all for their interest and their ideas, but to single out Martin Hyland and Glynn Winskel for special mention.

I would also like to mention the parallel work of Abramsky, who has given a treatment of just this set-up concentrating on the case of Scott domains (and so not including an account of the Plotkin power-domain). Abramsky has collected an extensive collection of applications in a series of manuscripts (*e.g.* Abramsky [1986]) and published a sketch of a more general account in Abramsky [1987].

1 *Denotational semantics of programming languages*

When one wants to give a denotational semantics to a programming language, which we assume to be context-free, the best way to think of it is as the free algebra over a many-sorted signature.

Let $G = (V, \Gamma, S, P)$ be a context-free grammar (V is the set of variables, Γ the terminal alphabet, S the start symbol, and P the set of production rules), then we define the signature Σ_G as follows:
Σ_G has V as its set of basic sorts, and given a rule

$$\rho : \quad X \longrightarrow \gamma = \alpha_0 Y_0 \alpha_1 Y_1 \ldots Y_{n-1}\alpha_n \qquad (\alpha_i \in \Gamma^*,\ Y_i \in V)$$

then Σ_G has an operation $f_\rho : Y_0 \times \ldots \times Y_{n-1} \to X$. Of course if $\gamma \in \Gamma^*$ we interpret this as a constant c_ρ of type X.

Take the free algebra over this signature. Any term of type S is given as a word in the f_ρ (and c_ρ). Dropping the function symbols, and retaining only the constants, we get a word of Γ^* which is necessarily in $L(G)$. Conversely, given a word w in $L(G)$, we can use G to obtain a parse-tree, and then observe that each parse-tree for w gives a term in the free algebra on Σ_G which represents w. Note, however, that it is impossible to recover G from Σ_G (in general it is not even possible to recover Γ). The best we can do is to recover from an arbitrary signature Σ a grammar G_Σ of which all grammars giving rise as above to Σ are interpretations (it is initial for g-interpretations, *cf.* Cremers & Ginsburg [1975]).

The algebraic approach does however give us the abstract syntax, and for our purposes this is what is important. It enables us to make the definition:

DEFINITION A denotational semantics for a programming language L given by a grammar G is an algebra in some category C for the signature Σ_G.

One of the main purposes of this definition is to avoid restricting ourselves to some particular category of "domains" (thus sidestepping the vexed question of which is the right one). Note, however, we do not even require that the category be a category of ordered sets, or that the maps involved be continuous. Another is to move problems such as those to do with the representation of non-termination away from the definition of the semantics and into the choice of the category in which it is to be valued (though this may necessitate the use of monoidal structures more general than cartesian product in order to get a satisfactory notion of algebra, if for example we want to use categories of partial maps *cf.* Robinson & Rosolini [1987]). The obvious drawback is that the definition seems to be drawn far too wide—many "semantics" will be nonsensical in computing terms. We have allowed in such "semantics" as the language itself (programs are equivalent iff textually identical) and the terminal semantics (all programs are the same). There seems to be no way to avoid this unless one is already provided with a semantics for the language, and a notion of relative correctness (or implementation) for different types of semantics (*cf.* the notion of full abstraction).

The idea behind our definition is that the elements of the carrier of an algebra should act like program fragments. There may not be an actual syntactic program which denotes them (for example, they might represent compiled modules written in a different or an extended language), but we should still be able to interpret the standard control structures of the language over them.

We should also say at once that although we have made this very general definition of semantics, the semantics with which we shall actually be concerned will either be conventional, in the sense of being valued in some category of cpo's, or else they will be valued in the category of locales.

We recall that a locale is a complete lattice which satisfies the infinite distributive law

$$a \wedge \bigvee b_\alpha = \bigvee (a \wedge b_\alpha).$$

An immediate consequence of completeness (which otherwise it would be necessary to state explicitly) is that every locale has a top element ($\top = \bigwedge \phi$) as well as a bottom ($\bot = \bigvee \phi$). Note that a locale is not necessarily a completely distributive lattice, but that the distributivity of $\wedge$ over colimits means (by the adjoint functor theorem) that $a \wedge (\ )$ has a right adjoint, which allows us to interpret implication (since the adjunction $a \wedge (\ ) \dashv a \rightarrow (\ )$ holds iff for all b and c, $a \wedge b \leq c$ iff $b \leq a \rightarrow c$).

Locales are thus the Lindenbaum algebras for infinitary intuitionistic propositional logic, but they can also be viewed as the algebraic counterparts of the open set lattices of topological spaces (the open sets of any topological space form a locale). For this reason we choose to define a map $f : A \to B$ in the category Loc as a map $f^* : B \to A$ of partially ordered sets, which preserves finite meets and arbitrary sups. Adopting the topological viewpoint, f^* is the counterpart of the inverse image map F^{-1} when F is a continuous function between topological spaces—but note that f^* can also be regarded as an interpretation of intuitionistic propositional theories. Since f^* preserves $\bigvee$'s, the adjoint functor theorem tells us that it has a right adjoint $f_* : A \to B$ (the direct image) which preserves arbitrary $\bigwedge$, and locale maps are often presented in terms of this pair.

Crucial to our theory is the existence of a pair of adjoint functors $\Omega \dashv \mathrm{pt} : \mathsf{Loc} \to \mathsf{Top}$ between the categories of locales and topological spaces.

Models of the theory represented by a locale A are given by the $\wedge\bigvee$-preserving maps $A \to 2 = \{\top = 1, \bot = 0\}$; or in more categorical terms, models are given by the representable functor $\mathrm{pt}(-) = \mathsf{Loc}(2, -)$. It is easy to check that if for $a \in A$ we define

$$U_a = \{p \in \mathrm{pt}\, A \mid p^*(a) = 1\},$$

then U_- is a locale map from the powerset of $\mathrm{pt}\, A$ to A. Translated into more conventional terms, this tells us that $\{U_a \mid a \in A\}$ forms a topology on $\mathrm{pt}\, A$, and so, after the easy verification that locale maps give rise to continuous functions we can treat $\mathrm{pt}(-)$ as a functor from Loc to Top, the category of topological spaces.

Conversely, given any topological space, X, its lattice of open sets forms a locale ΩX, and continuous maps of spaces give rise to locale maps between the corresponding topologies. This does not give an equivalence, but, as we claimed above, the two functors pt and Ω are adjoint.

1.1 LEMMA $\qquad \Omega \dashv \mathrm{pt} : \mathsf{Loc} \longrightarrow \mathsf{Top}.$

Proof. Given a continuous map $f : X \to \mathrm{pt}\, Y$ we must define an $\wedge\bigvee$-preserving map $f^* : Y \to \Omega X$, in order to give a locale map $\Omega X \to Y$. So, given $y \in Y$, we take $f^*(y)$ to be $f^{-1}\, U_y$, where U_y is defined as above.
Conversely, given $f^* : Y \to \Omega X$ we define a continuous map f on $x \in X$ by

$$f(x)\,(y) = \top \qquad \text{iff} \qquad x \in f^*(y).$$

We leave verification of the necessary equivalences to the reader (alternatively, see Johnstone [1982] II.1). $\qquad\square$

DEFINITION If the unit of this adjunction $\eta_X : X \to \mathrm{pt}(\Omega X)$ is an isomorphism, then we say that the space X is sober.

1.2 LEMMA A topological space X is sober if and only if each irreducible closed subset of X is the closure of a unique point. In particular, any sober space is T_0.

Proof. The irreducible closed subsets of X are the closed subets of X which cannot be expressed as the union of any finite number of proper closed subsets. Their complements therefore generate the prime principal ideals in $\Omega(X)$; but the prime principal ideals correspond naturally to the elements of $\mathrm{pt}(\Omega X)$. □

Furthermore

1.3 LEMMA If A is a locale, then $\mathrm{pt}\, A$ is sober.

Proof. Given a space X, η_X sends a point $x \in X$ to the point ηx of ΩX where $(\eta x)^*(U)$ is the truth-value of "$x \in U$". It is an immediate consequence of the triangle identities for adjunctions that η_{pt} is monic. Hence it suffices to show that it is also surjective.

Let p be a point of $\Omega(\mathrm{pt}\, A)$. Then $p^* : \Omega(\mathrm{pt}\, A) \to \mathbf{2}$. Consider now the counit map $\epsilon: \Omega\, \mathrm{pt}\, A \longrightarrow A$, and a moment's thought will show that this is surjective. We compose to obtain $\rho = \epsilon(p) \in \mathrm{pt}\, A$, and claim that $\eta \circ \rho = p$. We must show that for $U \in \Omega(\mathrm{pt}\, A)$,

$$(\eta \circ \rho)^* = p^* U;$$

but $(\eta \circ \rho)^* U$ is the truth value of $\rho \in U$, which is $\top$ iff $\rho^* a = \top$ (where $a = \bigvee\{\alpha \in A \mid \epsilon^* \alpha = U\}$), and this holds iff $p^* \epsilon^* a = \top$ iff $p^* U = \top$. □

We relate these spaces to the usual categories of cpo's by means of the specialisation ordering and Scott topologies.

We can define the specialisation ordering on the points of any space by $x \leq y$ if and only if $x \in \mathrm{cl}\{y\}$. If the space is T_0, in particular if it is sober, then this pre–order is a partial order. In the case of a sober space $X = \mathrm{pt}\, A$ we have $x \leq y$ iff $x^* \leq y^*$ in the usual functional ordering (since $x \in \mathrm{cl}\{y\}$ if and only if every open containing x also contains the point y).

1.4 LEMMA $X = \mathrm{pt}\, A$ has arbitrary directed sups in the specialisation ordering. The ordering has a bottom element iff X is irreducible.

Proof. Let S be a subset of X directed in the specialisation ordering. Then $\bigvee S$ is defined pointwise via $(\bigvee S)^*(a) = \bigvee\{p^* a \mid p \in S\}$ This clearly preserves $\bigvee$'s, and it also preserves $\wedge$, since $\wedge$ commutes with directed sups in $\mathbf{2}$.

The second statement is obvious. □

DEFINITION If X is a partially–ordered set, then the Scott topology $\Sigma(X)$ on X has as its open sets those subsets of X which are upwards closed in the partial order and inaccessible by directed joins.

When the order relation on a set X is given by the specialisation ordering for some topology, the Scott topology is not, of course, completely unrelated to the original.

1.5 LEMMA If $X = \mathrm{pt}\, A$ is sober, then the subsets of X which are open in its given topology also form open sets in the Scott topology on X.

Proof. It is easy to see that the open subsets of X are upwards closed in the specialisation preorder on X. Furthermore, since X is sober, any open of X is of the form U_a for some $a \in A$, so if $\bigvee S \in U_a$ for some directed set S, then $\bigvee\{p^*a \mid p \in S\}$ and the result follows from the compactness of 2. □

1.6 COROLLARY If X is sober, then ΣX is the finest topology on X compatible with the specialisation pre-order.

It is, however, important to note that this does not imply that the Scott topology on a poset X is necessarily sober.

Thus, we can obtain a predomain (a domain without $\perp$) from a locale. In order to go the other way, we look at the locale which is the Scott topology on the domain. Here we have:

1.7 LEMMA If D is an algebraic cpo, then it is sober in its Scott topology (or equivalently, it is the space of points of its Scott topology).

Proof. The Scott topology on D has as open sets the unions of sets of the form $x^\dagger$, where x is a finite (isolated) element of D (we shall call the collection of these isolated elements $\mathcal{B}[D]$). The Scott topology on D is therefore given by the same locale as the upwards closed topology on $\mathcal{B}[D]$. Now a point of a locale is given by a completely prime filter (the set of elements sent to $\top$). Suppose $\mathcal{F}$ is such a filter on $\Omega\, \mathcal{B}[D]$, then $\{x \mid \exists O \in \mathcal{F}.x^\dagger \subseteq O\}$ is upwards closed and downwards directed in $\mathcal{B}[D]$, and furthermore completely determines $\mathcal{F}$. It follows that the elements of $\mathrm{pt}(\Sigma\, D)$ correspond precisely to ideals in $\mathcal{B}[D]$, in other words to the elements of D. □

More generally, it is easy to show that continuous cpo's are sober in their Scott topologies. It is also known that there are other cpo's which possess this property. However, the question of finding an order-theoretic characterisation of the complete family remains open.

2 *Constructions on locales*

In the later sections of the paper we shall want both to use the category of locales as a semantic category, and to relate the semantics obtained there to semantics in categories of cpo's. In order to do this we need to give an interpretation of the basic operators for domain equations, which we take to be lifting, sum, product, function space and the three power domains. We intend that we should be able to solve domain equations in Loc, and that we should be able to obtain solutions in categories of cpo's, such as AlgCPO and SFP, as the points of solutions in Loc. It follows that our interpretations are constrained in such a way that taking points gives the usual interpretations in CPO. The one small exception to this general rule is the functor sum, which we shall interpret as the categorical sum, rather than the amalgamated sum (since at this stage we do not wish to restrict our domains to have $\perp$).

Note, however, that it would be perfectly plausible to make different choices for the interpretation of the basic operators, even for product, but more especically for lifting and for function space. The choice of operators constitutes a definite design decision.

In order to present the operations we make use of the fact that the category $\mathsf{Frm} = \mathsf{Loc}^{\mathsf{op}}$ is algebraic, and that locales can therefore be presented as the free frames generated by some family of tokens satisfying a given set of relations. The reader should be aware that this is a somewhat delicate property, and that it fails for complete boolean algebras. Intuitively, in Frm we can build terms using infinitary $\bigvee$, but only finitary $⩕$, and the distributivity available is sufficient for us to be able to put terms into a normal form consisting of an infinitary disjunction of a family of finite conjuncts. In a complete boolean algebra, on the other hand, we can use the anti-automorphism induced by $\neg$ to define an infinitary $\bigwedge$, but the lack of complete distributivity prevents us from being able to bound in advance the size of the algebra. In fact, it is possible to show that there are arbitrarily large countably generated complete boolean algebras (*cf.* Johnstone [1982]).

We illustrate this technique with the simplest operation—the lift $(-)_\perp$.

Given a cpo D (perhaps without $\perp$), its lift $D_\perp$ is obtained by adding a new bottom element. $\perp^\uparrow$ is then a new Scott open subset, strictly larger than all the old open subsets. The corresponding operation on locales is to add a new top element. Algebraically, the locale $A_\perp$ is generated by the tokens

$$\{\sigma(a) \mid a \in A\},$$

which satisfy the conditions

1. $\sigma(\perp_A) = \perp$
2. $\sigma(a) \wedge \sigma(b) = \sigma(a \wedge b)$
3. $\bigvee\{\sigma(s) \mid s \in S\} = \sigma(\bigvee S)$ provided $S \neq \emptyset$.

The condition that S be inhabited is needed to ensure that $\sigma(\top_A) = \sigma(\bigvee\emptyset) < \top_{A_\perp}$, where, confusingly, if $A = \Sigma D$, the Scott topology on an algebraic cpo D, then $\sigma(\top_A)$ is the open subset of $D_\perp$ containing just the elements of D, and $\top_{A_bot}$ is the whole of $D_\perp$ (the upwards closure of $\perp$).

The description of the categorical sum of two locales is even simpler: since the theory of frames is (infinitary) algebraic the sum of locales is their product as posets:

$$A + B = \{(a, b) \mid a \in A, b \in B\}$$

where $(a, b) \leq (a', b')$ iff $a \leq a'$ and $b \leq b'$. (Note that the amalgamated sum of two cpo's in which we identify the two bottom elements can be obtained from this coproduct by forcing as an additional requirement $(\top_A, b) = \top_{A+B} = (a, \top_B)$.)

The product is, however, slightly more complicated, even though it is interpreted as the categorical product in Loc. The product of two locales A and B is also generated by pairs of elements, though this time $\rho(a,b)$ is to be interpreted as the basic rectangle of the product space with sides a and b. $A \times B$ is the locale generated by $\{\rho(a,b) \mid a \in A,\ b \in B\}$, subject to the conditions:

1. $\rho(\bot_A, b) = \bot_{A\times B} = \rho(a, \bot_B)$
2. $\rho(\top_A, \top_B) = \top_{A\times B}$
3. $\rho(a_1, b_1) \wedge \rho(a_2, b_2) = \rho(a_1 \wedge a_2, b_1 \wedge b_2)$
4. $\bigvee\{\rho(a_i, b) \mid i \in I\} = \rho(\bigvee a_i, b)$
5. $\bigvee\{\rho(a, b_j) \mid j \in J\} = \rho(a, \bigvee b_j)$.

This presentation can be shortened somewhat if we note that the locale is generated by the $\rho(a, \top_B)$ together with the $\rho(\top_A, b)$ (which, from the logical point of view, are the interpretations of the propositions a and b in $A \times B$). Of course, since pt is a right adjoint it preserves cartesian product, and hence $\mathrm{pt}(A \times B) \simeq \mathrm{pt}(A) \times \mathrm{pt}(B)$ as spaces. However, it is not necessarily the case that $\Omega X \times \Omega Y = \Omega(X \times Y)$, unless at least one of X and Y is locally compact. Fortunately, this will be true for us.

Unfortunately Loc shares with the category of algebraic cpo's the property of not being cartesian closed. Say that a locale A is *exponentiable* if the functor $(-) \times A : \mathsf{Loc} \to \mathsf{Loc}$ has a right adjoint $[A \to -]$. Martin Hyland has shown (Hyland [1981]) that a locale is exponentiable iff it is locally compact (iff it is a continuous lattice in the sense of Scott). This includes all algebraic cpo's, not just the ones that are SFP, but in general the localic function space is not locally compact, even if both locales themselves are.

2.1 THEOREM If A, B are locales, with A locally compact, then the locale $[A \to B]$ is generated by symbols $w(a,b)$ ($a \in A, b \in B$) (which represent $\{f \mid a \ll f^* b\}$) subject to the following relations:

1. $w(\bot_A, b) = \top_{[A\to B]}$
2. $w(a_1 \vee a_2, b) = w(a_1, b) \wedge w(a_2, b)$
3. $w(a, b) \leq w(a, b') \qquad b \leq b'$
4. $w(a, \top_B) = \top_{[A\to B]} \qquad a \ll \top_A$
5. $w(a, b_1) \wedge w(a, b_2) \leq w(a', b_1 \wedge b_2) \qquad a' \ll a$
6. $w(a, b) = \bigvee\{w(a', b) \mid a \ll a'\}$
7. $w(a, \bigvee S) = \bigvee_S w(a, s)$ for S directed
8. $w(a, \bot_B) = \bot_{[A\to B]}$ unless $a = \bot_A$
9. $w(a, b_1 \vee b_2) = \bigvee\{w(a_1, b_1) \wedge w(a_2, b_2) \mid a \leq a_1 \vee a_2\}$.

In the case of the Scott topology on an algebraic cpo the relation $\ll$ is interpreted as

$$a \ll b \quad \text{iff} \quad \text{whenever } b = \bigvee b^{\uparrow}_{\alpha} \text{ then } a \text{ is contained in some finite union } b^{\uparrow}_{\alpha_1} \vee \ldots \vee b^{\uparrow}_{\alpha_n}.$$

Note that since the points of a locale are given by the representable functor $\mathsf{Loc}(\Omega, -)$ (where Ω is the terminal locale, the topology of the one-point space), if both A and B are spatial, then $\mathsf{pt}([A \to B]) = \mathsf{Loc}(\Omega \times A, B) = \mathsf{Loc}(A, B) = \mathsf{Top}(A, B)$, and so in the case of cpo's we inherit the space of Scott-continuous functions. It seems unlikely however that the localic function space is in general spatial; the exception unsurprisingly being in the case of SFP domains, where we can use the fact that the Scott topology is *strictly* locally compact to show that the function space (locale) is coherent, and thus spatial.

Suppose we are dealing with the function space of two algebraic cpo's ($[\Sigma X \to \Sigma Y]$). Then we have a basis for the two locales ΣX and ΣY, in the sense that any open is a filtered union of compact Scott opens. In this context the axioms 6 and 7 can be read as defining a general element of the function space in terms of the restricted family $\{w(a, b) \mid a, b \text{ finite}\}$. Note that if X is SFP, then any element is a directed join of these. Essentially the same axiom system can be used to present the function space in terms of the restricted basis, and in some cases we can even make simplifications. If $\top_{\Sigma X}$ is compact (certainly true if X has $\bot$) then axiom 4 is equivalent to "$w(\top, \top) = \top$" (and hence to "$w(a, \top) = \top$, all a"). Also, if B is a *stably locally compact* locale (*i.e.* $b \ll b_1$ and $b \ll b_2$ implies $b \ll b_1 \wedge b_2$), in particular if $B = \Sigma X$ where X is SFP, then axiom 5 can be strengthened to "5'. $w(a, b_1) \wedge w(a, b_2) = w(a, b_1 \wedge b_2)$. We shall need this strengthening later in section 4 when we come to consider the open set lattice of the domain of resumptions.

Finally, we come to the powerdomains. These can be given a "modal" description, which we sketch below; further details can be found in Robinson [1986]. We recall that if D is an algebraic cpo, then define a family of three different orderings on $\mathsf{M}[D]$, the set of non-empty finite subsets of $\mathcal{B}[D]$ as follows:

$$A \preceq_0 B \quad \text{iff} \quad \forall b \in B \; \exists a \in A \quad a \leq b$$
$$A \preceq_1 B \quad \text{iff} \quad \forall a \in A \; \exists b \in B \quad a \leq b$$
$$A \preceq_2 B \quad \text{iff} \quad A \preceq_0 B \text{ and } A \preceq_1 B.$$

The completions of $\mathsf{M}[D]$ with respect to these orderings give respectively the Smyth, Hoare and Plotkin powerdomains (which we write as $\mathsf{P}_0(D)$, $\mathsf{P}_1(D)$ and $\mathsf{P}_2(D)$).

The localic powerdomains are built on bases of tokens $\Box a$, $\Diamond a$ (where a runs over the elements of A). Although the notation is suggestive, we stress the fact that $\Box a$ is formally an indivisible token, and not the result of applying the modal operator $\Box$ to the (abstract) proposition a. Thus, although this approach seems very similar to that of Winskel [1985] it differs from it in that we do not allow the iteration of modal operators until they are needed for the solution of recursive domain equations.

The Plotkin powerdomain is very closely linked to the notion of Vietoris locale introduced by Peter Johnstone and differs from it essentially only in its treatment of the empty set.

Definition (*cf.* Johnstone [1985]) Given a locale A, the Vietoris locale $\mathsf{V}(A)$ of A is the free locale on the basis $\Box a$, $\Diamond a$ (where a runs through the elements of A) subject to the following relations:

1. if $a \leq b$ then $\Box a \leq \Box b$, and $\Diamond a \leq \Diamond b$

2. if S is directed in A then $\Box \bigvee S = \bigvee \{\Box s \mid s \in S\}$

3. (a) $\Box(a \wedge b) = \Box a \wedge \Box b$
   (b) $\Box \top = \top$

4. If S is any subset of A (including ϕ), then $\Diamond \bigvee S = \bigvee \{\Diamond s \mid s \in S\}$
   (and hence $\Diamond \bot = \bot$.)

5. $\Box(a \vee b) \leq \Box a \vee \Diamond b$

6. $\Diamond(a \wedge b) \geq \Diamond a \wedge \Box b$.

Roughly speaking, this corresponds to the Plotkin powerdomain together with the empty set as an isolated point. We shall actually need the strict sublocale $\mathsf{V}_2(A)$ of $\mathsf{V}(A)$ obtained by forcing in addition $\Diamond \top = \top$ (or equivalently $\Box \bot = \bot$).

In Robinson [1986] we prove the following

2.2 Theorem If D is an algebraic cpo, then $\mathsf{V}_2(\Sigma D)$ is the Scott topology on the Plotkin powerdomain of D.

We can give similar descriptions of the other two powerdomains:

2.3 Proposition

(i) Let $\mathsf{V}_0(A)$ be the locale generated by tokens $\Box a$ subject to the axioms 2 and 3 above, together with $\Box \bot = \bot$. Then $\mathsf{V}_0(\Sigma D) = \Sigma \mathsf{P}_0(D)$.

(ii) Let $\mathsf{V}_1(A)$ be the locale generated by the $\Diamond a$ subject to the axiom 4 above, together with $\Diamond \top = \top$. Then $\mathsf{V}_1(\Sigma D) = \Sigma \mathsf{P}_1(D)$.

3 *Axiomatic semantics of programming languages*

There are two stages to giving an axiomatic semantics for a programming language. The first is to find some set of propositions which the language may satisfy, which for simple imperative languages may be propositions relating input to output, while for more complex parallel languages they may be modal assertions about the necessary or possible eventual state of the execution. The second to give rules enabling us to deduce that a program satisfies some particular proposition. There is of course always more to it than this. The propositions are never simply an unstructured set—for a start they can always be pre-ordered if we define $a \leq a'$ whenever for all programs p

$$p \models a \Longrightarrow p \models a' .$$

Central to the philosophy of the axiomatic semantics is the idea that two programs are semantically equivalent iff they satisfy the same propositions, and hence that the propositions suffice to distinguish between programs. The converse, however, can not be so easily justified, and we shall not assume that there are necessarily enough actual syntactic programs to tell the propositions apart. This may result in our considering pre-orders finer than the one above.

It is useful to know that we can always consistently extend our propositions so that they interpret the full intuitionistic propositional calculus.

3.1 LEMMA Let P be a set of programs and A a partially ordered set of assertions such that

$$a \leq a' \text{ and } p \models a \quad \Longrightarrow \quad p \models a'.$$

Let $\tilde{A}$ be the free locale on A. Then we can extend the $\models$ relation to $P \times \tilde{A}$.

Proof. What I have called the free locale is of course given by the left adjoint to the forgetful functor $\mathsf{Loc}^{\mathsf{op}} \to \mathsf{Pos}$. For each $p \in P$ we have an order-preserving map $A \to P$ and so the result follows from the universal property. We can however give a more concrete definition of the extension. Any element of $\tilde{A}$ can be written formally as $\bigvee_\alpha(a_{\alpha 1} \wedge \ldots \wedge a_{\alpha n_\alpha})$, and we have p "$\models$" $\bigvee_\alpha(a_{\alpha 1} \wedge \ldots \wedge a_{\alpha n_\alpha})$ iff

$$\exists \alpha \; \forall i = 1, \ldots, n_\alpha \quad p \models a_{\alpha_i},$$

which is just what is demanded by a Tarskian definition of truth for the propositional connectives. □

Similar arguments show that if we can already interpret conjunction (disjunction) in the assertions, then we can manage the extension in such a way as to preserve them. From now on we assume that the assertions involved in an axiomatic semantics form a locale, and are thus closed under all intuitionistic propositional operations.

Recall that the notion of semantic equivalence is given by the propositions satisfied and not satisfied by the program. Thus the meaning of a program (or at least its semantic equivalence class) is given by a map from the locale to the Boolean algebra $\{\top, \bot\}$—a point of the locale. Again note that although the locale suffices to distinguish programs, it is not necessary that any point of the locale represents a program or for that matter that points suffice to distinguish assertions.

We must now consider the second component of the semantics—the proof rules which enable us to deduce properties of complete programs from properties of their component parts. Suppose we have a program p whose outer structure is

$$\text{operator} \quad \text{fragment1} \quad \text{fragment2}.$$

The principle of structurality implies that if we replace fragment1 and fragment2 by semantically equivalent fragments to get a new program p', then the semantics of p and p' are the same. Our axiomatic semantics has therefore to be strong enough to determine the semantics operator as a function

$$(\text{semantic equivalence classes})^2 \rightarrow (\text{semantic equivalence classes}).$$

Furthermore, since the semantics is axiomatic and structural, if we can deduce a property ρ of p, we have to be able to deduce it from properties ρ_1 and ρ_2 of fragment1 and fragment2. The map is therefore continuous for the topologies generated by the corresponding locales of assertions. This, however, is all we can say. Since the programs are not necessarily the whole of, or even dense in, the spaces of points of the locales of propositions, we cannot immediately deduce that this continuous map arises from a locale map on assertions. Nevertheless, I am going to claim that in a well-constructed axiomatic semantics this is precisely what happens. The argument is as follows: if the semantics did not give locale maps as semantics for some operator, then we could find essentially trivial extensions to the language which would necessitate a change in the semantics for the operator—all we have to do is to add new basic programs (viewed as atomic) which are declared to have particular properties, in other words to add programs representing points whose behaviour for the operator is not well-described by the semantics. By this means we see that the semantics of operator depends on the atomic programs available, which is clearly unsatisfactory.

To sum up, what the argument above attempts to show is that a satisfactory axiomatic semantics gives rise to a denotational semantics in the category of locales. Taking the points of these locales will give us a semantics in the category of cpo's (without $\perp$), though unfortunately, again we cannot say as much as we might like—there is no reason to suppose the cpo's are algebraic.

In the next section we examine the correspondence in the other direction, given a denotational semantics we shall extract a complete set of proof rules.

Before that however, we conclude this section with an essentially trivial but nevertheless messy example, that of the Hoare logic which treats the partial correctness of a simple imperative language. We steal the language IMP from Plotkin [1981]. The major syntactic category of IMP is the commands.

$$\begin{aligned} \text{comm} ::= a \ &\mid \text{skip} \mid c; c' \mid \text{if } b \text{ then } c \text{ else } c' \\ &\mid \text{while } b \text{ do } c \end{aligned}$$

where a is some "atomic" command such as

$$x := e$$

and b is a boolean expression. We remind the reader that a language of this form can be given a simple denotational semantics in which the commands denote endo-functions on a (flat) domain of *States*.

The axiomatic approach towards such a language is also very simple. It is conventional to write the basic tokens of the logic

$$\{p\}\ c\ \{p'\}$$

where p and p' are assertions about *States* (or from our point of view essentially open sets of *States*), and c is a command. We shall rewrite this as

$$c \models p \Rightarrow p',$$

for reasons which are readily apparent.

The semantics can be presented in the style of a system of natural deduction, for which the rules be found in *e.g.* de Bakker [1980]:

$$\frac{c \models p \Rightarrow p'' \quad c \models p' \Rightarrow p''}{c \models p \vee p' \Rightarrow p''}$$

$$\frac{c \models p \Rightarrow p' \quad c \models p \Rightarrow p''}{c \models p \Rightarrow p' \wedge p''}$$

$$\frac{c \models p \Rightarrow q \quad \text{and} \quad p' \leq p \quad q \leq q'}{c \models p' \Rightarrow q'}$$

$$\frac{}{\mathsf{skip} \models p \Rightarrow p}$$

$$\frac{c \models p \Rightarrow p' \quad c' \models p' \Rightarrow p''}{c; c' \models p \Rightarrow p''}$$

$$\frac{c \models p \wedge b \Rightarrow p' \quad c' \models p \wedge \neg b \Rightarrow p'}{\mathsf{if}\ b\ \mathsf{then}\ c\ \mathsf{else}\ c' \models p \Rightarrow p'}$$

$$\frac{c \models p \wedge b \Rightarrow p}{\mathsf{while}\ b\ \mathsf{do}\ c \models p \Rightarrow p \wedge \neg b}$$

We also have to be told which of the sequents $p \Rightarrow p'$ are satisfied by each of the atomic commands.

Our intuition is that this system of rules should present the classical denotational semantics in a localic form. In particular, the semantics of sequencing should turn out to be the composition operator for endo-functions. Unfortunately this is not quite the case. If we try to compare the two semantics, then we find that "$\{f \mid f \models p \Rightarrow p'\}$"is not in general a Scott open of $[States_\perp \to States_\perp]$. Part of the trouble is that p is not necessarily finite (there is something of an inconsistency of presentation here: the flat domain of states goes with the assumption that the value of a variable is represented by the contents of a particular storage location, and hence that either all variables are defined, or else none are—and this chimes badly with having predicates on the values of individual variables). We can however avoid this problem by using an ω-rule:

$$\frac{c \models p'' \Rightarrow p' \qquad \text{for all } p'' \ll p}{c \models p \Rightarrow p'}$$

More seriously, we can prove that the uniformly non-terminating program is hugely over-specified:

$$\text{while } \mathit{true} \text{ do skip} \models \mathit{true} \Rightarrow \mathit{true} \wedge \neg \mathit{true},$$

which contradicts the fact that in the domain $\mathit{States}_\perp$ the intepretation of $[\![\mathit{true} \wedge \neg\mathit{true}]\!]$ is the empty set. Intuitively, the more highly specified an object is (i.e. the more properties it has, or equivalently, the more open sets it is in) the higher up it is in the domain. But since we are dealing here only with partial correctness, the undefined state (representing non-termination) satisfies every predicate and so should come at the top of the domain. Not taking things too seriously, we shall suppose that the propositions are the Scott-open subsets of a flat domain $\mathit{States}^\top$, inverting the usual ordering. This does not of course affect the endo-functions as a set, but it does reverse the ordering, and so drastically affects the topology.

The Scott opens of $\mathit{States}^\top$ consist of the empty set and any subset of *States* unioned with the sets $\top$, with $\{a \cup \{\top\} \mid a \text{ finite}\}$ as the finite elements. In the semantics above, we take the predicates $p \Rightarrow p'$ with p finite to refer to elements of the locale, and interpret those where p is not finite by means of the ω-rule above. This gives us a sound interpretation, and the standard completeness theorem tells us that if we have enough atomic programs, then the locale presented really is $[\mathit{States}^\top \to \mathit{States}^\top]$.

4 *On denotational semantics*

The previous section of this paper was silent on one point. We claimed that it was possible to extract from an axiomatic semantics a denotational semantics valued in the category of locales, and hence, by taking points, in a category of cpo's, but gave no indication of how to find a domain equation which gave that semantics. Indeed I know of no way, other than the application of brute intelligence (or just guessing the right answer), in which this can be done. The rest of the paper is devoted to showing that the situation in the reverse direction is very different. Given a domain equation, and suitable presentations of the semantics maps for a language, it is possible effectively to extract a complete syntax-directed set of proof rules.

We already have all the tools we need to do this. For this section of the paper we shall restrict ourselves to working in SFP, where, as we have seen, the localic operators correspond precisely to the operations on cpo's. Since SFP has solutions to recursive domain equations with parameters from SFP, we can solve domain equations in Loc, provided the parameters come from Ω SFP, one solution being the Scott topology of some solution in SFP (this result is clearly not best possible, but at least it's cheap). The topology of the initial solution in SFP is what is going to provide us with our collection of propositions.

We recall how this solution is formed by successive approximation: as a zero'th order approximation we take the domain $\{\perp\}$, which is initial in the category of retractions contained in the identity (we write this SFP^{proj}). The domain equation itself gives us a functor on SFP^{proj}. We take as our example Plotkin's domain of resumptions

$$\mathit{Res} = [\mathit{States} \to_s \mathsf{P}_?(\mathit{States} + (\mathit{States} \otimes \mathit{Res}))],$$

where *States* is some domain of states, which is determined in advance, $\mathsf{P}_?(-)$ is one of the three power-domain functors, $[- \to_s +]$ the strict function space (obtained in this case by requiring

$w(a, \top) = \bot$ for all a not equal to $\top$), and $\otimes$ is the strict product. A solution to this equation is a fixed point of the functor

$$F(X) = [States \rightarrow_s \mathsf{P}_?(States + (States \otimes X))].$$

We obtain successive approximations to a fixed point by applying the functor to the previous approximation. The first approximation is thus $F\{\bot\}$. The fixed point is given as the colimit (limit) of the ω-sequence $(F^n\{\bot\} | n \in \mathsf{N})$ thus obtained. Now, the finite elements in the limit are obtained as the injections of the finite elements in the approximating domains. It is thus easy to see that the topology on the limit is the directed union of the inverse images of the topologies on the finite approximants, and, since we have the means of calculating each of these, can itself be calculated.

We return to our example of the domain of resumptions. In order to be definite we take $\mathsf{P}_?(-)$ to be the Plotkin power-domain $\mathsf{P}_2(-)$, the cases of the other power-domains are similar, though less complicated. Note that the strict product is obtained by adding the axiom

$$\rho(a, \top) = \rho(\top, \top) = \rho(\top, b) = \top$$

to the product space axioms, and adding the requirement that none of a_i, b_i be $\top$ to the intersection axiom:

$$\rho(a_1, b_1) \wedge \rho(a_2, b_2) = \rho(a_1 \wedge a_2 b_1 \wedge b_2.$$

Reading off from the definition we see that the topology of the first approximant $F\{\bot\}$ is generated by open sets of the form $w(p, \alpha)$,w) here p ranges over a basis of $\Omega(States)$, and α is a finite open of $\mathsf{P}_2(States + States \otimes \{\bot\})$, and is thus a finite positive combination of sets of the form

- $\Box \bigvee_i \mathrm{inj}_0\, p_i \wedge \mathrm{inj}_1\, p_i'$
- $\Diamond\ \mathrm{inj}_0\, p \wedge \mathrm{inj}_1\, p'$.

Further unpacking leads us to the result that the general Scott open subset of the domain of resumptions can be expressed as a directed join of elements of the form given by the following little grammar:

$$\begin{aligned}
\mathsf{Res} &::= \ \bot \ \mid \ \top \ \mid \ w(\mathsf{Sta}, \mathsf{Pow}) \\
\mathsf{Pow} &::= \ \bigvee \bigwedge \mathsf{Pow} \ \mid \ \Box \bigvee \mathsf{Sum} \ \mid \ \Diamond\ \mathsf{Sum} \\
\mathsf{Sum} &::= \ \mathsf{inj}_0\, \mathsf{Sta} \ \mid \ \mathsf{inj}_1\, (\mathsf{Sta}, \mathsf{Res}) \\
\mathsf{Sta} &::= \ p.
\end{aligned}$$

If $\rho \in \mathsf{Pow}$, we can read $w(p, \rho)$ as $p \rightarrow \rho$, and so obtain a modal logic in which we can iterate the modal operators. Note however that $\Box$ and $\Diamond$ are not dual (and hence this is not Hennessy-Milner logic).

If we replace the modal construction of power-domains by the original finite subset construction, then we get a system of assertions very similar to that in Brookes [1985] for the Owicki-Gries language. This has been investigated by Zhang, who has shown that Brookes' system can indeed be regarded as arising from an axiomatic presentation of the Plotkin powerdomain flavour of the domain of resumptions (Zhang [1987]).

REFERENCES

S. ABRAMSKY

[1986] *A Domain Equation for Bisimulation*, unpublished manuscript, 1986

[1987] *Domain Theory in Logical Form*, in **Proc. of the IEEE Symposium on Logic in Computer Science (LICS '87)**, IEEE Computer Society Press (1987) 47-53

S.D. BROOKES

[1985] *An Axiomatic Treatment of a Parallel Programming Language*, in **Logics of Programs** (edited by R. Parikh), Lecture Notes in Computer Science 193, Springer (1985) 41-60

J. DE BAKKER

[1980] **Mathematical Theory of Program Correctness**, Prentice Hall, New Jersey, 1980

A. CREMERS & S. GINSBURG

[1975] *Context-free grammar forms*, in J. Comput. System Sci. (1) (1975) 86-116

J.M.E. HYLAND

[1981] *Function spaces in the category of locales*, in **Continuous Lattices**, Lecture Notes in Mathematics 871, Springer-Verlag (1981) 264-281

P.T. JOHNSTONE

[1982] **Stone Spaces**, Cambridge University Press, 1982

[1983] *The Point of Pointless Topology*, in Bull. Amer. Math. Soc. (8) **1** (1983) 41-53

[1985] *Vietoris Locales and Localic Semilattices*, in **Continuous Lattices and Their Applications** (edited by R.-E. Hoffmann & K.H. Hoffmann), Pure & Applied Mathematics 101, Marcel Dekker (1985) 155-180

G.D. PLOTKIN

[1981] *"Pisa Notes"*, unpublished lecture notes, 1981

E.P. ROBINSON

[1986] *Power-domains, Modalities, and the Vietoris Monad*, Cambridge University Computer Laboratory technical report no. 98, 1986

D.S. SCOTT

[1982] *Domains for denotational semantics*, in **Automata, Languages and Programming, Proceedings of ICALP '82** (edited by M. Nielsen & E.M. Schmidt), Lecture Notes in Computer Science 140, Springer-Verlag (1982) 577-613

G. WINSKEL

[1985] *On Powerdomains and Modality*, in Theor. Comp. Sci. (36) (1985) 127-137

G.Q. ZHANG

[1987] *Towards Semantic Based Proof Systems*, unpublished manuscript, 1987

Connections between Partial Maps Categories and Tripos Theory

Maurizio Proietti
Istituto di Analisi dei Sistemi ed Informatica
Viale Manzoni 30, 00185 Roma (Italy)

Abstract

Categories of partial maps and triposes are abstract structures often used in the model-theoretic approach to computation theory. They have some common features: they are both able to model typed first-order logic, and in both structures one can build topoi. In this paper we compare the two structures and we show that, under some conditions, they give rise to equivalent topoi.

1. Introduction

Categories of partial maps have recently gained new interest as a useful tool in the model-theoretic approach to computation theory, recursiveness theory and universal algebra. The notion of partiality spontaneously arises in a model of computation where we would like to represent the possibility of a failure during the program execution.

Various categorical versions of the theory of partial maps have been proposed in the literature. In [Rosolini 86] *p-categories* are defined as categories with partial finite products; they are used to produce an abstract theory for computational processes.

A slightly different approach (but essentially equivalent, as shown in [Proietti 8?]), is the one presented in [Obtułowicz 82]. In that paper categories of partial maps are built in an equational form, that is, they are built by specifying some partial operations on the arrows and a set of suitable equations.

Categories of partial maps seem to be a good models for computation, because one can prove a soundness and completness theorem of a λ-calculus for partial functionals with respect to those structures [Moggi 85].

A fundamental feature of the categories of partial maps is that they provide a very natural framework for the interpretation of both first-order logic and effectiveness ([Obtułowicz 82],[Rosolini 86]).

Also triposes [Hyland,Johnstone,Pitts 80] have been proposed in the literature for representing partial computations. Using logic and partial combinatory algebras, triposes enable us to build various categorical models of recursive realizability and effectiveness ([Hyland,Johnstone,Pitts 80], [Pitts 81],[Hyland 82]).

With the following example, partly borrowed from [Poigné 86b], we try to outline some connections between the categories of partial maps and the triposes.Those connections will be analysed in the paper.

EXAMPLE.

Let $STACK$ be the specification defined as follows:

$SORTS$	$stack, \quad data$
$OPERATORS$	$empty\colon \longrightarrow stack$
	$push\colon stack \times data \longrightarrow stack$
	$pop\colon stack \longrightarrow stack$
	$top\colon stack \longrightarrow data$
VAR	$s\colon stack, \quad d\colon data$
$EQUATIONS$	$pop(push(s,d)) = s$
	$top(push(s,d)) = d$

This specification allows for partial operators, like *pop* and *top*. We may interpret a specification such as the above in a category of partial maps. Namely, we may consider a category $\mathcal{C}$ with finite products and then we may build the category $\mathcal{P}rt(\mathcal{C})$ of partial maps in $\mathcal{C}$. The objects of $\mathcal{P}rt(\mathcal{C})$ are those of $\mathcal{C}$, and its arrows are the pairs $\langle dom(f)\colon D_f \longrightarrow X, f\colon D_f \longrightarrow Y\rangle$. The domain-mono $dom(f)\colon D_f \longrightarrow X$ will be the identity map if f is total. The composition between two partial morphisms $\langle dom(f)\colon D_f \longrightarrow X, f\colon D_f \longrightarrow Y\rangle$, and $\langle dom(g)\colon D_g \longrightarrow Y, g\colon D_g \longrightarrow Y\rangle$, is defined by the diagram:

$$
\begin{array}{ccccc}
D_{g\circ f} & \longrightarrow & D_f & \overset{dom(f)}{\longrightarrow} & X \\
\downarrow & p.b. & \downarrow{\scriptstyle f} & & \\
D_g & \underset{dom(g)}{\longrightarrow} & Y & & \\
\downarrow{\scriptstyle g} & & & & \\
Z & & & &
\end{array}
$$

We shall write $f\colon X \rightharpoonup Y$ in place of $\langle dom(f)\colon D_f \longrightarrow X, f\colon D_f \longrightarrow Y\rangle$.

Let us suppose that $\mathcal{M}$ is a family of monomorphisms of $\mathcal{C}$ which is closed under identities and compositions, and pullbacks of elements of $\mathcal{M}$ always have representatives in $\mathcal{M}$. Thus we can restrict the choice of the domain-monos to the family $\mathcal{M}$, and we obtain the category $\mathcal{P}rt(\mathcal{C},\mathcal{M})$ *of partial maps in* $\mathcal{C}$ *with domains in* $\mathcal{M}$.

Thus, given a specification Σ, an interpretation of Σ in $\mathcal{P}rt(\mathcal{C},\mathcal{M})$ consists of an object $\underline{s_i}$ for each sort s_i, and a partial morphism $\underline{f}\colon \underline{s_1} \times \ldots \times \underline{s_n} \rightharpoonup \underline{s_{n+1}}$ for each operator $f\colon s_1 \ldots s_n \longrightarrow s_{n+1}$.

A category in which the partial morphisms of $\mathcal{P}rt(\mathcal{C},\mathcal{M})$ (that is, the interpretations of terms) could be thought as total morphisms, can be easily generated as follows. Let us take as objects of a new category the domain-monos in $\mathcal{P}rt(\mathcal{C},\mathcal{M})$ and as its arrows from $dom(f)\colon D_f \longrightarrow X$ to $dom(g)\colon D_g \longrightarrow Y$, the arrows $h\colon X \rightharpoonup Y$ in $\mathcal{P}rt(\mathcal{C},\mathcal{M})$, such that $dom(h) = dom(f) = dom(g)\circ h$ (that is, the domain of h is $dom(f)$ and the codomain of h is contained in $dom(g)$). Turning again to our $STACK$ example, the arrow $\underline{pop}$ which interprets the operator *pop*, will be a total arrow $\underline{pop}\colon dom(\underline{pop}) \longrightarrow \underline{stack}$.

There is another way of building a category in which we can interpret the operators of the signature $STACK$ as total arrows; it draws its inspiration from *tripos theory*. We will present that different approach using the same $STACK$ example, and we will show the connections with the approach presented above, based on $\mathcal{P}rt(\mathcal{C},\mathcal{M})$. It is indeed the main objective of our paper to study those connections and to establish the formal requirements for their existence.

We do not assume that the reader is familiar with tripos theory, and in the sequel we will recall the relevant definitions and properties.

If we want to interpret a many-sorted specificaton language with types in category theory, we need a category "of types", that is, a category with finite products and terminal object. In the $STACK$ example we will indeed construct a category $\mathcal{T}_{STACK}$ with finite products and terminal object. $\mathcal{T}_{STACK}$ will be the category $\mathcal{C}$ of the total arrows of $\mathcal{P}rt(\mathcal{C})$. Furthermore, we can describe the properties of the operators by means of first-order predicates. An intuitionistic approach will deal with partial operators in a more natural way. Thus, we will associate a category $\mathcal{P}(S)$ to each object S of $\mathcal{T}_{STACK}$. $\mathcal{P}(S)$ will enjoy the properties of an intuitionistic propositional calculus. The objects of $\mathcal{P}(S)$ are predicates $\phi(s)$ over a free variable of type S, and the arrows are the logical entailments $\phi(s) \vdash \psi(s)$. In order to deal with predicate calculus we have to define for any $f: S \longrightarrow R$ in $\mathcal{T}_{STACK}$, the *substitution* functor $\_\circ f: \mathcal{P}(R) \longrightarrow \mathcal{P}(S)$. We also need the universal and the existential quantification functors $\forall_f, \exists_f: \mathcal{P}(S) \longrightarrow \mathcal{P}(R)$, that is, the right and the left adjoint to $\_\circ f$, respectively.

A tripos over $\mathcal{T}_{STACK}$ is a generalization of the above structure.

To take into account the equational structure of $STACK$, one also need to define the equality predicates. For instance, we will define a predicate $Eq_{\underline{stack}}$ in $\mathcal{P}(\underline{stack} \times \underline{stack})$, for the object $\underline{stack}$ which corresponds to the sort *stack*. The predicate $Eq_{\underline{stack}}$ will be symmetric and transitive. The intended meaning of the chosen equality predicate is: "$Eq_{\underline{stack}}(s,t)$ iff both s and t are defined as elements of $\underline{stack}$ and they are equal". Therefore reflexivity is not a required property of equality predicates, and $Eq_{\underline{stack}}(s,s)$ means "s is defined as an element of $\underline{stack}$".

Now we are able to define a new category in which the (partial) operators of $STACK$ are intended as total morphisms. The objects of that category are $\mathcal{P}$*-objects* and the arrows are classes of logically equivalent *total functional relations.*

A $\mathcal{P}$*-object* is a pair (S, Eq_S). A (partial) functional relation F from (S_1, Eq_{S_1}) to (S_2, Eq_{S_2}), is an element of $\mathrm{P}(S_1 \times S_2)$ such that:

i) $F(x_1,y_1) \wedge Eq_{S_1}(x_1,x_2) \wedge Eq_{S_2}(y_1,y_2) \vdash F(x_2,y_2)$ (i.e. F respects equalities)

ii) $F(x,y) \vdash Eq_{S_1}(x,x) \wedge Eq_{S_2}(y,y)$ (i.e. F respects existence)

iii) $F(x,y_1) \wedge F(x,y_2) \vdash Eq_{s_2}(y_1,y_2)$ (i.e. F is single-valued)

Moreover, F is a *total* functional relation if the following holds:

iv) $Eq_{S_1}(x,x) \vdash \exists y.F(x,y)$

The above definitions generate a category, which is a topos if $\mathcal{P}$ is a tripos [Hyland,Johnstone,Pitts 80].

In our $STACK$ example we can define the interpretation of the operators as functional relations which verify further axioms like the following:

$$\vdash \forall s,t(\exists d.Push(s,d,t) \leftrightarrow Pop(t,s))$$
$$\vdash \forall d,t(\exists s.Push(s,d,t) \leftrightarrow Top(t,d))$$

Moreover, if we define a new equality predicate:

$$Eq_{nestack}(x_1,x_2) \dashv\vdash Eq_{\underline{stack}}(x_1,x_2) \wedge nonempty(x_1) \wedge nonempty(x_2)$$

where $nonempty(x) \dashv\vdash \exists t, d.Push(t, d, x)$, then Pop turns out to be a total functional relation from $(\underline{stack}, Eq_{nestack})$ to $(\underline{stack}, Eq_{\underline{stack}})$.

Therefore in the category of $\mathcal{P}$-objects we are able to interpret partial operators as total morphisms.

The above $STACK$ example shows that we can build a category in which the partial operators of the specification $STACK$ are interpreted as total morphisms using either categories of partial maps or triposes.

One of the differences between the two approaches is the meaning of equality.

In a category of partial maps two terms are equal iff they are identical or both undefined. In tripos theory two terms are equal iff they are identical and both defined. Therefore also the ways of generating total morphisms from partial operators are different: while in a category of partial maps we restrict the domains, in the tripos theory we restrict the extension of the equality predicate. In the sequel we will show that the two approaches may be viewed as equivalent.

In the following sections we outline the categorical theory of partial maps and the tripos theory. We will show how to build a tripos from a doctrine of functional relations [Obtułowicz 82]. We will also show the equivalence between the topos built from that tripos and the topos built from the doctrine of functional relations.

2. Categories of partial maps

In this section we describe the basic concepts of the theory of partial maps, essentially following [Obtułowicz 82].

The first step is an equational definition of the notion of partial maps in a category with finite products. We shall call those categories pc-categories. It is easy to check their equivalence to categories with ordered strict pre-cartesian structure which are described in [Obtułowicz 82], although we hope we have presented here a more readable and intuitive notion. A pc-category is also equivalent to a p-category with a terminal object for its subcategory of total morphisms (see [Rosolini 86]).

The second, third and fourth steps are equational definitions of the notion of partial maps in a category with finite limits, in a cartesian closed category and in a topos, respectively.

2.1. **Definition.** A *pc-category* is a category $\mathcal{C}$ equipped with:

i) a binary operation on the objects:

$$\otimes : Ob\mathcal{C} \times Ob\mathcal{C} \longrightarrow Ob\mathcal{C}$$

ii) a partial binary operation on the arrows:

$$\langle\!\langle \_, \_ \rangle\!\rangle : \frac{f: C \to A, g: C \to B}{\langle\!\langle f, g \rangle\!\rangle : C \to A \otimes B}$$

iii) a fixed object T (called the *domain classifier*)

iv) two arrows:

$$\pi_1^{A,B}: A \otimes B \to A \quad \text{and} \quad \pi_2^{A,B}: A \otimes B \to B$$

for any pair of objects A,B of $\mathcal{C}$ (upper indexes shall be omitted when there is not ambiguity)

v) an arrow:

$$t_A: A \to T$$

for any object A of $\mathcal{C}$.

Moreover, given $f: A \to B, g: B \to C, h: B \to D, \gamma: B \to T, \phi, \psi: A \to T, k: C \to A \otimes B$ the following equations hold:

(pc1)	$t_T = id_T$
(pc2)	$\pi_1 \circ \langle\langle f, t_A \rangle\rangle = f$
(pc3)	$\pi_1 \circ \langle\langle \pi_1 \circ \langle\langle f, \phi \rangle\rangle, \psi \rangle\rangle = \pi_1 \circ \langle\langle f, \pi_1 \circ \langle\langle \phi, \psi \rangle\rangle \rangle\rangle$
(pc4)	$\pi_1 \circ \langle\langle \phi, \psi \rangle\rangle = \pi_1 \circ \langle\langle \psi, \phi \rangle\rangle$
(pc5)	$\pi_1 \circ \langle\langle g \circ f, \gamma \circ f \rangle\rangle = \pi_1 \circ \langle\langle g, \gamma \rangle\rangle \circ f$
(pc6)	$\pi_1 \circ \langle\langle g \circ f, \phi \rangle\rangle = g \circ \pi_1 \circ \langle\langle f, \phi \rangle\rangle$
(pc7)	$\langle\langle \pi_1 \circ k, \pi_2 \circ k \rangle\rangle = k$
(pc8)	$\pi_1 \circ \langle\langle g, h \rangle\rangle = \pi_1 \circ \langle\langle g, t_D \circ h \rangle\rangle$
(pc9)	$\pi_2 \circ \langle\langle g, h \rangle\rangle = \pi_1 \circ \langle\langle h, t_C \circ g \rangle\rangle$

□

The intended meaning of the equations (pc1)...(pc9) becomes more clear if we write:

i) $f|_\phi$ (read: *f restricted to* ϕ) in place of $\pi_1 \circ \langle\langle f, \phi \rangle\rangle$,

ii) $dom(f)$ (read: *proper domain of* f) in place of $t_B \circ f$,

iii) $\phi \sqcap \psi$ (read: ϕ *intersection* ψ) in place of $\phi|_\psi$,

only if both ϕ and ψ have codomain the domain classifier T.

Using the new symbols, the above equations become:

(pc1)	$t_T = id_T$
(pc2)	$f|_{t_A} = f$
(pc3)	$(f|_\phi)|_\psi = f|_{(\phi \sqcap \psi)}$
(pc4)	$\phi \sqcap \psi = \psi \sqcap \phi$
(pc5)	$(g \circ f)|_{(\gamma \circ f)} = (g|_\gamma) \circ f$
(pc6)	$(g \circ f)|_\phi = g \circ (f|_\phi)$
(pc7)	$\langle\langle \pi_1 \circ k, \pi_2 \circ k \rangle\rangle = k$
(pc8)	$\pi_1 \circ \langle\langle g, h \rangle\rangle = g|_{dom(h)}$
(pc9)	$\pi_2 \circ \langle\langle g, h \rangle\rangle = h|_{dom(g)}$

Given two partial maps $f, g: A \to B$ it is natural to think that f *is contained in* g if $dom(f)$ is contained in $dom(g)$, and f is equal to g in the common domain. Thus we define:

$$f \sqsubseteq g \iff f = g|_{dom(f)}$$

It is easy to verify that $\sqsubseteq$ is a partial order in $C(A, B)$.

It should be clear that Definition 2.1 provides a generalization to categories of partial maps of the categorical concepts of terminal object and finite products. These concepts are expressed as intended, because in the subcategory C_T *of the total arrows* of the pc-category C (that is, those arrows $f: A \to B$ such that $dom(f) = t_A$), the domain classifier T is a terminal object and the other operations define indeed a product.

Obviously a partial map $\phi: A \to T$ can be thought as a representation of a subdomain of A. Thus, given $f: A \to B$, $f|_\phi$ is the restriction of f to the subdomain ϕ. The axioms (pc1)...(pc6) ensure that t_A is the top of $C(A, T)$ and that $\phi \sqcap \psi$ is the greatest lower bound of $\{\phi, \psi\}$ w.r.t. $\sqsubseteq$. Moreover, the composition preserves the greatest lower bounds. The axioms (pc7) ... (pc9) tells us that $\langle\langle f, g \rangle\rangle$ behaves as the pairing map restricted to the intersection of the proper domains of f and g.

Following this approach we can generalize to categories of partial maps, the other classical categorical constructions, such as the product between arrows. Given $f; A \to B$ and $g: C \to D$, then

$$f \otimes g = \langle\langle f \circ \pi_1, g \circ \pi_2 \rangle\rangle$$

moreover, given $\phi: A \to T$ and $\psi: B \to T$, then

$$\phi \times \psi = (\phi \circ \pi_1) \sqcap (\psi \circ \pi_2)$$

In order to give the extended version of pullbacks and equalizers for partial maps, we need, for each object A of C, an arrow $A \otimes A \to A$ which behaves like the restriction of the projections $\pi_1, \pi_2: A \otimes A \to A$ to the diagonal of $A \otimes A$.

2.2. **Definition.** An *equoidal category* is a pc-category C with an arrow

$$e_A: A \otimes A \to A$$

such that the following equations hold:

(e1) $$e_A \circ \langle\langle id_A, id_A \rangle\rangle = id_A$$

(e2) $$\pi_1|_{dom(e_A)} = \pi_2|_{dom(e_A)} = e_A$$

□

Equality arrows are indeed what we needed, since $e_B \circ \langle\langle f_1, f_2 \rangle\rangle$ is the greatest lower bound of $\{f_1, f_2\}$ w.r.t. the partial order $\sqsubseteq$ in $C(A, B)$. Thus, $dom(e_B \circ \langle\langle f_1, f_2 \rangle\rangle)$ is the greatest subdomain of A in which f_1 and f_2 are equal, that is, the domain of the equalizer of f_1 and f_2. Moreover, $dom(e_B \circ (f \otimes g))$ is the greatest subdomain of $A \otimes C$ in which $f \circ \pi_A$ is equal to $g \circ \pi_C$, that is, the domain of the pullback of f and g.

We use the following abbreviations:

$$f_1 \doteq f_2 = dom(e_B \circ \langle\langle f_1, f_2 \rangle\rangle)$$

$$f \asymp g = dom(e_B \circ (f \otimes g))$$

where $f_1, f_2, f: A \to B, g: C \to B$.

We can think of any partial map $f: A \to B$ in a pc-category C with proper domain $\phi: A \to T$ as a map which is total on ϕ. In such a way we naturally construct a new category ΔC whose objects are proper domains, and whose arrows have their proper domains as domains.

In ΔC it is very natural to derive finite limits in terms of the corresponding notions for partial maps. Thus, if C is a pc-category, ΔC has finite products and terminal object; if C is an equoidal category then ΔC will have equalizers too.

The category ΔC is very useful because it makes us able to formulate representation theorems. In fact one can show that every pc-category C is isomorphic to the category $Prt(C, \mathcal{M})$ of the partial maps on C with domains in a suitable family $\mathcal{M}$.

2.3. Definition. Given a pc-category C, the *category* ΔC *of domains of* C is defined as follows:

i) the objects of ΔC are the arrows of C with codomain the domain classifier T;

ii) given $f: A \to B, \phi: A \to T, \psi: B \to T$ such that $dom(f) = \phi$ and $\psi \circ f = \phi$, then f is an arrow of ΔC with domain ϕ and codomain ψ;

iii) given $f: \phi \to \psi, g: \psi \to \gamma$ their composition is $g \circ f: \phi \to \gamma$.

□

As it is natural to expect ΔC has all finite limits, provided that C is an equoidal category. In the following proposition those limits are explicitly constructed.

2.4. Proposition. Let C be an equoidal category and ΔC its category of domains. We have that:

i) The domain classifier T is a terminal object in ΔC ;

ii) given $\phi: A \to T$ and $\psi: B \to T$, $\phi \times \psi$ is a product in ΔC and $\pi_1|_{(\phi\times\psi)}, \pi_1|_{(\phi\times\psi)}$ are the projection arrows;

iii) given $f_1, f_2: \phi \to \psi$, $id_A|_{(f_1 \doteq f_2)}$ is an equalizer arrow for f_1 and f_2 in ΔC.

□

Hence ΔC has all finite limits. In particular, given $f: \phi \to \psi, h: \gamma \to \psi$, the following diagram is a pullback square:

$$\begin{array}{ccc} f \asymp h & \xrightarrow{\pi_1|_{(f \asymp h)}} & \phi \\ \downarrow{\scriptstyle \pi_1|_{f \asymp h}} & & \downarrow{\scriptstyle f} \\ \gamma & \xrightarrow[h]{} & \psi \end{array}$$

In order to get a category which behaves like the category of partial maps of a cartesian closed category, there should be, for each pair A, B of objects, an object $A \rightharpoonup B$ "of partial maps from A to B". The adjointness between the product and the exponential functors generalizes to a family of mappings $\lambda_{A,B,C}: C(A \otimes B, C) \longrightarrow C(A, B \rightharpoonup C)$, which defines a natural isomorphism between the functors $C(A \otimes \_, C)$ and $C_T(A, \_ \rightharpoonup C)$.

2.5. Definition. An *equoidal category with types* is an equoidal category C equipped with:

i) a binary operation on the objects :

$$\rightharpoonup: ObC \times ObC \longrightarrow ObC$$

ii) a unary partial operation on the arrows:

$$\lambda_{A,B,C}: \frac{f: A \otimes B \to C}{\lambda_{A,B,C}(f): A \to (B \rightharpoonup C)}$$

iii) an arrow:

$$ev_{A,B}: (A \rightharpoonup B) \otimes A \to B$$

for all pairs of objects A,B. (Indexes will be omitted, if there is not ambiguity.) Moreover the following equations hold:

(t1) $$dom(\lambda(f)) = t_A$$

(t2) $$ev \circ (\lambda(f) \otimes id_B) = f$$

(t3) $$\lambda(ev \circ (h \otimes id_B))|_{dom(h)} = h$$

where $f: A \otimes B \to C, h: A \to (B \rightharpoonup C)$.

□

The following step is an attempt to build a category which behaves like the category of partial maps of a topos. Therefore the notion of a *doctrine of functional relations* arises as an equoidal category with types with an additional family of arrows $\{\sigma_A: (A \rightharpoonup T) \to A | A \in ObC\}$. The intended meaning of these partial maps is to represent singleton maps, that is, σ_A is defined only on the partial maps $s \in (A \rightharpoonup T)$, whose proper domains are singletons and, σ_A applied to s is the single element in $dom(s)$.

The facts that the category $\Delta\mathcal{D}$ is a topos and $\mathcal{D}$ is fully embedded in $\mathcal{P}rt(\Delta\mathcal{D})$, confirm that a doctrine of functional relation $\mathcal{D}$ is what we wanted to build. Furthermore, the full embedding of $\mathcal{D}$ in $\mathcal{P}rt(\Delta\mathcal{D})$ allows us to interpret any first-order predicate language in $\mathcal{D}$.

2.6. **Definition.** A *doctrine of functional relations* is an equoidal category with types $\mathcal{D}$, equipped with an arrow:

$$\sigma_A: (A \rightharpoonup T) \to A$$

for each object A of $\mathcal{D}$. Moreover, given $f: A \to B, \phi: A \to T$, the following equations hold:

(d1) $$\sigma \circ \lambda(dom(e_A)) = id_A$$

(d2) $$f \circ \sigma_A \circ \lambda(id_B \asymp f) = id_B|_{dom(\sigma_A \circ \lambda(id_B \asymp f))}$$

(d3) $$\lambda(\phi \circ \pi_1) \asymp \lambda(t_{T \otimes T}) = \phi \circ \pi_1$$

□

The following technical lemmas will be useful in the proof of theorem 4.3

2.7. **Lemma.** Let $\mathcal{C}$ be an equoidal category. Given $f, g \in \mathcal{C}(A, B)$ we have:

$$f = g \qquad \text{iff} \qquad f \asymp id_B = g \asymp id_B$$

□

2.8. **Lemma.** Let $\mathcal{D}$ be doctrine of functional relations, T its domain classifier, and $\phi \in \mathcal{D}(A \otimes B, T)$. If

$$(\phi \circ \pi_1^{A \otimes B, B}) \sqcap (\phi \circ (\pi_1^{A,B} \otimes id_B)) \sqsubseteq \pi_2^{A,B} \asymp id_B$$

holds, then

$$\phi = (\pi_2^{A,B} \circ (\pi_1^{A,B}|_\phi)^*) \asymp id_B$$

where $f^* = \sigma_A \circ \lambda(id_B \asymp f)$ for $f \in \mathcal{D}(A, B)$. □

3. Triposes

Tripos theory arises as an attempt to model both the notion of recursive realizability and non-standard (intuitionistic) logic [Hyland,Johnstone,Pitts 80], [Pitts 81], [Hyland 82]. We can synthetize the definition of a tripos in the following four steps:

i) given a typed first-order predicate language without equality $\mathcal{L}$, we can think of its types as the objects of a category with finite products;

ii) for each type X we need to take care of the propositional logic for formulas over a free variable of type X; thus we associate to each type X a Heyting pre-algebra $\mathcal{P}X$, that is a set PX with a pre-order $\vdash$ (deduction is not antisymmetric), finite meets $\wedge$, finite joins $\vee$, implications $\rightarrow$, top $\top$ and bottom $\bot$. As usual, we can think of $\mathcal{P}X$ as a category (with at most one arrow for each pair of objects) with finite products ($\wedge$), finite coproducts ($\vee$), exponentials ($\phi \rightarrow (\_)$ is right adjoint to $(\_) \wedge \phi$), terminal object ($\top$) and initial object ($\bot$);

iii) in order to deal with predicate logic, for each $f: X \rightarrow Y$ we need the substitution functor $\mathcal{P}f: \mathcal{P}Y \rightarrow \mathcal{P}X$ (i.e. composition with f) and the quantification functors $\exists_f, \forall_f: \mathcal{P}X \rightarrow \mathcal{P}Y$, which are the left and right adjoints to $\mathcal{P}f$, respectively;

iv) in order to deal with higher-order logic it will be defined a membership predicate for each type X.

We will see later how the above structure might be thought of as codifying the internal logic of a topos; the leading examples are the localic and the realizability ones.

3.1. **Definition.** Let $\mathcal{C}$ be a category with finite limits. A *$\mathcal{C}$-tripos* $\mathcal{P}$ consists of:

i) for each object I of $\mathcal{C}$, a Heyting pre-algebra $(\mathcal{P}I, \vdash_I)$;

ii) for each arrow $f: I \rightarrow J$ in $\mathcal{C}$ there exist functors $\mathcal{P}f: \mathcal{P}J \longrightarrow \mathcal{P}I$, $\exists_f, \forall_f: \mathcal{P}I \longrightarrow \mathcal{P}J$, such that:

a) $\exists_f, \forall_f$ are, left and right adjoint to $\mathcal{P}f$, respectively;

b) the Beck condition holds, that is, for any pullback:

$$\begin{array}{ccc} I & \xrightarrow{h} & J \\ \downarrow{\scriptstyle k} & & \downarrow{\scriptstyle g} \\ H & \xrightarrow{f} & K \end{array}$$

in $\mathcal{C}$,we have:

$$\mathcal{P}g \cdot \forall_f \dashv\vdash \forall_h \cdot \mathcal{P}k \qquad \text{and} \qquad \mathcal{P}g \cdot \exists_f \dashv\vdash \exists_h \cdot \mathcal{P}k$$

c) $\mathcal{P}f$ preserves implications;

iii) for each object I of $\mathcal{C}$, an object Σ_I in $\mathcal{C}$ and an element $\in_I$ of $\mathcal{P}(I \times \Sigma_I)$, such that, given any ϕ in $\mathcal{P}(I \times J)$ there is an arrow $\{\phi\}: J \rightarrow \Sigma_I$ in $\mathcal{C}$ and $\mathcal{P}(id_I \times \{\phi\}) \in_I \dashv\vdash \phi$ holds in $\mathcal{P}(I \times J)$. $\in_I$ is called the *membership predicate* of I.

□

Given a many sorted first-order language $\mathcal{L}$ without equality and a $\mathcal{C}$-tripos $\mathcal{P}$, a *$\mathcal{P}$-interpretation* of $\mathcal{L}$ consists of:

a) for each type X of $\mathcal{L}$ an object X of $\mathcal{C}$;

b) for each function symbol f: $X_1 \cdots X_n \rightarrow X_{n+1}$ an arrow $f: X_1 \times \cdots \times X_n \rightarrow X_{n+1}$ in $\mathcal{C}$;

c) for each relation symbol $R \subset X_1 \ldots X_n$ an element R in $\mathcal{P}(X_1 \times \ldots \times X_n)$.

Now we can give an inductive interpretation of terms as arrows in $\mathcal{C}$, and the interpretation of a formula ϕ whose free variables are in $\bar{x} = (x_1, \cdots x_n)$, as an element $[\![\phi(\bar{x})]\!]$ of $\mathcal{P}(X_1, \cdots, X_n)$.

Given a finite collection Γ of formulas of $\mathcal{L}$, a formula ϕ and a string $\bar{x}$ containing all the free variables occurring in $\Gamma \cup \{\phi\}$ we define:

$$\Gamma \models_{\mathcal{P},\bar{x}} \phi \quad \text{iff} \quad \bigwedge_{\gamma \in \Gamma} [\![\gamma(\bar{x})]\!] \vdash [\![\phi(\bar{x})]\!] \quad in \quad \mathcal{P}(X_1 \times \ldots \times X_n)$$

The following *Soundness* lemma holds:

3.2. Lemma. For any $\mathcal{C}$-tripos $\mathcal{P}$ and any $\mathcal{P}$-interpretation of a language $\mathcal{L}$, if $\Gamma \vdash_{\bar{x}}$ holds in the intuitionistic predicate logic, then $\Gamma \models_{\mathcal{P},\bar{x}} \phi$ holds.

□

Given a $\mathcal{C}$-tripos $\mathcal{P}$, a $\mathcal{P}$*-object* is a pair $(X, =)$ where $=$ is a predicate in $\mathcal{P}(X \times X)$ such that:

$$\vdash \; x_1 = x_2 \;\rightarrow\; x_2 = x_1$$

$$\vdash \; x_1 = x_2 \;\wedge\; x_2 = x_3 \;\rightarrow\; x_1 = x_3$$

Reflexivity is not required because $=$ is intended to model an equality of partial elements. Thus, the predicate $[\![x \in X]\!] = [\![x = x]\!]$ represents the "extent to which x exists in X".

A *(partial) functional relation* F from $(X, =_X)$ to $(Y, =_Y)$, is an element of $\mathcal{P}(X \times Y)$ such that:

i) $F(x_1, y_1) \wedge \; x_1 =_X x_2 \;\wedge\; y_1 =_Y y_2 \vdash F(x_2, y_2)$ (i.e. F respects equalities)

ii) $F(x, y) \vdash \; x =_X x \;\wedge\; y =_Y y$ (i.e. F respects existence)

iii) $F(x, y_1) \wedge F(x, y_2) \vdash \; y_1 =_Y y_2$ (i.e. F is single-valued)

Morover, F is a *total* functional relation if the following holds:

iv) $x =_X x \; \vdash \exists y.F(x, y)$

3.3. Definition. The category $\mathcal{C}[\mathcal{P}]$ has $\mathcal{P}$-objects as objects, and equivalence classes w.r.t. $\dashv\vdash$ of total functional relations as arrows. Given two arrows $f: (X, =) \rightarrow (Y, =), g: (Y, =) \rightarrow (Z, =)$ represented by F and G respectively, their composition is represented by $\exists y(F(x, y) \wedge G(y, z))$. The identity arrow on $(X, =)$ is the equivalence class of $=$. □

3.4. Theorem. The category $\mathcal{C}[\mathcal{P}]$ of $\mathcal{P}$-objects is a topos. □

3.5. Lemma. Given a monomorphism $m: (Y, =) \rightarrow (X, =)$ in $\mathcal{C}[\mathcal{P}]$ represented by M, if we define $[\![R(x)]\!] = [\![\exists y M(x, y)]\!]$ and $[\![x_1 =_M x_2]\!] = [\![x_1 = x_2 \wedge R(x)]\!]$, then M represents an isomorphism $\tilde{m}$ from $(Y, =)$ to $(X, =_M)$. Furthermore, if $r: (X, =_M) \rightarrow (X, =)$ is represented by $=_M$ then $m = r \circ \tilde{m}$. □

4. Connections between tripos theory and categories of partial maps

In this section we show how to build a tripos $\mathcal{P}_{\mathcal{D}}$ from a given doctrine of functional relations $\mathcal{D}$. The category of types will be the subcategory of total maps. The logical structure will be the internal logic of the doctrine. The topos of $\mathcal{P}_{\mathcal{D}}$-objects actually has total functional relations as morphisms between objects with a partial membership predicate ($[\![x = x]\!]$ generally is not the top t_X). Thus it behaves as the topos $\Delta\mathcal{D}$ which has proper domains of $\mathcal{D}$ as objects and total arrows between proper domains as morphisms. We will show that they are equivalent topoi.

An immediate consequence of the full embedding of a doctrine $\mathcal{D}$ in $\mathcal{P}rt(\Delta\mathcal{D})$ is that we are able to interpret in it a first-order predicate language.

The following lemma will be a useful tool to recover such an interpretation from the one in the topos $\Delta\mathcal{D}$.

4.1. **Lemma.** Let us consider the functors:

$$\begin{array}{rccc} \mathcal{D}(\_,T): & \mathcal{D}_T^{op} & \longrightarrow & \mathcal{P.O.S.} \\ & I & & (\mathcal{D}(I,T),\sqsubseteq) \\ & \big\downarrow f & & \big\downarrow \_\circ f \\ & J & & (\mathcal{D}(J,T),\sqsubseteq bigr) \end{array}$$

$$\begin{array}{rccc} Sub_{\Delta D}(\_): & \Delta\mathcal{D}^{op} & \longrightarrow & \mathcal{P.O.S.} \\ & \phi & & (Sub_{\Delta D}(\phi),\leq) \\ & \big\downarrow f & & \big\downarrow f^{-1} \\ & \psi & & (Sub_{\Delta D}(\psi),\leq) \end{array}$$

$$\begin{array}{rccc} \kappa: & \mathcal{D}_T & \longrightarrow & \Delta\mathcal{D} \\ & I & & t_I \\ & \big\downarrow f & & \big\downarrow f \\ & J & & t_J \end{array}$$

then the family of morphisms in $\mathcal{P.O.S.}$:

$$\begin{array}{rccc} \eta_I: & (\mathcal{D}(I,T),\sqsubseteq) & \longrightarrow & (Sub_{\Delta D}(t_I),\leq) \\ & \phi & & id_I|_\phi: \phi \hookrightarrow t_I \\ & \big\downarrow \sqsubseteq & & \big\downarrow \leq \\ & \psi & & id_I|_\psi: \psi \hookrightarrow t_I \end{array}$$

defines a natural isomorphism:

$$\eta: \mathcal{D}(\_,T) \longrightarrow (Sub_{\Delta D}(\kappa(\_))$$

Proof. η_I *is a bijective function,* because it is the composition of the bijective function:

$$\begin{array}{ccc} \mathcal{D}(I,T) & \longrightarrow & \mathcal{P}rt(\Delta\mathcal{D})(t_I,t_T) \\ \phi & \longmapsto & \langle id_I|_\phi, 1_I\rangle \end{array}$$

with the bijective function:

$$\begin{array}{ccc} \mathcal{P}rt(\Delta\mathcal{D})(t_I, t_T) & \longrightarrow & Sub_{\Delta\mathcal{D}}(t_I) \\ \langle m, t_I\rangle & \longmapsto & m \end{array}$$

Both η_I and η_I^{-1} are order-preserving. In fact, given $\phi, \psi \in \mathcal{D}(I,T)$, we have that:

$$\begin{array}{llll} \phi \sqsubseteq \psi & \Rightarrow & (id_I|_\psi) \circ (id_I|_\phi) = (id_I|_\psi)|_\phi & \text{by (pc6)} \\ & & = id_I|_{(\psi \sqcap \phi)} & \text{by (pc3)} \\ & & = (id_I|_\phi) & \text{by hypothesis} \\ & \Rightarrow \eta_I(\phi) \leq \eta_I(\psi) & & \end{array}$$

vice-versa:

$$\begin{array}{llll} \eta_I(\phi) \leq \eta_I(\psi) & \Rightarrow & \text{there exists } f\colon \phi \to \psi \text{ such that } (id_I|_\psi) \circ f = (id_I|_\phi) & \\ & \Rightarrow & f = f|_\phi & \text{because } \phi = dom(f) \\ & & = f|_{(\psi \circ f)} & \text{because } f\colon \phi \to \psi \\ & & = (id_I|_\psi) \circ f & \text{by (pc6)} \\ & & = id_I|_\phi & \text{by hypothesis} \\ & \Rightarrow & \phi = \psi \circ f & \text{because } f\colon \phi \to \psi \\ & & = \psi \circ (id_I|_\phi) & \text{by the above} \\ & & = \psi \sqcap \phi & \text{by (pc 6)} \\ & \Rightarrow \phi \sqsubseteq \psi & & \end{array}$$

η is natural, because the following square commutes for all $f \in \mathcal{D}_T(J,I)$:

$$\begin{array}{ccc} (\mathcal{D}(I,T), \sqsubseteq) & \xrightarrow{\eta_I} & (Sub_{\Delta\mathcal{D}}(t_I), \leq) \\ \big\downarrow{\scriptstyle -\circ f} & & \big\downarrow{\scriptstyle f^{-1}} \\ (\mathcal{D}(J,T), \sqsubseteq) & \xrightarrow{\eta_J} & (Sub_{\Delta\mathcal{D}}(t_J), \leq) \end{array}$$

fact the following diagram is a pullback for all $\phi \in \mathcal{D}(I,T)$:

$$\begin{array}{ccc} \phi \circ f & \xrightarrow{id_J|_{(\phi \circ f)}} & t_J \\ \big\downarrow{\scriptstyle f|_{\phi \circ f}} & & \big\downarrow{\scriptstyle f} \\ \phi & \xrightarrow[id_I|_\phi]{} & t_J \end{array}$$

t is:

$$f^{-1}(\eta_I(\phi)) = \eta_J(\phi \circ f)$$

□

Using Lemma 4.1 we can recover the interpretation of first-order predicate logic in $\mathcal{D}$ from one in $\Delta\mathcal{D}$. As an example we define an operator of join by the following square:

$$\begin{array}{ccc} \mathcal{D}(I,T) \times \mathcal{D}(I,T) & \xrightarrow{\sqcup_I} & \mathcal{D}(I,T) \\ \big\downarrow{\scriptstyle \eta_I \times \eta_I} & & \big\uparrow{\scriptstyle \eta_I^{-1}} \\ Sub_{\Delta\mathcal{D}}(t_I) \times Sub_{\Delta\mathcal{D}}(t_I) & \xrightarrow{\vee_I} & Sub_{\Delta\mathcal{D}}(t_I) \end{array}$$

In a similar way we define in $\mathcal{D}$(I,T) meet $\sqcap_I$, implication $\rightarrow_I$, top $\top_I$, and bottom $\bot_I$ se definitions of meet and top are equivalent to the ones in section 2).

Furthermore, we define a universal quantification functor $\forall_{J,I}\colon \mathcal{D}(J \otimes I, T) \longrightarrow \mathcal{D}(J,T)$ the following square:

$$\begin{array}{ccc} \mathcal{D}(J \otimes I, T) & \xrightarrow{\forall_{J,I}} & \mathcal{D}(J,T) \\ \big\downarrow{\scriptstyle \eta_{J \otimes I}} & & \big\uparrow{\scriptstyle \eta_J^{-1}} \\ Sub_{\Delta\mathcal{D}}(t_J \otimes t_I) & \xrightarrow{\mathcal{U}_{\pi_1}} & Sub_{\Delta\mathcal{D}}(t_J) \end{array}$$

where $\mathcal{U}_{\pi_1}$ is the universal quantification in $\Delta\mathcal{D}$.

Similarly for the existential quantification functor.

4.2. **Theorem.** Let $\mathcal{D}$ be a doctrine of functional relations and $\mathcal{D}_T$ its subcategory of total arrows. A tripos over $\mathcal{D}_T$ is given by:

i) if I is an object of $\mathcal{D}_T$ then:
$\mathcal{P}(I) = (\mathcal{D}(I,T), \sqsubseteq_I, \sqcap_I, \sqcup_I, \rightarrow_I, t_I, \perp_I)$

ii) if $f \in \mathcal{D}_T(I,J)$, $\phi \in \mathcal{D}(I,T)$, $\psi \in \mathcal{D}(J,T)$ then:
$\mathcal{P}f(\psi) = \psi \circ f$
$\forall_f(\phi) = \forall_{J,I}((id_J \asymp f) \rightarrow (\phi \circ \pi_2^{J,I}))$
$\exists_f(\phi) = \exists_{J,I}(id_J \asymp (f|_\phi))$
where $\forall_{J,I}, \exists_{J,I}$ are defined as above;

iii) if $\phi \in \mathcal{P}(I \times J)$ then:
$\Sigma_I = I \rightharpoonup T$
$\in_I = ev_{I,T} \circ \langle\langle \pi_2, \pi_1 \rangle\rangle$
$\{\phi\} = \lambda_{J,I,T}(\phi)$

Proof. It follows immediately from Lemma 4.1 that (i) defines a Heyting algebra. Using Lemma 4.1. it is easy to prove that

$$\mathcal{P}f = \eta_J^{-1} \circ f^{-1} \circ \eta_I$$
$$\forall_f = \eta_I^{-1} \circ \mathcal{U}_f \circ \eta_J$$
$$\exists_f = \eta_I^{-1} \circ \mathcal{E}_f \circ \eta_J$$

where $\mathcal{U}_f, \mathcal{E}_f$ are the quantifier functors in $\Delta\mathcal{D}$. Thus, the functors defined in (ii), satisfy conditions (ii) a),b),c) of Definition 3.1.

Finally, it follows from the definition 2.5 of an equoidal category with types, that $\in_I$ actually is a membership predicate for I.

□

4.3. **Theorem.** Let $\mathcal{D}$ be a doctrine of functional relations and $\mathcal{P}$ the tripos defined as in 4.1., then there exists an equivalence from the topos of $\mathcal{P}$-objects $\mathcal{D}_T[\mathcal{P}]$ to $\Delta\mathcal{D}$, which is injective on objects.

Proof. Let α be the functor defined as follows:

$$\begin{array}{cccc} \alpha: & \Delta\mathcal{D} & \longrightarrow & \mathcal{D}_T[\mathcal{P}] \\ & \phi: A \rightarrow T & & (A, =_\phi) \\ & \downarrow h & & \downarrow h \asymp id_B \\ & \psi: B \rightarrow T & & (B, =_\psi) \end{array}$$

where the equality predicate $=_\phi$ is $id_A|_\phi \asymp id_A$.

Notice first that α is indeed a functor. In fact, for any $\phi \in \mathcal{D}(A,T)$, the subobject represented by $id_A|_{(=_\phi)}$ is the interpretation of the formula of the language $\mathcal{L}_{\Delta\mathcal{D}}$ of $\Delta\mathcal{D}$, $a = a' \wedge (a \in \phi)$, thus it is symmetric and transitive. It follows from Lemma 4.1, that also $=_\phi \in \mathcal{P}(A \otimes A)$ is symmetric and transitive.

Furthermore $id_{A \otimes B}|_{(h \asymp id_B)}$ is the interpretation of the formula of $\mathcal{L}_{\Delta\mathcal{D}}$, $h(a) = b \wedge a \in \phi \wedge b \in \psi$, thus, by Lemma 4.1, $h \asymp id_B \in \mathcal{P}(A \otimes B)$ is a functional relation from $(A, =_\phi)$ to $(B, =_\psi)$.

Now we will prove that α is an equivalence.

α is faithful. Let's suppose $h, k \in \Delta\mathcal{D}(\phi, \psi)$ and $\alpha(h) = \alpha(k)$, that is $h \asymp id_B = k \asymp id_B$; thus, by lemma 2.7., we get $h = k$.

α is full. Let $F \in \mathcal{P}(A \otimes B)$ be a functional relation from $\alpha(\phi) = (A, =_\phi)$ to $\alpha(\psi) = (B, =_\psi)$, then

$$\mathcal{P} \models F(a,b) \wedge F(a,b') \rightarrow b =_\psi b'$$

Therefore:

$$F \circ \pi_1^{A\otimes B,B} \sqcap F \circ (\pi_1^{A,B} \otimes id_B) \sqsubseteq ((id_B|_\psi) \asymp id_B) \circ (\pi_2^{A,B} \otimes id_B)$$

by the $\mathcal{P}$-interpretation definition

$$\sqsubseteq t_B \circ e_B \circ (\pi_2^{A,B} \otimes id_B)$$

because $(id_B|_\psi) \asymp id_B \sqsubseteq t_B \circ e_B$

$$= \pi_2^{A,B} \asymp id_B$$

Hence, by Lemma 2.8, we have:

$$\begin{aligned} F &= (\pi_2^{A,B} \circ (\pi_1^{A,B}|_F)^*) \asymp id_B \\ &= \alpha(\pi_2^{A,B} \circ (\pi_1^{A,B}|_F)^*) \end{aligned}$$

Every $\mathcal{P}$-object is isomorphic to $\alpha(\phi)$ for some object ϕ of $\Delta\mathcal{D}$. Let (A, eq) be a $\mathcal{P}$-object, that is, A is an object of $\mathcal{D}_T$ and eq is a symmetric and transitive predicate in $\mathcal{D}(A \otimes A, T)$. The arrow $\phi: A \rightarrow T$, equal to $eq \circ \langle\langle id_A, id_A\rangle\rangle$, is an object of $\Delta\mathcal{D}$ such that:

$$\mathcal{P} \models eq(a,a) \leftrightarrow a =_\phi a$$

where $=_\phi$ is defined as above. In fact:

$$\begin{aligned} [\![a =_\phi a]\!] &= ((id_A|_{eq\circ\langle\langle id_A, id_A\rangle\rangle}) \asymp id_A) \circ \langle\langle \pi_1^{A,A}, \pi_1^{A,A}\rangle\rangle \\ &= \pi_1^{A,A}|_{eq\circ\langle\langle \pi_1^{A,A}, \pi_1^{A,A}\rangle\rangle} \doteq \pi_1^{A,A} \\ &= eq \circ \langle\langle \pi_1^{A,A}, \pi_1^{A,A}\rangle\rangle \\ &= [\![eq(a,a)]\!] \end{aligned}$$

Thus, the subobjects of $(A, t_{A\otimes A})$ represented by $eq(a,a')$ and $a =_\phi a'$ have isomorphic canonical monomorphisms, and therefore they are isomorphic.

Moreover *α is injective on objects.* In fact, given two objects ϕ, ψ in $\Delta\mathcal{D}$ such that $\alpha(\phi) = \alpha(\psi)$, then $(id_A|_\phi) \asymp id_A = (id_A|_\psi) \asymp id_A$; by Lemma 2.7, it turns out that $id_A|_\phi = id_A|_\psi$, and hence $\phi = \psi$.

□

4.4. **Remark.** If we require that equality predicates enjoy a *consistence property*, that is, for any two equality predicates $=_1$ and $=_2$ in $\mathcal{P}(A \times A)$, we have:

$$\mathcal{P} \models a \in_1 A \ \wedge\ a' \in_1 A \ \wedge\ a \in_2 A \ \wedge\ a' \in_2 A \ \rightarrow\ (a =_1 a' \ \leftrightarrow\ a =_2 a')$$

then $\mathcal{D}_T[\mathcal{P}]$ and $\Delta\mathcal{D}$ are isomorphic. In that case, in the proof of 4.3. it holds that:

$$\mathcal{P} \models eq(a,a) \leftrightarrow a =_\phi a$$

Using the consistence property we deduce:

$$\mathcal{P} \models eq(a,a') \leftrightarrow a =_\phi a'$$

In the tripos $\mathcal{P}$ the relation $\models$ is antisymmetric, and therefore, eq and $=_\phi$ are equal.

□

5. Conclusions

We compared two structures used in the model-theoretic approach to computation theory: the categories of partial maps and the triposes.

We showed how to build a tripos from a doctrine of functional relations. We also proved the equivalence between the topos built from that tripos and the one built directly from the doctrine of functional relations.

This equivalence formally establishes the corrispondence between the two theories and it also allows us to transfer many results from a theory to the other.

6. Acknowledgements

I would like to thank Anna Labella and Alberto Pettorossi for their encouragement and helpful suggestions. Their comments on a previous version of this paper were highly appreciated.

This research was supported by an ESPRIT scholarship of the Enidata S.p.A. (Italy) and IASI Institute of the Italian National Research Council.

References

[Hyland 82]
Hyland, J.M.E.: *The Effective Topos,* in Troelstra and van Dalen editors, The L.E.J. Brouwer Centenary Symposium, North Holland, 1982, pp 165-216.

[Hyland,Johnstone,Pitts 80]
Hyland, J.M.E., Johnstone, P.T. and Pitts A.M.: *Tripos Theory,* in Math. Proc. Camb. Phil. Soc. (88), 1980, pp 205-232.

[Lambek & Scott 86]
Lambek, J. and Scott, P.: *Introduction to Higher Order Categorical Logic*, Cambridge University Press, 1986.

[MacLane 71]
MacLane, S.: *Categories for the Working Mathematician,* Springer-Verlag, 1971.

[Makkai & Reyes 77]
Makkai, M. and Reyes, G.: *First Order Categorical Logic*, in Lecture Notes in Mathematics 611 Springer Verlag, 1977.

[Moggi 85]
Moggi, E.: *Lambda Calculus and Categories,* Thesis Proposal, Computer Science Departement, University of Edinburgh, 1985.

[Obtułowicz 82]
Obtułowicz, A.: *The Logic of Categories of Partial Functions and Its Applications,* PhD Thesis, Institute of Mathematics, Polish Academy of Science, Warsaw, 1982.

[Pitts 81]
Pitts, A..M.: *Tripos Theory in general and Realizability Toposes in particular*, Dissertation for Research Fellowship, St. John's College, Cambridge, 1981.

[Poigné 86a]
Poigné, A.: *Algebra categorically*, Proceedings of Category Theory and Computer Science 1985, in Lecture Notes in Computer Science, 240, Springer-Verlag, 1986, pp. 76-102.

[Poigné 86b]
Poigné, A.: *Category Theory and Logic*, Proceedings of Category Theory and Computer Science 1985, in Lecture Notes in Computer Science, 240, Springer-Verlag, 1986, pp. 103-142.

[Proietti 8?]
Proietti, M.: *Comparison between pc-categories and p-categories*, to be published, 198?.

[Rosolini 86]
Rosolini, G.: *Continuity and Effectiveness in Topoi*, PhD Thesis, Computer Science Departement, Carnegie-Mellon University, Pittsburgh, 1986.

[Scott 77]
Scott, D.S.: *Identity and Existence in Intuitionistic Logic*, in Fourman, M.P., Mulvey, C.J. and Scott, D.S.(eds.) Application of Sheaves, Lecture Notes in Mathematics, 753, 1977, pp. 660-694.

A Fixpoint Construction of the p-adic Domain

Steven Vickers

Department of Computing,
Imperial College of Science and Technology,
180 Queen's Gate,
London,
SW7 2BZ.

Abstract The Kahn domain on p symbols can be given an arithmetic structure so that its maximal elements are isomorphic to the p-adic integers. This is described as a fixpoint of a functor in a category of sheaves of rings.

1. Introduction

In a companion paper, "An Algorithmic Approach to the p-adic Integers" [Vickers 87], we show how the Kahn domain on p symbols – the set of finite and infinite lists of symbols, ordered by the prefix order – can be used to give an algorithmic setting for p-adic analysis. The object is to test out our formal methods of program verification on a rich body of known mathematical constructions and theorems.

The methods there rely very heavily on the arithmetic structure of the domain: it is essentially a ring (with addition, subtraction, zero, multiplication and unit). It is, of course, possible to define the arithmetic operations in the same way as any other continuous functions, using the standard domain equation for the Kahn domain. However, we have also explored the idea of defining the domain *with the arithmetic already built in*. It is now a fixpoint of an endofunctor on a category whose objects already have arithmetic structure.

The construction uses sheaves, and relies on a number of observations.

First, a sheaf gives rise to a cpo. This is described in Fourman and Scott [79] and is done by collecting together all the local sections of the sheaf and ordering them by $x \sqsubseteq y$ iff x is a restriction of y. Not all cpos can be obtained this way, but it doesn't matter here. Our domain of p-adics can be considered a sheaf over the frame (cHa) $\omega+1$.

Second, given that sheaves and cpos can in some circumstances be considered the same thing, their morphisms are still very different. The sheaf morphisms must be very strongly strict. Not only must they map $\perp$ to $\perp$, but they must also preserve "extents of definition" (in our context, lengths of lists) at all higher levels. The arithmetic operations are compatible with this, and so can be considered as sheaf morphisms.

Third, the theory of sheaves of algebras is well worked out, and used successfully in a number of branches of mathematics.

1.1 The p-adic domain

This is described in detail in Vickers [87], but, briefly, consider single-length arithmetic (addition, subtraction and multiplication) in an n-bit computer. This follows the usual methods of arithmetic, but throws away overflow. Thus 2^n is equivalent to 0, and the system is really doing arithmetic modulo 2^n. This puts a ring structure on the finite lists of n bits, the possible contents of a single register. (The head of the list is to be the *least* significant bit.)

In an ∞-bit computer we do arithmetic in exactly the same way. Now we can handle all natural numbers without overflow, and also the generalized natural numbers, the *2-adic integers,* that have infinitely many bits set. This puts a ring structure on the infinite lists of bits.

Given two lists of unequal length, we can operate on them by first truncating the longer one: we cut off (at the most significant end) what in the shorter context would count as overflow. This puts an arithmetic structure on the entire Kahn domain on two letters (0 and 1).

We've been doing arithmetic to base 2, but the same process works to any base p, giving an arithmetic structure to the Kahn domain on p letters (0, 1, …, p–1). Its maximal (infinite) elements are the *p-adic integers,* its finite elements are the integers modulo p^n for varying n.

1.2 Sheaves

The most helpful account for our purposes is Fourman and Scott [79]. The Crash Course in Sheaf Theory in Johnstone [82] summarizes also the aspects of sheaf theory that we don't use, and would be useful in relating our view of sheaves with some others.

An intuition behind a sheaf is that it is a "continuously varying set", with the variation indexed by points in a topological space. For each point there is a set, and continuity is formalized by a technical *local homeomorphism* property. This is the *display space* or *bundle* view of sheaves.

An "element" of this continuously varying set gives for each point an element of the corresponding set, again subject to continuity conditions. These "elements" are called *global sections*. However, it is also useful to consider "elements" that are not defined at all points, but only at those in some open set. These "partial elements" are *local sections* over that open set.

If one open set contains another, then a local section over the larger can be *restricted* to a local section over the smaller. If we think of the restriction, the less defined section, as being an approximation to the more defined section, then the local sections form a cpo under the approximation ordering.

An alternative view of the sheaf is now via its local sections: for each open set we have the set of local sections over it, and for each inclusion between open sets we have the corresponding restriction map. This is the *presheaf* viewpoint (sheaves satisfy an extra *pasting* condition). It does not rely on the open sets being sets of anything, so it really defines the notion of sheaf over a *frame*, i.e. a complete Heyting algebra (see Johnstone [82]).

In our context –

- The frame is the ordinal $\omega+1$ whose elements are the finite natural numbers together with ω (∞). This is in fact isomorphic to the frame of open sets for a topology on the natural numbers.
- The local sections over a finite ordinal n are the lists of length n.
- The global sections (local sections over ω) are the infinite lists.
- Restriction is truncation (at the tail end).

1.3 Notation and Conventions

Throughout, a *ring* is a commutative ring with a multiplicative identity, written usually as 1, but sometimes as *e* to avoid confusion with the integer 1.

p and q will denote integers, *not necessarily prime*. All the theory of p-adic integers that is touched on here works for composite p.

$\mathbb{N}$ is the set of natural numbers, including 0.

$\mathbb{Z}$, $\mathbb{Z}/n$ and $\mathbb{Z}_p$ are the rings of integers, integers modulo n and the p-adic integers. A useful introduction to the p-adics can be found in Cohn [77].

Functions and functors are composed with *;* (diagramatic order) or $\circ$ (applicational order). $\circ$ may be elided. e.g. if $f: A \rightarrow B$ and $g: B \rightarrow C$, then

$$f;g = g\circ f = gf: A \rightarrow C$$

Categories are usually written as emboldened descriptions of their objects, e.g. **Rings, Sets.**

A *frame* is, as in Johnstone [82], a complete Heyting algebra; frame morphisms preserve finite meets and arbitrary joins.

$\bigvee^{\uparrow}$ and $\bigsqcup^{\uparrow}$ are *directed* joins.

2. p-maps of rings

The notion we describe here is presented not for its general interest, which is not clear, but in order that the main construction of the next paragraph can be specified by a universal property.

Definition 2.1 Let R, S be rings, and f: R → S a function. Then f is a *p-map* iff

(i) f is an additive group homomorphism

(ii) $p\, f(xy) = f(x)\, f(y)$ for all $x, y \in R$

(iii) $f(1) = p$

Proposition 2.2

(i) In 2.1, (ii) and (iii) may be generalized to

$$p^{n-1} f(\Pi_{i=1}^{n} x_i) = \Pi_{i=1}^{n} f(x_i) \qquad n \geq 0, x_i \in R$$

provided we are careful to interpret the case n=0 as clause (iii).

(ii) The 1-maps are precisely the ring homomorphisms.

(iii) A p-map composed with a q-map gives a pq-map.

(iv) Defining qf by $(qf)(x) = q\, f(x)$, if f is a p-map then

$$qf = (q\, Id_R);f = f;(q\, Id_S)$$

is a pq-map.]

3. p-adic integers as a fixpoint

We first define $\mathbf{Z}_p$ as the final fixpoint of an endofunctor $F = F_p$ of the category of rings. Since each p-adic integer can be represented by an infinite list of base p digits, we see that as sets,

$$\mathbf{Z}_p \cong \mathbf{Z}_p \times p \qquad d{::}x \leftrightarrow \langle x, d\rangle$$

This describes $\mathbb{Z}_p$, in a well-known way, as a fixpoint of an endofunctor of **Sets**. However, to do this in **Rings** we have to be more clever because it is not true that as rings $\mathbb{Z}_p \cong \mathbb{Z}_p \times \mathbb{Z}/p$: operations on the first digits can generate carries to the rest, which is not the case in the direct product. The method used, which was suggested by Mike Smyth, is to go first to $\mathbb{Z}_p \times \mathbb{Z}$ and then take a quotient ring.

Given a ring R, we define F(R) and a p-map $\eta_R: R \to F(R)$ in three stages.

First, let $(R_1, +, *)$ be the ring-without-1 with underlying additive group R, and multiplication

$$x*y \qquad = \qquad pxy$$

The original 1 in R we now write as e in R_1, since it is no longer a multiplicative identity.

Define also a function $\eta_1: R \to R_1$ by $\eta_1(x) = x$. This satisfies clauses (i) and (ii) of the definition of a p-map.

Next, define R_2 as the ring R_1 with a 1 freely adjoined. Its additive structure is that of $R_1 \oplus \mathbb{Z}$, and multiplication is

$$\langle x, m\rangle\langle y, n\rangle = \langle x*y+nx+my, mn\rangle = \langle pxy+nx+my, mn\rangle$$

Let η_2 be the injection of generators, defined by $\eta_2(x) = \langle x, 0\rangle$. This is a ring-without-1 homomorphism.

Finally, let

$$F(R) = R_3 = R_2/(\langle e, -p\rangle) \qquad \text{i.e. } R_2 \text{ factored out by the ideal generated by } \langle e, -p\rangle$$

and let $\eta_3: R_2 \to R_3$ be the natural map, a ring homomorphism. Let $\eta = \eta_R = \eta_1;\eta_2;\eta_3$.

It is easily checked that η is a p-map. The three stages cooperate to give this. R_1 gives clause (ii) of the definition of p-maps, R_2 takes us back to rings-with-1, and R_3 gives clause (iii).

Definition 3.1 If $x \in R$ and $n \in \mathbb{Z}$, then we write

$$[x, n] = \eta(x)+n = \eta_3(\langle x, n\rangle) \in F(R)$$

Proposition 3.2

(i) $[x, m]+[y, n] = [x+y, m+n]$

(ii) $-[x, m] = [-x, -m]$

(iii) $n = [0, n]$ in $F(R)$

(iv) $[x, m].[y, n] = [pxy+nx+my, mn]$

(v) $[x, m] = [y, n]$ iff $m \equiv n \pmod p$ and $x+(m-n)/p = y$

(vi) η is 1-1.

Proof

Most of this is obvious.

(v) First, suppose $[x, m] = 0$. Then in R_2 we must have, for some $[y, n]$,

$$\langle x, m\rangle = \langle y, n\rangle\langle e, -p\rangle = \langle ne, -np\rangle$$

$\therefore$ $p \mid m$ (with $n = -m/p$) and $x+m/p = 0$.

Now if $[x, m] = [y, n]$ then $[x-y, m-n] = 0$ and we can apply the above observation.]

Theorem 3.3

Let R and S be rings, and let $f: R \to S$ be a pq-map. Then there is a unique q-map $f': F_p(R) \to S$ such that $f = \eta;f'$.

Proof

We must have

$$f'([x, m]) = f'([x, 0]) + f'([0, m]) = f'\eta(x) + f'(m) = f(x) + qm$$

so that f', if it exists, is unique.

Suppose $[x, m] = [y, n]$. Then

$$f(y) + qn = f(x+(m-n)/p) + qn = f(x) + pq.(m-n)/p + qn = f(x) + qm$$

Thus f' is well-defined; it is easily shown to be a q-map.]

Theorem 3.4

F is functorial. More precisely, if f: R → S is a q-map then there is a unique q-map F(f): F(R) → F(S) such that

$$\eta;F(f) = f;\eta$$

and if moreover g: S → T is an r-map then F(f;g) = F(f);F(g).

On 1-maps, this restricts to a functor F: **Rings** → **Rings**.]

Note also the ring homomorphisms

ε_R: F(R) → R, [x, m] ⟼ px+m, corresponding to p Id: R → R

ε'_R: F(R) → $\mathbb{Z}/p$, [x, m] ⟼ m (mod p), corresponding to 0: R → $\mathbb{Z}/p$

Lemma 3.5
F preserves codirected limits in **Rings**.
Proof

Let C be a directed set, and (R_i) ($i \in C$) a C^{op} diagram in **Rings**. If $i \leq j$ in C, we write the corresponding map from R_j to R_i in restriction form, $r \mapsto r|i$.

Let R be the limit of this diagram. An element of R is a sequence (r_i) ($i \in C$) which is coherent in the sense that if $i \leq j$ then $r_j|i = r_i$.

Let S be the limit of $(F(R_i))$, and let g: F(R) → S be the induced map, $[(r_i), m] \mapsto ([r_i, m])$. We show that g is an isomorphism.

First, suppose $([r_i, m]) = 0$. Then for all i, $[r_i, m] = 0$, i.e. $p|m$ and $r_i+(m/p)e_i = 0$.

$$\therefore \quad [(r_i), m] = [(r_i) + (m/p)(e_i), 0] = [(r_i + (m/p)e_i), 0] = 0$$

Now take $([r_i, m_i])$ in S. If i and j are in C, let k be an upper bound for them. Then

$$[r_i, m_i] = F(-|i)([r_k, m_k]) = [r_k|i, m_k], \text{ so } m_i \equiv m_k \pmod p$$

Similarly $m_j \equiv m_k \pmod p$, so we can choose m so that all $m_i \equiv m \pmod p$

$\therefore \qquad ([r_i, m_i]) = ([r_i, m + p((m_i - m)/p)]) = ([r_i + ((m_i - m)/p)e_i, m])$

$\qquad\qquad = g([(r_i + ((m_i - m)/p)e_i), m])$

and g is onto. Thus g is an isomorphism, and F preserves such limits.]

Actually, the same argument shows that F preserves quite a lot of limits, including all filtered or cofiltered limits.

Proposition 3.6

F_{pq} is naturally isomorphic to $F_p;F_q$.

After composition with the appropriate isomorphisms, $\eta^{(pq)} = \eta^{(p)};\eta^{(q)}$ and $\varepsilon^{(pq)} = \varepsilon^{(q)};\varepsilon^{(p)}$.

Proof

Essentially, pqr-maps from R to S factor uniquely, first, via qr-maps from F_pR to S, and, then, via r-maps from F_qF_pR to S.]

Proposition 3.7

$F_p0 \cong \mathbb{Z}/p$

Proof

A p-map f from 0 to S must (by additivity) be the zero map; but 1 = 0 in the zero ring, so

$$0 = f(0) = f(1) = p \qquad \text{in } S$$

This holds iff the unique map from $\mathbb{Z}$ to S factors via $\mathbb{Z}/p$.]

Corollary 3.8 $F^i(0) = F_p{}^i(0) \cong F_{p^i}(0) \cong \mathbb{Z}/p^i$]

Theorem 3.9

The ring $\mathbb{Z}_p$ of p-adic integers is the final fixed point of F in the category of commutative rings with 1.

Proof

Apply the Tarski construction, starting with the zero ring (final in the category of rings). We get the diagram

$$0 \leftarrow \mathbb{Z}/p \leftarrow \mathbb{Z}/p^2 \leftarrow \mathbb{Z}/p^3 \leftarrow \ldots$$

whose limit, given lemma 3.5, is the final fixed point. But it is well-known that $\mathbb{Z}_p$ is the limit of this diagram.]

4. cpos and sheaves

We present a simple theory that makes sense of algebraic structure on a wide class of cpos, including the ones used in constructing our domain of p-adic integers. It applies to cpos in which there is not only the relative notion of one (less defined) element approximating another (more defined) one, but also an absolute measure of how much an element is defined. This measure takes its values in a frame.

Essentially what we do is to use the ideas of Fourman and Scott to get our cpos into a context where we can use the theory of sheaves of algebras.

Definition 4.1 Let A be a frame and S a sheaf over it. We define D(A, S), the *domain* of (A, S), to be the disjoint union of the sets S(a) for a in A, so that its elements are all the local sections of the sheaf.

We define the *extent* function E: D(A, S) $\rightarrow$ A by writing Ex = a if x is in S(a). This is the measure of definition.

Proposition 4.2

(i) D(A, S) is a bounded complete cpo.
(ii) An element x of D(A, S) is finite iff Ex is finite.
(iii) D(A, S) is algebraic iff A is (when considered as a cpo).

Proof

(i) We define an order $\sqsubseteq$ on D(A, S) by writing $x \sqsubseteq y$ if x is the restriction of y to a smaller area of definition [element of A], i.e. iff

$$Ex \leq Ey \text{ and } x = y|Ex$$

(If $a \in A$ and f is the unique morphism in A from $a \wedge Ey$ to Ey then $y|a$, the restriction of y to a, is defined as S(f)(y).)

Reflexivity and transitivity come directly from the fact that S is a functor. As for antisymmetry, suppose that $x \sqsubseteq y$ and $y \sqsubseteq x$. Then clearly Ex = Ey, and $x = y|E(x) = y|E(y) = y$. (Thus a presheaf S gives rise to a poset D(A, S).)

Any subset of D(A, S) that is bounded above is compatible (look at the restrictions of the bound), and so can be pasted together to give the join of the subset. Note that $\bot$ is got by pasting together the empty set. Similarly, any directed set is compatible.

(ii) $\Rightarrow$: Let $Ex = a = \bigvee B$ in A. By taking all finite joins within B, we may assume wlog that B is directed. Let $Y = \{x \mid b : b \text{ in } B\}$. Y is directed, and x is its sup, so for some b in B $x \sqsubseteq x \mid b$ and so $b = a$.

$\Leftarrow$: Suppose $x \sqsubseteq \bigsqcup Y$ where Y is directed. Let $B = \{Ey: y \text{ in } Y\}$, directed in A, and let a be its join. $Ex \leq a$, so $Ex \leq Ey$ for some y in Y.

Therefore, $x = (\bigsqcup Y) \mid Ex = (\bigsqcup Y) \mid Ey \mid Ex = y \mid Ex$ and $x \sqsubseteq y$.

(iii) Easy corollary of (ii).]

Since we shall be dealing freely with changes of the base frame A, we work in the category **Sheaves**. Its objects are pairs (A, S) with A a frame and S a sheaf over it, and a morphism from (A, S) to (B, T) is a pair (f, θ) where f: A→B is a *frame* morphism and θ: S→f;T is a natural transformation. Although not immediately obvious, it is the case that f;T is a sheaf.

Proposition 4.3
D is the object part of a functor from **Sheaves** to **cpos**.
Proof

If $(f, \theta): (A, S) \to (B, T)$ and $x \in D(A, S)$ – so $x \in S(Ex)$ – then

$$D(f, \theta)(x) = \theta_{Ex}(x) \in T(f(Ex)) \qquad]$$

5. Algebraic structure on cpos

Now let $\mathbb{T}$ be a finitary algebraic theory, defined by operators with given arities and identities. If A is a frame, then, as is well-known (see Johnstone [82]), the $\mathbb{T}$-algebra objects in the category Sh-A of sheaves over A are equivalent to the functors from A^{op} to **$\mathbb{T}$-algebras** whose (presheaves of) carriers are sheaves. Thus these can conveniently be thought of as sheaves of algebras.

Definition 5.1 The category **$\mathbb{T}$-frames** has –

- objects are pairs (A, S) with A a frame and S a sheaf of $\mathbb{T}$-algebras over A.
- a morphism from (A, S) to (B, T) is a pair (f, θ) such that f: A → B is a frame morphism and

θ: $S \to f;T$, a natural transformation, is a homomorphism of sheaves of algebras.

There is a standard terminology of ringed spaces, locales and topoi. Putting $\mathbb{T}$ the theory of rings, as we shall in our applications, our **Ringed-frames** is $(\textbf{Ringed-locales})^{op}$.

Suppose (A, S) is a $\mathbb{T}$-frame, and let D = D(A, S). If ω is an n-ary operator in $\mathbb{T}$, we define an operation ω: $D^n \to D$ by

$$\omega(x_1, \ldots, x_n) = \omega(x_1|a, \ldots, x_n|a)$$

where a is the meet $\bigwedge_i Ex_i$ (ω is already defined in the algebra S(a)).

Proposition 5.2

ω is continuous.

Proof

We use a bold-face vector notation for tuples in D^n.

Suppose $\mathbf{x} \sqsubseteq \mathbf{y}$ in D^n, and let $a = \bigwedge_i Ex_i$, $b = \bigwedge_i Ey_i$. $a \le b$.

$$\omega(\mathbf{x}) = \omega(\mathbf{x}|a) = \omega(\mathbf{y}|a) = \omega(\mathbf{y}|b|a) = \omega(\mathbf{y}|b)|a \text{ (the restriction maps are homomorphisms)}$$

$$= \omega(\mathbf{y})|a$$

$$\therefore \quad \omega(\mathbf{x}) \sqsubseteq \omega(\mathbf{y})$$

Now suppose $\mathbf{x} = \bigsqcup^{\uparrow} \mathbf{x}^\lambda$. Since we already know that $\omega(\mathbf{x}^\lambda) \sqsubseteq \omega(\mathbf{x})$, to show $\omega(\mathbf{x}) = \bigsqcup^{\uparrow} \omega(\mathbf{x}^\lambda)$ it suffices to show that they have the same extent.

$$E\omega(\mathbf{x}) = \bigwedge_i Ex_i = \bigwedge_i \bigvee_\lambda Ex^\lambda_i = \bigvee_\lambda \bigwedge_i Ex^\lambda_i \text{ (using directedness)}$$

$$= \bigvee_\lambda E\omega(\mathbf{x}^\lambda) = E\bigsqcup^{\uparrow} \omega(\mathbf{x}^\lambda) \qquad]$$

Proposition 5.3

If (f, θ): $(A, S) \to (B, T)$ is a morphism of $\mathbb{T}$-frames, then $D(f, \theta)$ commutes with the operations ω.

Proof

Suppose $\mathbf{x} \in D(A, S)^n$.

$$\omega(D(f, \theta)(\mathbf{x})) = \omega(\theta(x_i)) = \omega(\theta(x_i) |\bigwedge_i E\theta(x_i)) = \omega(\theta(x_i) |\bigwedge_i f(Ex_i))$$

$$= \omega(\theta(x_i) |f(\bigwedge_i Ex_i))$$

$$= \omega(\theta(x_i|\bigwedge_i Ex_i)) \qquad \text{because } \theta \text{ is natural}$$

$$= \theta(\omega(x_i|\bigwedge_i Ex_i)) \qquad \text{because } \theta \text{ gives homomorphisms}$$

$$= D(f, \theta)(\omega(\mathbf{x})) \qquad]$$

Proposition 5.4

Let e = e' be an identity in $\mathbb{T}$, in which e contains the variables x_i ($1 \leq i \leq m$) and e' contains the variables y_j ($1 \leq j \leq n$). Then when the variables are interpreted in D,

$$e\,|a = e'\,|a \qquad \text{where } a = \bigwedge_i Ex_i \wedge \bigwedge_j Ey_j$$

Proof

This follows immediately from the fact that the restriction maps are $\mathbb{T}$-algebra homomorphisms between the algebras of sections.]

Note that if both sides of an identity have the same variables, no restriction need be made: it will have been done already in the process of the calculation. But if the two sides have different variables, more care needs to be taken. For example, if $\mathbb{T}$ is the theory of groups, the identity

$$x.x^{-1} = 1 \quad \text{must be interpreted as} \quad x.x^{-1} = 1\,|Ex$$

Thus we get a theory of $\mathbb{T}$ structure on certain cpos as a by-product of the known theory of sheaves of $\mathbb{T}$-algebras.

6. The domain of p-adic integers

We have from section 3 an endofunctor F of **Rings**; we define an analogous endofunctor F of **Ringed-frames** (as in section 5, with $\mathbb{T}$ the theory of rings – still commutative with 1).

Let (A, R) be a ringed frame. We define

$$F(A, R) = (A_\perp, R')$$

where $A_\perp$ is A with a new bottom adjoined, and

$$R'(\perp) = 0 \qquad \text{(the zero ring, final in } \mathbf{Rings}\text{)}$$

$$R'|_A = R;F: A^{op} \to \mathbf{Rings} \to \mathbf{Rings}$$

The restriction maps to $\perp$ are the unique maps to 0.

Lemma 6.1

R' is a sheaf of rings over $A_\perp$.

Proof

Let X be a subset of $A_\perp$, with least upper bound x, and for each $a \in X$ let $s_a \in R'(a)$ with $s_a |a \wedge b = s_b |a \wedge b$. If X is empty or $\{ \perp \}$ these paste together trivially, and otherwise we can omit $\perp$ from X.

$$\therefore \qquad s_a = [r_a, m_a] \qquad \text{with } r_a \in R(a).$$

$$[r_a |a\wedge b, m_a] = s_a |a\wedge b = s_b |a\wedge b = [r_b |a\wedge b, m_b]$$

Therefore $m_a \equiv m_b$ (mod p) and we can assume the m_a s are all equal:

$$s_a = [r_a, m] \qquad \text{with } r_a |a\wedge b = r_b |a\wedge b$$

Now pasting the s_a s together is seen to be equivalent to pasting the r_a s together, which is possible because R is a sheaf.]

F as defined so far is the object part of a functor; for suppose $(f, \theta): (A, R) \to (B, S)$ is a morphism of ringed frames. We define

$$F(f, \theta) = (f_\perp, \theta_1): (A_\perp, R') \to (B_\perp, S')$$

as follows:

$$f_\perp(\perp) = \perp$$

$$f_\perp(a) = f(a) \qquad \text{if } a \in A$$

If $a \in A$ then $\theta_a: R(a) \to Sf(a) = Sf_\perp(a)$ and we define

$\theta_{1a} = F(\theta_a): R'(a) = F(R(a)) \rightarrow F(S(f_\perp(a))) = S'(f_\perp(a))$

$\theta_{1\perp} = 0: R'(\perp) = 0 \rightarrow 0 = S'(f_\perp(\perp))$

This makes θ_1 a natural transformation from R' to $S'f_\perp$, so $(f_\perp, \theta_1)$ is a morphism of ringed frames. F thus defined is easily shown to be functorial.

Lemma 6.2

Suppose we are given a linear inverse limit diagram in **Ringed-frames**:

$$(A_0, R_0) \xleftarrow[(f_0, \theta_0)]{} (A_1, R_1) \xleftarrow[(f_1, \theta_1)]{} (A_2, R_2) \xleftarrow[(f_2, \theta_2)]{} \ldots$$

Then the limit (A, R) of this has the following structure.

An element of A is a sequence (a_i) where $a_i \in A_i$ and $f_i(a_{i+1}) = a_i$.

An element of $R((a_i))$ is a sequence (r_i) where $r_i \in R_i(a_i)$ and

$\theta_i(r_{i+1}) = r_i$.]

Lemma 6.3

F preserves ω-limits in **Ringed-frames**.

Proof

Suppose we are given an ω-sequence as in lemma 6.2.

First, clearly $A_\perp = \lim A_{i\perp}$. For the coherent sequences (a_i) in $\lim A_{i\perp}$ are either in A or are constant ⊥.

Let us write

$F(A, R) = (A_\perp, R')$ as before,

and $\lim F(A_i, R_i) = \lim (A_{i\perp}, R'_i) = (A_\perp, R'')$

If $(a_i) \in A$, then a typical element of $R'((a_i)) = F(R((a_i))$ can be written

$[(r_i), m]$ where $r_i \in R_i(a_i)$ and $\theta_i(r_{i+1}) = r_i$

and a typical element of $R''((a_i))$ can be written

$$([r_i, m_i]) \qquad \text{where } r_i \in R_i(a_i) \text{ and } [\theta_i(r_{i+1}), m_{i+1}] = [r_i, m_i] .$$

The natural homomorphism from $F(A, R)$ to $\lim F(A_i, R_i)$ takes $[(r_i), m]$ to $([r_i, m_i])$. The proof that this is an isomorphism is formally the same as in **Rings**.]

We can now apply the Tarski construction to F, starting from the final ringed frame(*, 0) where * is the one-element locale and 0 takes that one element to the zero ring.

Theorem 6.4
Let (A, R) be the final fixed point of F, and let $D_p = D(A, R)$. This is the *domain of p-adic integers* . Then

$A = \mathbf{N} \cup \{\infty\}$ (the ordinal $\omega+1$)

$R(n) = \mathbf{Z}/p^n$

$R(\infty) = \mathbf{Z}_p$

The restriction maps are as in the limit construction of $\mathbf{Z}_p$. In fact, they are the only maps possible between these rings.

D_p is algebraic.

Proof
Let –

$(A_n, R_n) = F^n (*, 0)$

$(f_0, g_0)\colon (A_1, R_1) \to (A_0, R_0)$ be the unique map

$(f_n, g_n) = F^n(f_0, g_0)$.

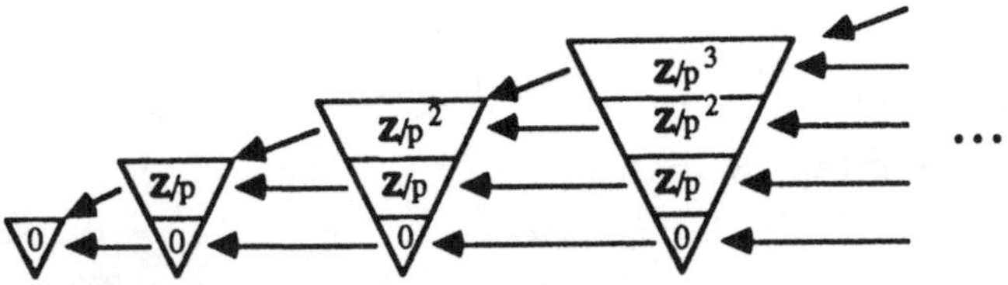

First, A_i has i+1 elements, linearly ordered; this is clear from the construction $A_{i+1} = A_{i\,\perp}$. Thus we can identify A_i with {0, 1, ..., i}. (However, we must be careful in this because the construction

$A_{i+1} = A_{i\,\perp}$ implicitly identifies A_i with the top i+1 elements of A_{i+1}, and the two identifications are not compatible.)

Next, $f_i: A_{i+1} \to A_i$ maps the bottom i+1 elements of A_{i+1} in order to the i+1 elements of A_i, and the top element i+1 of A_{i+1} to the top element i of A_i. For this is true at i = 0 and remains true when new bottoms are adjoined.

Now an element of A is a sequence (a_i) where $a_i \in A_i$ and $f_i(a_{i+1}) = a_i$. If, for some i, $a_i = n < i$, then $a_j = n$ in all A_j with $j \geq i$. Such elements of A correspond to $\mathbf{N}$. The only remaining case is when $a_i =$ i for all i, and this corresponds to ∞.

Next, if $n \leq i$ then $R_i(n) = \mathbf{Z}/p^n$. For this is certainly true at i = 0; and if true at i then

$$R_{i+1}(0) = 0 = \mathbf{Z}/p^0$$

while for $0 < n \leq i+1$

$$R_{i+1}(n) = F(R_i(n-1)) = F(\mathbf{Z}/p^{n-1}) = \mathbf{Z}/p^n \text{ by corollary 3.8.}$$

Now consider R(a) for $a \in A$. If $a = n \in \mathbf{N}$, then an element of R(n) is a coherent sequence (r_i) with $r_i \in R_i(n) = \mathbf{Z}/p^n$ $(i \geq n)$, so it can be considered to be just an element of $\mathbf{Z}/p^n$. If $a = \infty$, then an element of R(a) is a coherent sequence (r_i) with $r_i \in R_i(i) = \mathbf{Z}/p^i$, i.e. an element of $\mathbf{Z}_p$.

Thus $R(n) = \mathbf{Z}/p^n$, and $R(\infty) = \mathbf{Z}_p$.

The statements about the uniqueness of the ring maps is easy, and the fact that D_p is algebraic follows from proposition 4.2 (iii).]

We now define a [..., ...] notation analogous to that used in the rings.

Let –

(A, R) be a ringed frame,

$(A_\perp, R') = F(A, R)$

$D_1 = D(A, R)$

$D_2 = D(F(A, R))$

We define a function from $D_1 \times \mathbf{Z} \to D_2$, $\langle r, l\rangle \mapsto [r, l]$, by

$$[r, m] = [r, m] \in F(R(a)) = R'(a)$$

where $a = Er \in A$.

Note a subtlety about how this works in the domain of p-adics. It identifies an element $n \in A$ with the element $n+1 \in A \cong A_\perp$, so if
$Er = n$ then $E[r, m] = n+1$.

Proposition 6.5
Let A, R, etc. be as above.

(i) The function $\langle r, m\rangle \mapsto [r, m]$ is continuous in r.

(ii) Every element of D_2 is either $\perp$ or $[r, m]$ for some r and m; it cannot be both.

(iii) $[r, m] = [s, n]$ iff $m \equiv n$ (modulo p) and $r = s + ((n-m)/p)e$.

Proof

(i) Suppose $r \sqsubseteq s \in D_1$. Let $a = Er$, $b = Es$.

$E[r, m] = a \leq b = E[s, m]$ (a and b here considered as elements of $A_\perp$)

and $[s, m] |a = [s |a, m] = [r, m]$, so $[r, m] \sqsubseteq [s, m]$

Now suppose X is a directed set in D_1. The system $(R(Ex))$ $(x \in X)$ forms a codirected limit diagram in **Rings**, and from the fact (lemma 3.5) that F preserves this limit it follows that

$$[\bigsqcup^{\uparrow} X, m] = \bigsqcup^{\uparrow} \{[x, m] : x \in X\}$$

(ii) This follows immediately from the definition.

(iii) Suppose $[r, m] \sqsubseteq [s, n]$. Then $a = E[r, m] = Er \leq b = E[s, n] = Es$.

$$[r, m] = [s, n] |a = [s |a, n] \quad (\text{so } m \equiv n \ (\text{mod } p))$$
$$= [s |a, m + p((n-m)/p)] = [s |a + ((n-m)/p)e, m]$$

If [r, m] = [s, n] then a = b , so s |a = s and r = s + ((n–m)/p)e.]

Corollary 6.6 Let (A, R) be the domain of p-adic integers, D = D(A, R) . Then [..., ...] maps D× $\mathbb{Z}$ to D , and every element of D is either ⊥ or of the form [r, m].]

Finally, we show that our domain of p-adics is indeed the Kahn domain.

Theorem 6.7

Let D_p be the domain of p-adic integers and K the Kahn domain using as symbols the p natural numbers 0, 1, ..., p–1. Then as cpos, $D_p \cong K$.

Proof

The elements of K are the finite and infinite lists of the p symbols. We define strict functions $g:K \rightarrow D_p$ and $h: D_p \rightarrow K$ recursively by

g(m::x) = [g(x), m]

h([r, m]) = m'::h(r') where $0 \le m' < p$, $m' \equiv m \pmod p$ and r' = r + ((m–m')/p)e

Note that in the definition of h, [r', m'] = [r, m] . Then h is well-defined, for suppose [r, m] = [s, n], implying that [r', m'] = [s', n']. By proposition 6.5 $m' \equiv n' \pmod p$, so m' = n' and r' = s'.

We show that g and h are mutually inverse isomorphisms.

If x is a finite element of K, we show by induction on its length that h(g(x)) = x; then since K is algebraic hg must be the identity on the whole of K .

If x has length 0, then it is ⊥ and h(g(⊥)) = ⊥ by strictness.

Otherwise, x = m::x' and

h(g(x)) = hg(m::x') = h([g(x'), m]) = m::h(g(x')) = m::x' by induction

= x

Similarly, we prove by induction on Er that for finite $r \in D_p$, g(h(r)) = r .

If Er = 0 then r = ⊥ and g(h(⊥) = ⊥, again by strictness.

Otherwise, $r = [s, m]$, and we can choose m between 0 and $p-1$ so that $m' = m$ and $s' = s$ in the definition of h. Note that $Er = Es+1$, finite.

$$g(h(r)) = gh([s, m]) = g(m::h(s)) = [g(h(s)), m] = [s, m] \text{ by induction}$$
$$= r \qquad]$$

7. Conclusion

We have achieved our original aim of presenting the domain of p-adics *together with its arithmetic structure* as a fixpoint of a functor F. What is still not clear is how this functor can clarify our view of the domain.

In the usual fixpoint definition of a domain, via a domain equation

$$D \cong G(D)$$

the functor G often describes in a good intuitive way the internal structure of D, and F ought to do the same for D_p.

The structure $F^n 0$ can be seen as representing a calculating machine that can add, subtract and multiply integers, subject to a restriction to n digits: that is all the internal registers can handle. The limit D_p is then the infinite limit of these finite machines. The functor F in some sense ought then to connect up an extra digit to all the internal registers. However, I have not managed to formalize this idea even for these basic machines, let alone for the entire category **Ringed-frames**.

Even granted that the correct way to handle the arithmetic structure is via sheaves, there might be grounds for using other categories or functors. For instance, **Ringed-frames** might be thought too big – we actually only use a rather small part of it; or too small – whatever works in **Ringed-frames** ought also to work in (the opposite of) **Ringed-topoi**. This raises the possibility – exemplified in the methods of Tierney [76] – that D_p is a free some-kind-of-ring-or-other that can't exist in **Sets** and has to go to a different topos, that of sheaves over $\omega+1$.

Again, although the functor $A \mapsto A_\perp$ gives the right fixpoint $\omega+1$, so does the Smyth (or t or $\Box$) part of the Vietoris functor. This takes A to a frame generated by symbols $\Box a$ ($a \in A$) subject to

$$\Box \bigwedge_{i=1}^{n} a_i = \bigwedge_{i=1}^{n} \Box a_i$$

and $$\Box \bigsqcup^{\uparrow}_{\lambda} a_\lambda = \bigsqcup^{\uparrow}_{\lambda} \Box a_\lambda$$

Every element of this is a join of elements $\Box a$. When the functor is applied to the linearly ordered frame with n elements, the only join that is not already some □a is the empty join **false**, so its effect is just to adjoin a new bottom. Thus our functor F might well be better defined with a different frame part.

Of course, there is no point in proving in detail that the construction can be made to work in all other plausible categories or with all other plausible functors. The present paper shows that the construction exists, but not really what it does. What is still required is a presentation of the construction in a context where there is a good intuition for what it does. I hope that work on this will elucidate connections between topos theory and computer science.

8. Bibliography

P.M. Cohn, 1977: "Algebra", vol. 2, Wiley.

M.P. Fourman and D.S. Scott, 1979: "Sheaves and Logic", in "Applications of Sheaves", Springer LNM 753.

P.T. Johnstone, 1982: "Stone Spaces", Cambridge University Press.

M. Tierney, 1976: "On the Spectrum of a Ringed Topos", in "Algebra, Topology and Category Theory: a collection of papers in honor of Samuel Eilenberg", Academic Press.

S.J. Vickers, 1987: "An Algorithmic Approach to the p-adic Integers", in the Third Workshop on the Mathematical Foundations of Programming Language Semantics, held at Tulane University; Springer.

A CATEGORY OF GALOIS CONNECTIONS

J. M. McDill
Department of Mathematics
California Polytechnic
State University
San Luis Obispo, CA 93407

A. C. Melton*
Department of Computing
and Information Sciences
Kansas State University
Manhattan, KS 66506

G. E. Strecker*
Department of
Mathematics
Kansas State University
Manhattan, KS 66506

§0. Abstract

We study Galois connections by examining the properties of three categories. The objects in each category are Galois connections. The categories differ in their hom-sets; in the most general category the morphisms are pairs of functions which commute with the maps of the domain and codomain Galois connections. One of our main results is that one of the categories--the one which is the most closely related to the closed and open elements of the Galois connections--is Cartesian-closed.

§1. Introduction

Galois connections, which have been the subject of study for some time, occur naturally in many areas of computer science. For example, they are present in the Scott inverse limit construction [S]; some data flow analyses can be formulated as Galois connections [N]; and in some cases the correctness of a compiler may be verified by showing that a Galois connection exists [MSS]. As is evident from the compiler example, the presence of a Galois connection can prove to be useful. In order that we can both recognize the existence of Galois connections and use them effectively, we need a good understanding of Galois connections themselves. Galois connections have been studied by many researchers; see, for example, [BJ], [G], [O], [Sch] and [HH]. In this paper we approach Galois connections from a new perspective.

We begin by defining a category **GAL** of Galois connections. During our investigation of **GAL**, we find it useful to introduce two related categories: $\mathbf{GAL_P}$ and **GAL***. Some of the results of this study are known and some are new. Perhaps the most interesting new results concern Cartesian-closedness and **GAL**. Although we are not able to show that **GAL** itself is Cartesian-closed, we do show that **GAL*** is Cartesian-closed. The exponential objects in **GAL*** are built, as is the case in many other Cartesian-closed concrete categories over **SET**, on the respective hom-sets. However, an example is given which shows that in some Cartesian-closed concrete categories over **SET** exponential objects need not be built on the hom-sets. In our concluding section, we mention the work from [HH] which shows the "proper" generalization from Galois connections to categories and functors.

§2. Definition of the Category GAL

Galois connections, as originally defined by Ore [O] in 1944, are certain pairs of order-reversing mappings $P \xrightarrow{f} Q \xrightarrow{g} P$ between partially-ordered sets. An equivalent concept is a

*This research was partially funded by the National Science Foundation under grant DCR-8604080.

residuated-residual pair of mappings [BJ]. A residuated-residual pair of mappings is obtained from a Galois connection pair by replacing the partial ordering on Q with its dual ordering; thus, residuated-residual mappings are order-preserving. However, the term "Galois connection" has been used in both the order-reversing context and the order-preserving context. The first use of Galois connections for the order-preserving pairs of mappings seems to have been in 1953 in [Sch], where they were called *Galois connections of mixed type*. More recently this usage has been promoted in [G] and [HH]. Since in computer science we are often interested in "information-preserving" mappings, order-preserving mappings seem more natural than order-reversing ones. Hence, in this paper Galois connections are defined in the order-preserving context. Results analogous to ours hold, of course, for "order-reversing" Galois connections.

2.01 Definition

Let $P \xrightarrow{f} Q \xrightarrow{g} P$ be order-preserving maps on posets. (P, Q, f, g) is called a **Galois connection** provided that:

$$\forall\, a \in P,\ a \leqslant gfa \text{ and } \forall\, c \in Q,\ fgc \leqslant c$$

(The map f is said to be **residuated**, and the map g is said to be **residual**.)

In any Galois connection (P, Q, f, g), f uniquely determines g and g uniquely determines f.

2.02 Definition

When defining a category of Galois connections, one could let the objects be partially ordered sets and let the morphisms be Galois connection mappings [P]; or one could let the objects be Galois connections. We have chosen the latter approach in our definition of **GAL**. The natural choice for morphisms in **GAL** is pairs of mappings; if $A = (P_A, Q_A, f_A, g_A)$ and $B = (P_B, Q_B, f_B, g_B)$ are objects in **GAL**, then a morphism from A to B is naturally a pair of functions (h, k) with $h: P_A \to P_B$ and $k: Q_A \to Q_B$. (See the diagrams.) The question is what properties should h and k have. From an earlier paper [MSS], we know that the essential structure of a Galois connection (P, Q, f, g) is determined by the closed elements, the open elements, and the isomorphic levels created in P and Q by f and g. (See Definition 2.03.) As Proposition 2.04 shows, requiring the morphisms in **GAL** to form commuting diagrams is what is needed to preserve the essential structure from A to B. Thus, we define **GAL** to be the category whose objects are Galois connections and whose morphisms are defined as follows: (h, k) is a morphism from (P_A, Q_A, f_A, g_A) to (P_B, Q_B, f_B, g_B) if and only if $h: P_A \to P_B$, $k: Q_A \to Q_B$, $kf_A = f_B h$, and $hg_A = g_B k$, (i.e., both of the diagrams below commute).

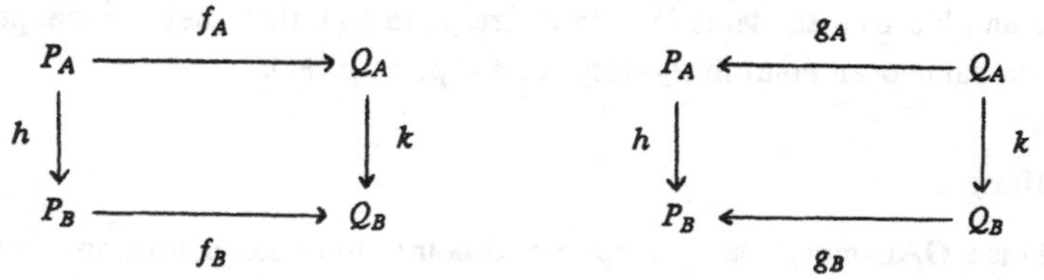

For any Galois connection (P,Q,f,g), gf is a closure operator on P (i.e., it is idempotent and $gfp \geqslant p \; \forall \; p \in P$), and fg is an interior operator on Q (i.e., it is idempotent and $fgq \leqslant q \; \forall \; q \in Q$).

Definition 2.02 has no requirement on h and k other than commutation of diagrams. It is natural to consider the case where h and k are required to be order-preserving. We denote by **GAL$_P$** the subcategory of **GAL** whose objects are Galois connections and whose morphisms are pairs of order-preserving functions. Note that **POS**, the category of partially ordered sets and order-preserving functions, is isomorphic to a full subcategory of **GAL$_P$**.

Many of the results obtained for GAL hold also for **GAL$_P$**. In our defining of special objects in GAL, we have constructed objects which also work in **GAL$_P$**; for example, the coseparator in GAL given in Proposition 4.03 is also a coseparator in **GAL$_P$**. However, not all the results obtained for GAL hold for **GAL$_P$**. For example, GAL is a balanced category, but **GAL$_P$** is not (see Propositions 3.05 and 306 and the comment following Proposition 3.06).

2.03 Definition

Let (P,Q,f,g) be a Galois connection. An element $p \in P$ is called **closed** iff $gfp = p$; an element $q \in Q$ is called **open** iff $fgq = q$. Open points and closed points are called **skeletal** points. A **level** in P (resp., a level in Q) is an equivalence class obtained by the relation: p and p' are equivalent if $gfp = gfp'$ (resp., q and q' are equivalent if $fgq = fgp'$).

There is a natural isomorphism between the levels of P and the levels of Q [MSS]. Levels which correspond under this isomorphism are said to be **associated**. Skeletal points are said to be **associated** if they are in associated levels.

2.04 Proposition

If (P_A,Q_A,f_A,g_A) and (P_B,Q_B,f_B,g_B) are Galois connections and $h:P_A \rightarrow P_B$, $k:Q_A \rightarrow Q_B$ are functions, then the following are equivalent:

1. (h,k) is a GAL-morphism.
2. (i) h maps closed points in P_A to closed points in P_B, and k maps open points in Q_A to open points in Q_B.
   (ii) h and k are level preserving, i.e., if a and b are on the same level in P_A, then h(a) and h(b) are the same level in P_B; if c and d are on the same level in Q_A, then k(c) and k(d) are on the same level in Q_B.
   (iii) If a in P_A and c in Q_A are in associated levels in A, then h(a) and k(c) are also in associated levels in B.
3. If a and b are on the same level in P_A (resp., in Q_A), then they are mapped by either path to the same open point in Q_B (resp., closed point in P_B).

2.05 Corollary

If (h,k) is a GAL-morphism, h maps closed points to closed points, and k maps open points to open points.

§3. Special Morphisms in GAL

The proofs of the next two propositions are simplified by the lemmas which precede them. We determine what special morphisms do to the closed points and open points before looking at the actions of the special morphisms on general elements in the posets.

3.01 Lemma

If (h,k) is a GAL–monomorphism, then h must be injective on closed points, and k must be injective on open points.

3.02 Proposition

If (h,k) is a GAL–morphism, then (h,k) is a GAL–monomorphism iff h and k are injective maps.

3.03 Lemma

If (h,k) is a GAL–epimorphism, then h must be surjective on closed points, and k must be surjective on open points.

3.04 Proposition

If (h,k) is a GAL–morphism, then (h,k) is a GAL–epimorphism iff h and k are surjective maps.

3.05 Proposition

GAL is a balanced category; that is, isomorphisms in GAL are precisely bimorphisms (i.e., morphisms which are both monomorphisms and epimorphisms).

3.06 Proposition

If (h,k) is a $\mathbf{GAL_P}$–morphism, then (h,k) is a $\mathbf{GAL_P}$–isomorphism iff
h and k are bijective and
h^{-1} and k^{-1} are order-preserving.

Since there are morphisms (h,k) in $\mathbf{GAL_P}$ where h and k are bijections but h^{-1} and k^{-1} are not order-preserving, then there are bimorphisms in $\mathbf{GAL_P}$ which are not $\mathbf{GAL_P}$–isomorphisms. Hence, $\mathbf{GAL_P}$ is not balanced as GAL is.

§4. Limits and Colimits in GAL

4.01 Proposition

The category GAL has an initial object and terminal objects.

The initial object is $\varnothing \overset{\varnothing}{\rightarrow} \varnothing \overset{\varnothing}{\rightarrow} \varnothing$. The terminal objects are formed from singleton sets and constant maps between them, $\{\cdot\} \overset{f}{\rightarrow} \{\cdot\} \overset{g}{\rightarrow} \{\cdot\}$

The following proposition shows that GAL is complete (i.e., all limits of small functors exist, including multiple pullbacks, multiple equalizers, terminal objects, products and inverse limits [HS]). The products in GAL are formed from the products in POS in a straightforward fashion.

4.02 Proposition

The category GAL has products and equalizers. (Hence, the category GAL is complete [HS].)

The next results show that GAL is co-well-powered (i.e., it has a representative class of quotient objects) and that GAL is cocomplete (i.e., all colimits of small functors exist in GAL, such as coproducts, multiple coequalizers, multiple pushouts and direct limits [HS]).

4.03 Proposition

The category GAL has a coseparator.

A coseparator C is the Galois connection (X, X, f, g) illustrated below:

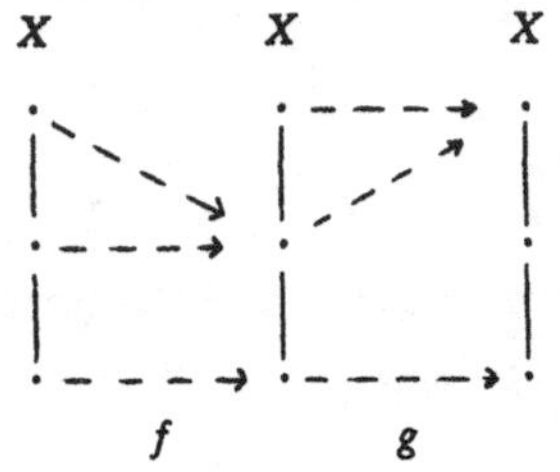

4.04 Proposition

The category GAL is well-powered.

4.05 Corollary

GAL is cocomplete and co-well-powered.

4.06 Proposition

The category GAL has a separator.

A separator S is the Galois connection (B, B, f, g) shown below:

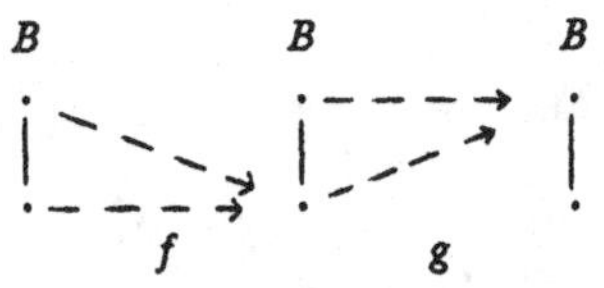

The existence of a separator is an important smallness condition for the category GAL, as can be seen in the following corollary.

4.07 Corollary

GAL is concretizable, i.e., there exists a faithful functor $hom(S,\_)$ from GAL to SET.

§5. Exponential Objects and Cartesian-Closedness

It is often the case that one wishes to form exponential objects in a category that has products of pairs. In many concrete categories an exponential object B^A, associated with objects A and B, is formed by endowing the set of all morphisms from A to B with an appropriate structure. One should note, however, that it is not always the case that the members of B^A correspond with the morphisms from A to B. The category REL whose objects are pairs (X,ρ) with $\rho \subseteq X \times X$ and whose morphisms $f:(X,\rho) \rightarrow (Y,\sigma)$ are relation-preserving functions (i.e., if $x\rho x'$, then $f(x)\,\sigma\, f(x')$) is Cartesian-closed with

$$(Y,\sigma)^{(X,\rho)} = (Y^X,\omega)$$

where $f\,\omega\, g$ means that

$$\forall (x,x') \in X \times X, \text{ if } x\rho x' \text{ then } f(x)\,\sigma\, g(x').$$

The exponential object (Y^X,ω) is built on the set of all functions from X to Y and not on the set of all morphisms from X to Y.

In an arbitrary category, an object A is said to be **exponential** provided that exponentiation $(\_)^A$ is a functor which is right-adjoint to $A \times \_$, the Cartesian-product-by-A functor. Specifically, A is exponential provided that for each object B, there is an evaluation morphism $eval_B : A \times B^A \rightarrow B$ and for each object C and each morphism $f:A \times C \rightarrow B$, there is a unique morphism $\bar{f}:C \rightarrow B^A$ such that $eval_B \circ (id \times \bar{f}) = f$.

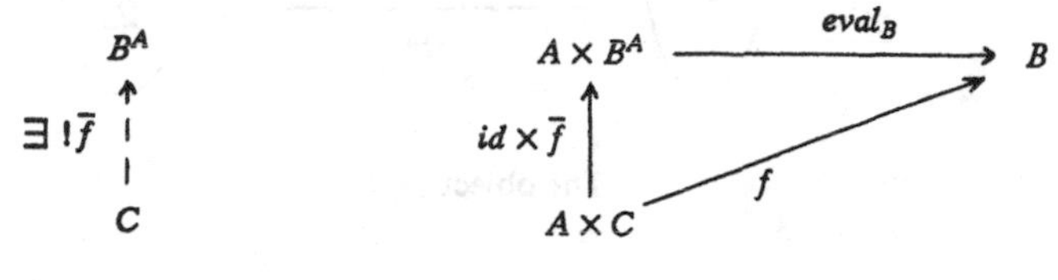

If each object A of a category is exponential, then the category is called **Cartesian-closed.**

In this section, we investigate a natural structure $[A,B]$ which yields a Galois connection associated with the set of GAL-morphisms from A to B and show that it comes very close to providing exponential objects (Proposition 5.13). We then define a very natural category GAL* whose objects are the same as those of GAL and whose morphisms are precisely the associated pairs of skeletal elements in the Galois connections of the form $[A,B]$ mentioned above, and we show that GAL* is Cartesian-closed.

5.01 Definition

Let $A = (P_A, Q_A, f_A, g_A)$ and $B = (P_B, Q_B, f_B, g_B)$ be Galois connections. Define $[A,B] = (P^*, Q^*, f^*, g^*)$ as follows:

$$P^* = \{h : P_A \rightarrow P_B \mid \exists\, k : Q_A \rightarrow Q_B \text{ such that } (h,k) \in Mor\ \text{GAL}\}$$

$Q^* = \{k: Q_A \to Q_B \mid \exists\, h: P_A \to P_B$ such that $(h, k) \in Mor$ **GAL**$\}$

$f^*(h) = f_B h g_A \qquad \forall\, h \in P^*$

$g^*(k) = g_B k f_A \qquad \forall\, k \in Q^*$

The partial orders in P^* and Q^* result from the pointwise-ordering of the functions (i.e., $h \leqslant h'$ iff $h(a) \leqslant h'(a)\ \forall\, a \in P_A$ and $k \leqslant k'$ iff $k(c) \leqslant k'(c)\ \forall\, c \in Q_A$).

The set P^*, which for clarity is denoted by $[P_A.P_B]$, is the set of all functions from P_A to P_B which are the "left halves" of GAL-morphisms from A to B. Likewise Q^*, which is denoted by $[Q_A.Q_B]$, is the set of all functions from Q_A to Q_B which are the "right halves" of GAL-morphisms from A to B. The function $f^*: [P_A.P_B] \to [Q_A.Q_B]$ represents a natural way to convert a mapping from P_A to P_B to a mapping from Q_A to Q_B. f^* takes an $h: P_A \to P_B$ and composes it with the appropriate mappings from the Galois connections A and B; $f^*(h) = Q_A \overset{g_A}{\to} P_A \overset{h}{\to} P_B \overset{f_B}{\to} Q_B$. Likewise, g^* is a natural mapping from $[Q_A.Q_B]$ to $[P_A.P_B]$.

5.02 Proposition

$[A.B]$ is an object in GAL.

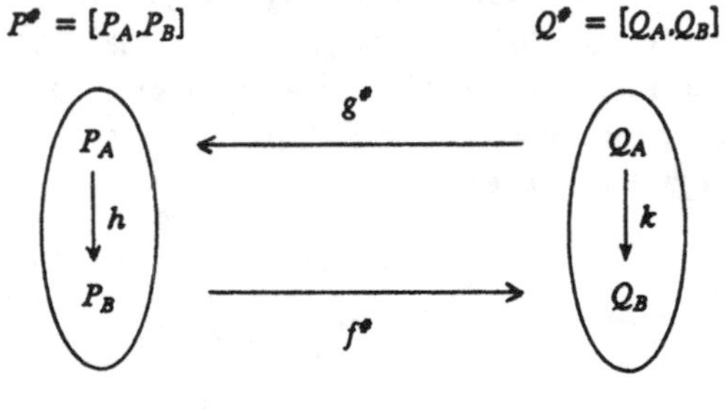

The object $[A.B]$

5.03 Notation

Let $h \in P^*$. Then $\bar{h}$ denotes the closure of h in P^*, i.e., $\bar{h} = g^* f^*(h) = g_B(f_B h g_A) f_A = h g_A f_A = g_B f_B h$. Similarly, k° denotes the interior of k in Q^*, i.e., $k^\circ = f^* g^*(k) = f_B(g_B k f_A) g_A = f_B g_B k = k f_A g_A$.

5.04 Proposition

The following statements are equivalent:

1. h is a closed element of P^*
2. $h[P_A] \subseteq g_B[Q_B]$, (i.e., the image of P_A under h consists of closed elements)
3. $h(p_A) = g_B f_B h(p_A) \qquad \forall\, p_A \in P_A$
4. $h(p_A) = h g_A f_A(p_A) \qquad \forall\, p_A \in P_A$ (i.e., the image of each level is a closed point)
5. $|h[f_A^{-1}[q_A]]| \leqslant 1 \qquad \forall\, q_A \in Q_A$

6. $|h[f_A^{-1}[f_A(p_A)]]| = 1 \qquad \forall\, p_A \in P_A$ (i.e., each level is mapped by h to a singleton)

5.05 Proposition

Let $h \in P^*$. Then for all closed points p_A in P_A, $h(p_A) = \bar{h}(p_A)$; equivalently,

$$h|_{g_A[Q_A]} = \bar{h}|_{g_A[Q_A]}$$

5.06 Proposition

If $h \in P^*$, then $(h, f^*(h))$ is a GAL-morphism from A to B.

Propositions analogous to 5.04, 5.05 and 5.06 can be given for the case when $k \in Q^*$.

5.07 Corollary

If (h,k) is a GAL-morphism from A to B, then $(\bar{h},k)$, (h,k°) and $(\bar{h},k^{\circ})$ are also GAL-morphisms from A to B.

5.08 Lemma

If (h,k) is a GAL-morphism, then $\bar{h} = g^*(k)$ and $k^{\circ} = f^*(h)$.

5.09 Proposition

$(h,k): A \to B$ is a GAL-morphism iff h and k are in associated levels in $[A,B]$.

5.10 Notation

If $A = (P_A, Q_A, f_A, g_A)$ and $B = (P_B, Q_B, f_B, g_B)$ are objects in GAL, then let

$$e_P : P_A \times P^* \to P_B \text{ and}$$
$$e_Q : Q_A \times Q^* \to Q_B \text{ be given by:}$$
$$e_P(p_A, h) = h(p_A) \qquad \forall\, p_A \in P_A \text{ and}$$
$$e_Q(q_A, k) = k(q_A) \qquad \forall\, q_A \in Q_A.$$

and let $eval = (e_P, e_Q)$

$$\begin{array}{ccc}
P_A \times [P_A, P_B] & \overset{g_A \times g^*}{\underset{f_A \times f^*}{\rightleftarrows}} & Q_A \times [Q_A, Q_B] \\
\downarrow e_P & & \downarrow e_Q \\
P_B & \overset{g_B}{\underset{f_B}{\rightleftarrows}} & Q_B
\end{array}$$

5.11 Proposition

1. $eval : A \times [A, B] \to B$ is a GAL-morphism.
2. e_P and e_Q are order-preserving functions.

5.12 Notation

Let C be any object in GAL.

1. Define $\psi : [C, [A, B]] \to [A \times C, B]$ by
   $\psi(h, k) = eval \circ (id \times (h, k))$ for each GAL-morphism $(h, k) : C \to [A, B]$
   More conveniently:
   $\psi(h, k) = (\psi_P(h), \psi_Q(k)) = (e_P \circ (id \times h), e_Q \circ (id \times k))$
2. Define $\phi : [A \times C, B] \to [C, [A, B]]$ as follows:
   for each GAL-morphism $(h', k') : A \times C \to B$,
   $\phi(h', k') = (\phi_P(h'), \phi_Q(k'))$ where
   $\phi_P(h') : P_C \to [P_A, P_B]$ is given by $(\phi_P(h')(p_C))(p_A) =$
   $g_B f_B h'(p_A, p_C)$ for each $p_C \in P_C$ and each $p_A \in P_A$, and
   $\phi_Q(k') : Q_C \to [Q_A, Q_C]$ is given by $(\phi_Q(k')(q_C))(q_A) =$
   $f_B g_B k'(q_A, q_C)$ for each $q_A \in Q_A$ and $q_C \in Q_C$

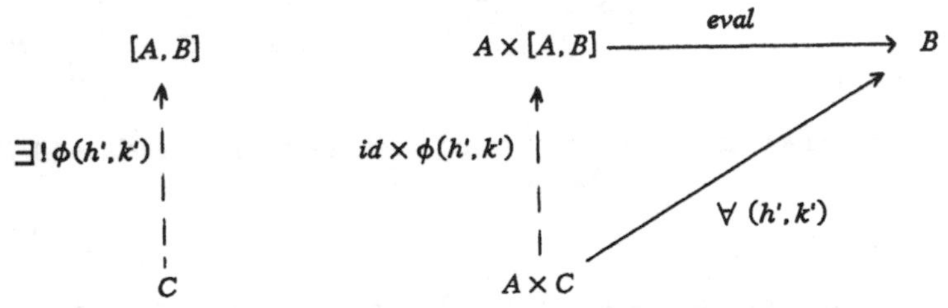

5.13 Proposition

1. ψ and ϕ are GAL-morphisms.
2. ψ and ϕ preserve order.
3. $\phi\psi\phi = \phi$ and $\psi\phi\psi \geqslant \psi$.
4. $([A \times C, B], [C, [A, B]], \phi, \psi)$ and $([C, [A, B]], [A \times C, B], \psi, \phi)$ need not be Galois connections.

Since in many applications of Galois connections the emphasis is on the closed points and the open points, it is often the case that (h, k) could be replaced by $(\bar{h}, k^{\circ})$ because these maps agree on the skeletal points in their respective domains. (See Proposition 5.05.) With this replacement in mind, we define the category GAL*.

5.14 Definition

Let GAL* be the category defined as follows:

Each object in GAL is also in GAL*.
A GAL-morphism (h, k) is a (GAL)*-morphism iff $(h, k) = (\bar{h}, k^{\circ})$.

Looking back at Proposition 5.09, we can see that $(h,k):A \to B$ is a **GAL***-morphism if and only if h and k are associated skeletal points in $[A,B]$. Thus h and k map levels to skeletal points in P_B and Q_B, respectively.

Identities in **GAL*** do not generally consist of pairs of identity functions. If $A = (P,Q,f,g)$ is in **GAL*** and if P or Q has a non-skeletal point (i.e., if f and g are not bijections), then the **GAL***-identity 1_A does not consist of $(1_P,1_Q)$. However, the identities in **GAL*** are easily determined; the **GAL***-identity on A is (gf,fg). The existence of these "interesting" identities in **GAL*** means that **GAL*** is not a subcategory of **GAL**. However, based on our work in **GAL** and on the definition of morphism in **GAL***, we have the following theorem.

5.15 Theorem

The category **GAL*** is Cartesian-closed.

At this point, it is an open question as to whether **GAL** is Cartesian-closed.

§6. Generalizing Galois Connections to Categories and Functors

It is generally assumed that a proper generalization of (order-preserving) Galois connections is adjoint situations. However, in [HH] Galois connections are generalized to concrete categories and concrete functors over a common base category. This generalization appears to be the correct one since virtually all results concerning Galois connections have categorical analogies in this setting whereas there are only a few analogous results in the setting of adjoint situations.

6.01 Definition

Let $\mathbf{X}$ be a category. A concrete category over $\mathbf{X}$ is a pair $(\mathbf{A},U)$ where $\mathbf{A}$ is a category and $U:\mathbf{A}\to\mathbf{X}$ is a forgetful functor. A concrete functor $G:(\mathbf{A},U)\to(\mathbf{B},V)$ between concrete categories over $\mathbf{X}$ is a functor $G:\mathbf{A}\to\mathbf{B}$ which commutes with the forgetful functors (i.e., $VG = U$).

If $(\mathbf{A},U)$ is a concrete category, then there exists a natural order on the class of $\mathbf{A}$-objects.

$$A \leqslant B \text{ iff } (UA = UB \text{ and } 1_{UA}:A\to B \in Mor\,\mathbf{A})$$

6.02 Definition

Let $G:(\mathbf{A},U)\to(\mathbf{B},V)$ and $F:(\mathbf{B},V)\to(\mathbf{A},U)$ be concrete functors between concrete categories over $\mathbf{X}$. $(\mathbf{A},\mathbf{B},F,G)$ is called a Galois connection between concrete categories over $\mathbf{X}$ if and only if

$$f:FB\to A \in Mor\,\mathbf{A} \text{ iff } f:B\to GA \in Mor\,\mathbf{B} \qquad (*)$$

for A in $\mathbf{A}$, B in $\mathbf{B}$, and f an $\mathbf{X}$-morphism. (In $(*)$ $f:FB\to A$ is the unique $\mathbf{A}$-morphism from FB to A whose image under U is the $\mathbf{X}$-morphism f, and $f:B\to GA$ is the unique $\mathbf{B}$-morphism from B to GA whose image under V is the $\mathbf{X}$-morphism f.)

Let $\mathbf{T}$ be a category with one object and one morphism. A partially ordered set P can be considered to be a concrete category $(\mathbf{P},U)$ over $\mathbf{T}$ where $\mathbf{P}$ is formed in the standard way (i.e., the elements in P are the objects in $\mathbf{P}$ and there is a $\mathbf{P}$-morphism from a to b iff $a \leqslant b$ in P) and U is the unique constant functor over $\mathbf{T}$.

6.03 Proposition

Let P and Q be partially ordered sets, and let $f: P \to Q$ and $g: Q \to P$ be functions. (P, Q, f, g) is a Galois connection between partially ordered sets if and only if $(\mathbf{P}, \mathbf{Q}, f, g)$ is a Galois connection between categories over **T**.

The material in this section comes from [HH]. Our Galois connection between categories is called a Galois connection of the third kind in [HH]. The interested reader is encouraged to see [HH].

§7. References

[BJ] Blyth, T. S., and Janowitz, M. F. *Residuation Theory.* Pergammon Press, Oxford, 1972.

[G] Gierz, G., Hofmann, K. H., Keimel, K., Lawson, J. D., Mislove, M., and Scott, D.S. *A Compendium of Continuous Lattices.* Springer-Verlag, Berlin, 1980.

[HH] Herrlich, H. and Husek, M. Galois connections. *Springer-Verlag Lecture Notes in Computer Science,* 239(1986), 122-134.

[HS] Herrlich, H. and Strecker, G. E. *Category Theory.* Allyn and Bacon, Boston, 1973; second edition Helderman Verlag, Berlin, 1979.

[MSS] Melton, A., Schmidt, D. A., and Strecker, G. E. Galois connections and computer science applications. *Springer-Verlag Lecture Notes in Computer Science,* 240(1986), 299-312.

[N] Nielson, F. A denotational framework for data flow analysis. *Acta Informatica,* 18(1982), 265-287.

[O] Ore, O. Galois connections. *Trans. Amer. Math. Soc.* 55(1944), 493-513.

[P] Plotkin, G. The category of complete partial orders. Postgraduate course notes, Computer Science Dept., Edinburgh University, Edinburgh, Scotland, 1982.

[Sch] Schmidt, J. Beiträge zur Filtertheorie. II. *Math. Nachr.* 10(1953), 197-232.

[S] Scott, D. Continuous Lattices. *Springer-Verlag Lecture Notes in Math.* 274(1972), 97-136.

Vol. 245: H.F. de Groote, Lectures on the Complexity of Bilinear Problems. V, 135 pages. 1987.

Vol. 246: Graph-Theoretic Concepts in Computer Science. Proceedings, 1986. Edited by G. Tinhofer and G. Schmidt. VII, 307 pages. 1987.

Vol. 247: STACS 87. Proceedings, 1987. Edited by F.J. Brandenburg, G. Vidal-Naquet and M. Wirsing. X, 484 pages. 1987.

Vol. 248: Networking in Open Systems. Proceedings, 1986. Edited by G. Müller and R.P. Blanc. VI, 441 pages. 1987.

Vol. 249: TAPSOFT '87. Volume 1. Proceedings, 1987. Edited by H. Ehrig, R. Kowalski, G. Levi and U. Montanari. XIV, 289 pages. 1987.

Vol. 250: TAPSOFT '87. Volume 2. Proceedings, 1987. Edited by H. Ehrig, R. Kowalski, G. Levi and U. Montanari. XIV, 336 pages. 1987.

Vol. 251: V. Akman, Unobstructed Shortest Paths in Polyhedral nvironments. VII, 103 pages. 1987.

Vol. 252: VDM '87. VDM – A Formal Method at Work. Proceedings, 1987. Edited by D. Bjørner, C.B. Jones, M. Mac an Airchinnigh and .J. Neuhold. IX, 422 pages. 1987.

ol. 253: J.D. Becker, I. Eisele (Eds.), WOPPLOT 86. Parallel rocessing: Logic, Organization, and Technology. Proceedings, 986. V, 226 pages. 1987.

ol. 254: Petri Nets: Central Models and Their Properties. Advances Petri Nets 1986, Part I. Proceedings, 1986. Edited by W. Brauer, /. Reisig and G. Rozenberg. X, 480 pages. 1987.

ol. 255: Petri Nets: Applications and Relationships to Other Models f Concurrency. Advances in Petri Nets 1986, Part II. Proceedings, 986. Edited by W. Brauer, W. Reisig and G. Rozenberg. X, 516 ages. 1987.

ol. 256: Rewriting Techniques and Applications. Proceedings, 987. Edited by P. Lescanne. VI, 285 pages. 1987.

ol. 257: Database Machine Performance: Modeling Methodologies nd Evaluation Strategies. Edited by F. Cesarini and S. Salza. X, 250 ages. 1987.

ol. 258: PARLE, Parallel Architectures and Languages Europe. olume I. Proceedings, 1987. Edited by J.W. de Bakker, A.J. Nijman d P.C. Treleaven. XII, 480 pages. 1987.

ol. 259: PARLE, Parallel Architectures and Languages Europe. olume II. Proceedings, 1987. Edited by J.W. de Bakker, A.J. Nijman d P.C. Treleaven. XII, 464 pages. 1987.

l. 260: D.C. Luckham, F.W. von Henke, B. Krieg-Brückner, Owe, ANNA, A Language for Annotating Ada Programs. V, 143 ges. 1987.

l. 261: J. Ch. Freytag, Translating Relational Queries into Iterative ograms. XI, 131 pages. 1987.

l. 262: A. Burns, A.M. Lister, A.J. Wellings, A Review of Ada sking. VIII, 141 pages. 1987.

l. 263: A.M. Odlyzko (Ed.), Advances in Cryptology – CRYPTO '86. oceedings. XI, 489 pages. 1987.

. 264: E. Wada (Ed.), Logic Programming '86. Proceedings, 86. VI, 179 pages. 1987.

. 265: K.P. Jantke (Ed.), Analogical and Inductive Inference. oceedings, 1986. VI, 227 pages. 1987.

. 266: G. Rozenberg (Ed.), Advances in Petri Nets 1987. VI, 451 es. 1987.

. 267: Th. Ottmann (Ed.), Automata, Languages and Programng. Proceedings, 1987. X, 565 pages. 1987.

. 268: P.M. Pardalos, J.B. Rosen, Constrained Global Optimiza: Algorithms and Applications. VII, 143 pages. 1987.

. 269: A. Albrecht, H. Jung, K. Mehlhorn (Eds.), Parallel Algons and Architectures. Proceedings, 1987. Approx. 205 pages. 7.

. 270: E. Börger (Ed.), Computation Theory and Logic. IX, 442 es. 1987.

Vol. 271: D. Snyers, A. Thayse, From Logic Design to Logic Programming. IV, 125 pages. 1987.

Vol. 272: P. Treleaven, M. Vanneschi (Eds.), Future Parallel Computers. Proceedings, 1986. V, 492 pages. 1987.

Vol. 273: J.S. Royer, A Connotational Theory of Program Structure. V, 186 pages. 1987.

Vol. 274: G. Kahn (Ed.), Functional Programming Languages and Computer Architecture. Proceedings. VI, 470 pages. 1987.

Vol. 275: A.N. Habermann, U. Montanari (Eds.), System Development and Ada. Proceedings, 1986. V, 305 pages. 1987.

Vol. 276: J. Bézivin, J.-M. Hullot, P. Cointe, H. Lieberman (Eds.), ECOOP '87. European Conference on Object-Oriented Programming. Proceedings. VI, 273 pages. 1987.

Vol. 279: J.H. Fasel, R.M. Keller (Eds.), Graph Reduction. Proceedings, 1986. XVI, 450 pages. 1987.

Vol. 280: M. Venturini Zilli (Ed.), Mathematical Models for the Semantics of Parallelism. Proceedings, 1986. V, 231 pages. 1987.

Vol. 281: A. Kelemenová, J. Kelemen (Eds.), Trends, Techniques, and Problems in Theoretical Computer Science. Proceedings, 1986. VI, 213 pages. 1987.

Vol. 282: P. Gorny, M.J. Tauber (Eds.), Visualization in Programming. Proceedings, 1986. VII, 210 pages. 1987.

Vol. 283: D.H. Pitt, A. Poigné, D.E. Rydeheard (Eds.), Category Theory and Computer Science. Proceedings, 1987. V, 300 pages. 1987.